CENTURIONS OF A CENTURY.

Off to the Crimea.

FRONTISPIECE.

Centurions of a Century

AMONG WHICH ARE MANY WHO HAVE
SOLDIERED IN

THE TWELFTH, or THE SUFFOLK REGIMENT OF FOOT,

DISTINGUISHED BY ITS SERVICES AT

DETTINGEN, June 27th, 1743; MINDEN, August 1st, 1759;

AND IN

The Glorious DEFENCE OF GIBRALTAR from the year 1779 to 1782,

OBTAINING FOR ITS BADGE AND MOTTO

"Gibraltar," Castle and Key—"Montis Insignia Calpe;"

FOR THE

STORMING AND CAPTURE OF SERINGAPATAM, May 4th, 1799;

And for its Gallant Conduct on many duties in

INDIA from 1798 to 1807;

SOUTH AFRICA, 1851-2-3; NEW ZEALAND;

AFGHANISTAN, 1878-80; SOUTH AFRICA, 1899-1902.

This book (10/6) can be obtained from the Editor, 4, Montpellier Crescent, Brighton; or from Messrs. Cox & Co. (Shipping Agency), 16, Charing Cross, S.W.

DEDICATION.

This Book is dedicated, first, to the 12th (now Suffolk) Regiment, in memory of twenty-seven most happy years passed with that distinguished Corps; secondly, to the youth (male and female) of His Majesty's dominions, in hopes that they may train themselves to do, in time of need, decisive things;

By the Author,

C. H. GARDINER.

CONTENTS.

CHAPTER XXI.

CHAPTER XXII.

CHAPTER XXIII.

CHAPTER XXIV.

CHAPTER XXV.

CHAPTER XXVI.

CHAPTER XXVII.

CHAPTER XXVIII.

CHAPTER XXIX.

CHAPTER XXX.

CHAPTER XXXI.

CHAPTER XXXII.

CHAPTER XXXIII.

CHAPTER XXXIV.

CHAPTER XXXV.

CHAPTER XXXVI.

CHAPTER XXXVII.

CHAPTER XXXVIII.

CHAPTER XXXIX.

CHAPTER XL.

CHAPTER XLI.

CHAPTER XLII.

CHAPTER XLIII.

CHAPTER XLIV.

CHAPTER XLV.

APPENDIX.

LIST OF ILLUSTRATIONS.

Errata, Addenda and Omissions.

p. 13. Created Sir John Ligonier (K.B.), on field of Dettingen, by George II., in 1743.

Created Irish Viscount, 1757.

Created Earl and Field-Marshal, 1766.

p. 43. Monckton, Murray, Guy Carleton, Townshend were Brigadiers to Wolfe. Monckton was dangerously wounded with Wolfe at Quebec. Guy Carleton successfully defended both town and province of Quebec against all inroads of Americans in 1775-76.

p. 51. A 74 line-of-Battleship required for building 2,000 well-grown trees, each tree weighed 2 tons—hence 4,000 tonnage.

p. 54. British Regiments with Burgoyne, on surrender at Saratoga, were 9th, 20th, 21st, 53rd, 62nd.

p. 154. The body of Sir Thomas Picton was exhumed in 1859 and reburied in crypt of St. Paul's.

p. 161. The first fight between Air, Surface, and Submarine craft took place on Xmas Day, 1914. Commander Francis Hewlett, son of Maurice Hewlett, a brave young pilot, was the one not to return. A New Year's joy to all was to learn Commander Hewlett had been saved and was now ashore in Holland.

p. 387. The 4th Foot was also in Abyssinia.

p. 466. For 13th read 15th Sikhs.

Introduction.

"CENTURIONS."

"To do in time of need decisive things."

THE 1st August, 1759, is a date well known to soldiers, when the 12th, 20th, 23rd, 25th, 37th, and 51st, the "Minden" Regiments, all cover themselves with roses in memory of that Battle.

On 1st August, 1914, England had to realize that her Army would be required about Minden once again, for War had been forced upon her by the Hapsburghs and the Hohenzollerns, who had set Europe in a blaze—the rappel for England, France, Belgium, Russia at once sounded, to extinguish first the flame, then next, we hope, the wicked hands who had planned and fanned it.

At this moment the mighty fleets and armies of all Europe are scheduled out for War—England must be with France, Russia, and Belgium as they fight towards Berlin.

Pages 60 to 250 tell us of Actors and their doings in those grand scenes of War 100 years ago—pages from the pen of Field-Marshal French and Admiral Jellicoe have again told us the oft-quoted tale, that the British Officer, the Soldier, and the Sailor for coolness, courage, decision, energy, and gentleness remain as usual entirely unsurpassed.

The terrible chill that comes to one in reading the daily "Roll of Honour" is like to that of a benumbed or frozen limb, but, thank God, slowly and surely comes the warm sensation of the returning Life Blood, telling one to "carry on" if only in Memory of the Brave and Young, the very best of Britain's sons.

The rivers Marne*, Seine, Aube, and Aisne have all been reddened once again, in conflicts so similar for skill and bravery as upheld the genius of Napoleon in 1814. In 1914 we have the siding different, but our Regiments just the same.

"French" rides up to the prostrate 12th, or Suffolks, when, on 26th August, those critical hours of fighting had reduced them to a shadow, and says:—

> "I want to tell you Suffolks how pleased I am with your work. You cannot realize how valuable it was, and in two or three days England will know of it, the Empire will know of it, and ring with your Glory."

Could the hum-drum English but realize that the Empire Sir John French spoke of is now represented within 80 miles of our shores, surely all

*A house freshly battered on the banks of the Marne in 1914 bore a tablet to mark the damage to it of shot and shell in 1814.

should be like greyhounds at the leash, straining every nerve to be with them in fight.

"Shall we not through good and ill
Cleave to one another still?
Britons' myriad voices call
Sons, be welded each and all
Into one Imperial whole;
One with Britain, heart and soul!
One life, one flag, one fleet, one Throne!
Britons hold your own! against this infernal Hun."

This Imperial whole is now beautifully welded into our Great Seal, for in December, 1899, when England was at War, and wanted help, our very best from North, South, East, West came forth and gave it.

Then our late Gracious Sovereign, King Edward VII., to honour this loyalty and zeal, on his accession to the throne, in 1901, proclaimed himself by the Grace of God, King of the United Kingdom of Great Britain and Ireland, and of the British Dominions beyond the Seas, and gave to each Dominion a scroll of honour on his Great Seal, to which his gracious son, King George, has now added Egypt.

"There is a great force hidden in a sweet command."

C. H. GARDINER.

6th Dec., 1914.

CENTURIONS OF A CENTURY.

CHAPTER I.*

A centurion said: "Doubtless this man was the Son of God." A soldier said: "Speak one word only, and thy servant shall be whole."

THE national character shone grandly one hundred and odd years ago, and was able to produce, from a small population, a proportion of great men and women, perhaps unsurpassed before or since. Their calling was various—soldiers, sailors, statesmen, poets, men of science, etc. This was due, perhaps, to the grit instilled in the age of Elizabeth (d. 1603), Cromwell (d. 1658), and the Pitts, for Elizabeth never claimed any power which she could not exercise effectually, and in Continental policy her voice was listened to with full attention. How different with James I. and Charles I., who carried no weight in European politics, and deadened the aspirations of their people. The latter monarch so excelled in domestic tyranny that England sought a Protector in Cromwell (the soldier yeoman), who saved the British Empire from partition, and maintained all civil liberties; whilst with our great Commoners, the Pitts, England breathed and breathed again. When patriotism is shown, patriots soon will follow, for—

> " We are all sprung from men who fought at Crecy;
> We all were Englishmen when Shakespeare wrote;
> And we are still compatriots "*in essè*,
> Though called Australians, Yankees, and what not."

* Readers are asked kindly to allow chapters 1-2 to pass as an Introduction.

Only by refreshing our knowledge of past events can the memory of those who have gone before be kept evergreen; only by rekindling the same imagination, sympathy, and admiration which they created, can encouragement be given to those now toiling up the hill, forging their links to the past, in times we call our own.

The roll of soldiers and sailors now presented is of distinguished men who held principal command in camp and ship from the month of May, 1660, when London was *en fete* to welcome Charles II. back again as King. He had been met at Dover by Monk, the first Captain-General, afterwards Duke of Albemarle; his crown, however, was to be reset with well-defined restrictions—no standing army, no financial independence, no unparliamentary taxes.

The titles ran as — Captain-General, Commander-in-Chief, General-Commanding-in-Chief, Field-Marshal on the Staff, and General on the Staff. The Duke of Wellington was Commander-in-Chief, but his successors were only appointed as "Commanding-in-Chief," a distinction which places this high office under direct Ministerial control.

1660 George Monk, Duke of Albemarle, Captain-General.

1674 James, Duke of Monmouth, Captain-General; beheaded 1685.

1690 John, Earl of Marlborough.

1691 Meinhart, Duke of Leinster and Schomberg.

1702 John, Duke of Marlborough, Captain-General.

1711 James, Duke of Ormond, Captain-General.

1714 John, Duke of Marlborough, Captain-General, ceased to hold office in 1717.

1744 John, Earl of Stair.

1748 H.R.H. Duke of Cumberland, Captain-General.

1757 John, Earl Ligonier [see note, page 13].

1766 John, Marquis of Granby; office vacant from 1769.

1778 Jeffrey, Lord Amherst.

1782 Hon. Henry Seymour Conway.

1793 Jeffrey, Lord Amherst.

1795 H.R.H. Frederick, Duke of York, Captain-General.

1809 Sir David Dundas, Bart.
1811 H.R.H. Frederick, Duke of York.
1827 Arthur, Duke of Wellington.
1828 Rowland, Viscount Hill.
1842 Arthur, Duke of Wellington.
1852 Henry, Viscount Hardinge.
1856 H.R.H. George, Duke of Cambridge.
1895–1900 Lord Wolseley.
Sir Evelyn Wood in interim.
1901–1904 Lord Roberts (Army Council).

In the Navy the First Lords have been :—

1751 Lord Anson.
1762 Earl of Halifax.
1762 George Grenville.
1763 Earl of Sandwich.
1763 Earl of Egmont.
1766 Sir Charles Saunders.
1766 Sir Edward (afterwards Lord) Hawke.
1771 Earl of Sandwich.
1782 Viscount Keppel (one of three brothers).
1783 Viscount (afterwards Earl) Howe.
1783 Viscount Keppel.
1783 Viscount Howe (one of three brothers).
1788 Earl of Chatham (eldest brother to Pitt).
1794 Earl Spencer.

Now we pass from the centurions to the men who—

"Gallant in strife, and noble in their ire,
The battle is their pastime, They go forth
Gay in the morning, as to summer's sport;
When evening comes, the glory of the morn—
The youthful soldier—is a clod of clay."

"It is, I believe, a common mistake that since the Revolution we have no such thing as impressment or conscription for the Army. Of course, no permanent law provided for it, because there was no permanent law for the Army. Also it is true that this means of raising a force was only made lawful in times of war, and was applied in a limited way. But in the first place it was at times applied to insolvent debtors. Imprisoned debtors were discharged on condition of their enlisting or finding a substitute. This seems to have been done on many occasions during the eighteenth century. Then again, convicted criminals were released upon condition of their enlisting. This was, I believe, done until the end of the Peninsular War. Thirdly, conscription was applied to the pauper class. In 1703 'Justices are to raise and levy such able-bodied men as have not any lawful calling or employment, or visible means for their maintenance or subsistence, and hand them over to the officers of the Queen's forces.' Similar Acts were passed during the reigns of George II. and George III.; the persons liable to be impressed were 'all such able-bodied, idle and disorderly persons, who cannot upon examination prove themselves to exercise and industriously follow some lawful trade or employment, or to have some substance sufficient for their support and maintenance.' I believe that clauses directing the impressment of able-bodied paupers were in force until 1780. A British army of the eighteenth century must have been largely composed of bad characters, insolvent debtors, criminals, idle and disorderly persons. The Army was never popular; the soldiers, as a class, were despised."

And it was so. The soldier on enlistment had to be impeached as one *de profundis*. Virtuous and vicious every man must be, whether tinker, tailor, soldier, sailor; but certainly the soldier and the sailor, when he puts on his red or blue jacket for the first time, also becomes a new man, ready to pour out his life blood like rain for his king and country.

"'Halt! Shoulder Arms! Recover! As you were! Right wheel! Eyes left! Attention! Stand at ease!' O Britain! O, my country! words like these have made thy name a terror and a fear to all the nations."

Were these words not a "terror" when the Governor of Southampton, on hearing that his old regiment had mutinied on the beach by positively refusing to embark for India, set off at a gallop down to the scene of action? The moment the unruly set caught sight of their old chief, a thrill took hold of all, and when his well-known voice exclaimed: "Grenadiers! Recover arms! Shoulder arms! To the right wheel! March!" to their boats and ships they went. All grievances, needless to say, were listened to, and if acceptable at once redressed.

There is no need to call these incidents other than slight mishaps. Something similar, in 1869, happened to the writer when in command of time-expired men homeward bound from India. They were unarmed, for their service was over; consequently their arms and accoutrements had been returned to store for fresh regimental use. Affairs went merrily until reaching Meerut, when three prisoners (infantry) on long sentence for imprisonment at home were added to my numerical strength. Prisoners require a guard, so with them some stands of arms were sent. The hour of day was early morning, when the trusty Sergeant-Major entered my tent, and with a grave salute informed me my men had mutinied and would not go on guard. I hurried through my toilet and attended the parade, where I saw two horse gunners and a Lancer man treating my accoutrements with contempt. The situation appealed to me as ludicrous, for how could one expect these men to handle arms and struggle into belts not made for them at all. However, I took up a bayonet and said: "This bit of steel must be your sabre or your lance, and on sentry you must go." I need not say this mutiny died out in less time than it has taken me to write.

To return, the ruling class, or lawgivers, had but little more to boast about than the class they ruled, for their ways and means of getting into power were quaint—family preserves made an election a farce. The laws they gave made it a crime to be an Irishman, and for a theft of forty shillings it meant a sentence requiring a black cap for the Judge.

"Never go to Law!" is even now the best of all advice, for if you win, it is but scanty justice, and a roaring bill to pay; and take the Borough Bench, where men of no legal training pronounce most solemn judgment from bye-laws now so obsolete that Lycurgus might have framed them! Never go to Law.

———:———

> "I did pluck allegiance from men's hearts,
> Loud shouts and salutations from their mouths,
> Even in the presence of the crowned king."

James, Duke of Monmouth, born 1649, succeeded the Duke of Albemarle (Monk) as Captain-General of the Forces. He was the natural and favourite son of Charles II., and at Bothwell Bridge had scored a victory for his father in 1678, which victory, coupled with the affection of the king, somewhat raised his hopes that his legitimacy might be established, and the succession secured to him. In this he was deceived, for on the somewhat sudden death of Charles in 1684, James II. was declared King.

Monmouth now rebelled, which gave reason for King James to increase his Army. Thus the 12th Regiment, with some few other corps, were amongst the newly raised in 1685. If now we make the 12th a centrepiece for one to pirouette around, it is from no wish to specialise, for are not all battalions, batteries and ships best described in the words *Nulli Secundus?* So apart from the reason given in the Preface, it becomes to the writer an easier task to tread the way the Suffolks take in their pilgrimage of war.

The regiment, with the Duke of Norfolk as their Colonel, gained their first "Cross Swords" under Marlborough (that man of extraordinary genius, the greatest lustre in the reign of Anne, whose splendid qualities were mixed with an alloy of great defects); for in the above miserable rebellion, Marlborough found himself pitted against his former commander—the unhappy Duke. At the head of a small body of troops, he harassed the rebels on their march from Bridgewater, and by some skilful dispositions at Sedgemoor decided the fate of the day, which at a time seemed very doubtful, for the followers of Monmouth saw in him a hero fit to conduct the worship of their cause—the Protestant Faith; and like the Highlanders for Prince Charlie, and the Vendeans of 1793, no devotion could be greater than that of the peasants and yeomen for their young Duke, as he was called. Tales were told of him at nightfall, and his return was expected long after he was dead. After Sedgemoor he was captured, and beheaded on July 15th, 1685.

The cruel severities which followed the suppression of this feeble rebellion greatly lowered the character of the King in the estimation of his subjects, for Marlborough, in urging the claim of a poor suppliant at Whitehall, struck the chimney-piece and said: "This marble is not harder than the King." And we know that the Lord Chancellor of England was then a man of brutal nature, a sensualist, a sot; and in connection with this rebellion his heart became saturated with the very worst of actions, which he fully gave vent to in his bloody Black Assize. Judge Jeffreys was his name; he goes down to history as an ogre of inhuman dye, for even to-day in Devonshire children play a game called "Judge Jeffreys' Ground" (elsewhere styled "Tom Tiddler's Ground"), and look upon the man as a sanguinary and ferocious beast. His time soon came, for on the flight of his royal master, James II., he was discovered in a cellar disguised as a sailor, and promptly sent to the Tower, where he died in 1689.

England now came under Romanist domination; so when the 12th Regiment, with others, were paraded on Hounslow Heath, 1688, and were asked to subscribe to the tenets held by James, they quietly laid down their pikes and toed the line with the feeling of the day.

Seven bishops, sheet anchors of our faith, were imprisoned in the Tower; seven statesmen, of whom Edward Russell* was one, crossed the sea to invite, persuade, and to return, if possible, with

---

* Earl Russell was a nephew of the Duke of Bedford, a worthy gentleman who lived to the age of eighty-seven, dying in 1700. His life, however, had been sadly overshadowed by the execution of his son Lord William (for high treason), who in some way became mixed up in the Rye House Plot to assassinate the two brothers, Charles II. and James II. He accepted the patent of Duke from William of Orange, on condition that in the preamble should appear: "To solace his excellent father for so great a loss, to celebrate the memory of so noble a son, and to excite his worthy grandson to do like his father, this high dignity is entailed upon the Earl and his posterity."

William of Orange (who had married Mary, daughter of James) as our future king and queen.

The Earl of Lichfield had succeeded Norfolk as Colonel, and Lord Hunsden the former; but on this occasion of removing Sovereigns the latter would not subscribe, so the regiment fell to one John Wharton, an honest, bluff, good soul.

William of Orange reached Torbay on November 5th, 1689. On his Staff were three men—Count Schomberg, Earl Russell, and Van Keppel. The former had a reputation of being the greatest living master of the art of war. This, with other eminent virtues and polished manners, made him a great favourite with the English people. The latter was head of a family soon to be known as the Albemarles or Keppels, who have fought by sea and land for England ever since. The eldest son, Lord Bury St. Edmunds, takes his title from the town, now the military homestead of the 12th (Suffolks).

Count Schomberg was at once appointed to command an expedition sent to resist the forces of King James in Ireland; the 12th went with him. He landed at Antrim in August, took Carrickfergus, and marched towards Dublin; but his army was ill-trained and worse appointed, and considerably below the strength of the troops under James. He, therefore, entrenched himself near Dundalk, and resolved to make a stand there until his men were well disciplined, and reinforcements and supplies should arrive. Heavy rains, however, turned his camp into a marsh; disease broke out among his troops, and swept them away in shoals. Amongst them was poor Colonel Wharton, a man beloved by all.

To make matters worse, a formidable conspiracy broke out among the French emigrants in his camp; but the brave veteran, with his indomitable spirit, scouted the word "impossible." He dismissed the conspirators by shooting six and deporting all the rest, and his masterly dispositions soon brought back the health and courage of his men. Although he could only muster 5000, he kept at bay an army four times as numerous, and brought them back to Ulster without the loss of a flag or gun. For his services he was made a Duke and a Knight of the Garter, and received a grant of £100,000.

Sic transit gloria mundi, for one year later, July 1st, 1691, the same general, the same troops, were on the brink of the Boyne. The drums stopped beating, and the drummers, ten abreast, descended into the stream. Next came the main body of the English infantry, struggling through the river up to their armpits in water.

In a few minutes the Boyne was alive with muskets; a wild shout of defiance rose from the opposite shore. For a moment the result seemed doubtful; but the Protestants pressed forward, and in another moment the whole Franco-Irish line gave way. In the melée Schomberg, leading on his troops, was seen to fall. When his friends raised him, he was already dead; two sabre wounds were on his head, and a bullet in his neck. Meinhart Schomberg, his son, in command of the right wing, soon heard of his father's death, which left him in command. This son lived to become Captain-General of our Army.

King William, who was present at the Boyne, soon entered Dublin with his troops. There he took the Mint and the specie called the "gun money;" for James, whilst there in 1689, finding himself short of cash, established mints and debased the coinage by using copper from the guns and stamping same as half-crowns, shillings and sixpences; the same at Limerick; so when the latter city fell, roughly speaking, one hundred tons of scrap metal represented in nominal value over a million and a quarter sterling. When William made his proclamation that all nominal value was to cease, and the coppers to be taken at their intrinsic worth, it seemed rough on the poor Irish who had adhered to James throughout. James, not wishing for the fate of Charles I., fled to France.

Marlborough had no liking for King William; neither had Earl Russell, his other counsellor. His attachment seemed to be for the Princess Anne. He had refused, when asked by William (on Schomberg's death) to come over and take command, not wishing to be in Ireland whilst her father was there in person; but on the latter's escape to France, he joined King William for a while, and put the troubled country in order once again. Then both left for Holland, where some English forces were still employed against the French; the 12th Regiment also went with them, and became engaged in one or two minor operations.

Marlborough and the regiment were back again in 1692, the latter little thinking that the former had in view a project of handing over the English forces in Flanders to the French camp, if James should so wish. When this was discovered, Queen Mary rushed at her sister, Princess Anne, and demanded full explanation. When this painful interview was over, Marlborough* was recalled, dismissed from all employment, and forbidden to appear at Court again.† This incident over, William continued his operations against the French, and the 12th Regiment were again despatched to Flanders with other reinforcements; for William had met with heavy reverses both at Steinkerk and in the great and sanguinary battle of Landen on July 19th, 1693, where the 14th Regiment alone lost five officers killed and five wounded, and 40 per cent. of men.

Luxemburg was the general pitted against William. In early days he had been aide-de-camp to the great Prince Condé, and had earned the reputation of being the greatest general of his age. He died in 1695. His mantle fell on General Villeroy. The 12th and some other regiments fell foul of him at Dixmude, for here our force surrendered. The 12th Regiment remonstrated at this, and broke their arms and destroyed their colours, sooner than surrender them to the French.

General Ellemberg, who sanctioned this surrender, was executed at Ghent; whilst Colonel Brewer, of the 12th, and other command-

---

* Perhaps, after all, Marlborough was not so bad as he was painted, for at the same time Earl Russell had treasonably offered to carry over the fleet. Strong colour is given to this intrigue of Marlborough, for in the same year Louis XIV. prepared a numerous fleet for the invasion of England. Our navy (ready, aye! ready) under Earl Russell and Sir George Rooke, in concert with the Dutch, completely defeated them off La Hogue, destroying seventeen vessels. At this time Queen Mary died, December 28th, 1694, of smallpox. Her husband was with her at the time. A portion of the park and palace at Greenwich was given by the King to become a hospital for disabled seamen. [† See note, page 13].

ing officers who remonstrated, were fully acquitted. This regrettable incident in no way affected the siege of Namur, although Villeroy had boasted of its certain fall. Namur was perhaps the strongest fortress in Europe, being fortified by Vauban, and defended by Marshal Boufflers. Its surrender to William on August 26th, 1695, was hailed as a great triumph.

Villeroy, in his rage at this reverse, barbarously bombarded the city of Brussels. The 12th, Royals, and 15th Foot were soon on his trail, but he marched and countermarched, and would not fight. The Peace of Ryswick, 1697, terminated the war, and the 12th Regiment returned to England.

One provision of this peace was to make Louis abandon the Stuart cause and recognise William as King, with a view to closing the conspiracy continued for so long to turn England into a Roman Catholic country and a dependency of France. Events moved quickly; James II. died at St. Germains in September, 1701, leaving one son by his second marriage, to be called the Old Pretender, his son being known as the Young Pretender.

Fresh foreign complications soon arose, for Louis had seized the Netherlands, and put his grandson on the throne of Spain; and at St. Germains, whilst James II. was dying, he promised to acknowledge his son as King of our three kingdoms. This meant another war.

William sent Lord Albemarle to Holland to make plans for the ensuing campaign. He soon, however, received the news that his royal master was *in extremis*, caused by a fall from his horse. Albemarle came back to England in time to hear from the dying king that his plans were much approved, and that he had asked Anne to make Marlborough her adviser. He gave the keys of his closet and private drawers to Albemarle; then turning on his side, just said: "Je tire vers ma fin." He died on March 8th, 1702.

"War is honourable,
Bring then these blessings to a strict account.'

If from grim-visaged war a blessing can be wrung, the following affords pleasure to relate. Steele, the great essayist, and friend of

Addison, was also a keen politician; so much so, that he determined to go into the Army and fight for King William. Unable to procure a commission, he entered as a private, soon obtained an ensigncy, and became secretary to his Colonel, who shortly after procured him a Company in Lord Lucas' Fusiliers. On his return home, after the fall of Namur, he wrote his first essay, "The Christian Hero," he having witnessed the following scene:—

At the siege of Namur by the Allies, there were in the ranks of the Company commanded by Captain Vincent in Colonel Frederick Hamilton's regiment, one Unnion, a corporal, and a private sentinel. There happened between these two men a dispute about an affair of love, which upon some aggravation grew to an irreconcilable hatred. Unnion, being the officer of Valentine, took all the opportunities, even to strike his rival, and profess the spite and revenge which moved him to it. The sentinel bore it without resistance, but frequently said he would die to be revenged of that tyrant. They had spent whole months in this manner—the one injuring, the other complaining—when, in the midst of their rage towards each other, they were commanded upon the attack of the castle, where the corporal received a shot in the thigh, and fell. The French pressing on, and he expecting to be trampled to death, called out to his enemy: "Ah, Valentine! can you leave me here?" Valentine immediately ran back, and in the midst of a thick fire of the French, took the corporal on his back and brought him through all that danger as far as the Abbey of Salsine, when a cannon ball took off his head, and his body fell under his enemy, whom he was carrying off. Unnion immediately forgot his wound, rose up tearing his hair, and threw himself upon the bleeding carcass, crying: "Ah, Valentine! was it for me who have so barbarously used thee?" He was not by any means to be forced from the body, but was removed with it bleeding in his arms, and attended with tears by all their comrades, who knew their enmity. When he was brought to a tent his wounds were dressed by force; but the next day, still calling upon Valentine, and lamenting his cruelties to him, he died in the pangs of his remorse.

The same story greatly touched the heart of Hugh Miller, for in the story of his life he comments that often as men who have loved do often learn to hate the object of their affections, and men who have hated, sometimes learn to love. But the portion of the anecdote specially worthy of remark appears to be that which, dwelling on the over-mastering remorse and sorrow of the rescued soldier, shows how effectually his poor dead comrade had, by dying for him "while he was yet his enemy," heaped coals of fire upon his head; it seemed to him the scheme on which our redemption has been framed, "in that while we were yet sinners, He died for us."

Self-abnegation is so often witnessed on every field of battle. We anticipate our story by giving for example: Baird, when he denied himself for Wellesley, and later, when Outram did the same for Havelock.

---

* [Page 2] Ligonier was of noble French descent; he preferred being a Protestant in England to a Catholic in France; was a brilliant cavalry general under Marlborough. Queen Anne made him Field Marshal, and also made him Earl. Ligonier had for his secretary a Mr. Richard Cox, who appointed him agent to the 1st Regiment of Foot Guards. Mr. Cox's office was in Albemarle Street, where he carried on his business till 1765, in which year he removed to Craig's Court. We all have been that quivering subaltern who, with his account overdrawn, still had the courage to enter that office for another £5 note or so (time, the Ascot Week in town). Cox has always been so good!

† [Page 10] The Duchess Sarah did not fare much better, for Anne (Mrs. Freeman) wrote to Sarah (Mrs. Morley) thus: "Be assured, my M——, that although I write to you with almost the same madness of affection, I will ever imitate her example for all its royalty, and exchange you for a mushroom of your own raising" (Mrs. Masham).

CHAPTER II.

"As I think most things are governed by destiny, having done all things, we should submit with patience."—MARLBOROUGH.

PRINCE GEORGE, the last survivor of Anne's thirteen children, died shortly before King William; hence the new Queen deserved great sympathy, for these sorrows must have negatived her character, and robbed her of all cheer. Luckily she had Marlborough to give lustre to her reign by his victories on land, and Sir George Rooke and Sir Cloudesley Shovel by victories at sea; for in October, 1702, Rooke, by a swoop, captured seventeen vessels and about four million dollars off Vigo (which town was afterwards stormed and taken by the Duke of Ormond). The two admirals then joined forces, and with their combined fleets attacked and took Gibraltar. The dollars were recoined and stamped with "Vigo" under the bust of Anne.

Thus commenced the war known in history as that of the Spanish Succession; now we will call upon old Pierre, the drummer, to enliven us by thumping on the skin, singing as we march:—

"And at Blenheim, in spite of our valiance,
The victory lay with Malbrook;
The news it was brought to King Louis;
Corbleu! how his Majesty swore,
When he heard they had taken my grandsire,
And twelve thousand gentlemen more!

"At Namur, Ramillies, and Malplaquet
Were we posted, on plain or in trench;
Malbrook only need to attack it,
And away from him scampered we French.

Cheer up! 'tis no use to be glum, boys,
'Tis written, since fighting begun,
That sometimes we fight and we conquer,
And sometimes we fight and we run.

"To fight and to run was our fate;
Our fortune and fame had departed,
And so perished Louis the Great,
Old, lonely, and half broken-hearted.
His coffin they pelted with mud,
His body they tried to lay hands on;
And so, having buried King Louis,*
They loyally served his great grandson."

In a short seven years Marlborough had gained for England those splendid victories, and at an age approaching sixty. After Malplaquet, August 31st, 1709—the last of his great victories—charges of peculation were cast upon him; and in 1712, imprisonment awaited him; so he took the hint and quitted England—to return, however, on the day Queen Anne ceased to reign, and George became King.

His death was sad. Mrs. Godfrey had been for some time at variance with the Duchess Sarah, for reason of her rude exposure and neglect of her brother—for now that he was quite in his second childhood, and completely crippled with paralysis, sans teeth, sans everything, she drove the great man about in her coach wherever she went, and left him exposed at times to the gaze and jokes of the populace. It was when his death came (1722) Mrs. Godfrey besought of her husband and niece to visit the lying-in-state of England's greatest soldier; they returned with the report that within the death chamber there was not even a single attendant or a glimmering candle, and all other surroundings were equally out of character. This so depressed Mrs. Godfrey that next morning (Sunday) when dressing for Chapel, she suddenly fell into a faint or trance. The doctors reported her dead, and it was with the greatest difficulty Colonel Godfrey, her husband, could persuade them to

* Louis XIV. died on September 1st, 1715.

delay the funeral. Mrs. Godfrey lay thus apparently lifeless from Sunday until the same hour of the following Sunday, when she awoke, and upbraided her servants for not having dressed her for Chapel. Mrs. Godfrey was rightly never told of this sad and thrilling narrative, in which she so nearly escaped being buried alive.

Marlborough received burial in Westminster Abbey, but not at the expense of the nation. He now rests at Blenheim.

The clash of arms died out when the Peace of Utrecht was signed in 1713; by this treaty Gibraltar was confirmed to the English in the most thorough and complete way, viz.: "The Catholic King (*i.e.*, of Spain) doth hereby, for himself, his heirs and successors, yield to the Crown of Great Britain the full and entire property of the town and castle of Gibraltar, together with the port, fortifications, and forts thereunto belonging; and he gives up the said property to be held and enjoyed absolutely, with all manner of right, for ever, without any exception or impediment whatsoever." The integrity of which was so gloriously maintained by Elliot and his garrison seventy years later, when France and Spain did their best to seize Gibraltar.

Queen Anne died in 1714. The passing of the Crown to Hanover was only carried by a majority of one—Joseph Paice, M.P. for Lyme, Dorset. The choice of who was to be successor was uppermost in all minds.

Steele, for writing the "Crisis," which told the nation that the Protestant Succession in the House of Hanover was in danger from the then Ministry, was expelled from the House. Dr. Sacheverell, for preaching on the same subject, was suspended from preaching for three years. And when this matter was settled for George to be King, off set a member of the Stock Exchange—John Carter, by name—on a walk to Portsmouth, to communicate the same. The Governor of the garrison was one General Gibson, who, not being of Mr. Carter's way of thinking, promptly put him into prison and kept him there for three days—for it took that time for official news to reach Portsmouth from London.

Marlborough had been well served throughout his campaigns by his aide-de-camp, John Dalrymple, Earl of Stair; but when the career of Marlborough was stopped in 1711, Lord Stair retired from the Army and spent the next twenty-two years of his life, with occasional breaks, in agricultural pursuits. Although a most distinguished soldier, he was somewhat a recluse in private life, for in early youth he had the misfortune to kill his elder brother by the accidental discharge of a pistol. It was, however, on the threat of a French invasion that the Earl again tendered his services, and was appointed Commander-in-Chief, 1742. He was the directing General in the battle of Dettingen, June 26th, 1743.

In 1742, over a dispute between Maria Theresa and the Elector of Bavaria for the throne of Spain, in which England took the part of the former and France the latter, the dogs of war again commenced to howl. England sent 16,000 men and the same number of Hanoverians, under Stair, later to be joined by the King and his son Cumberland. They encamped on the river Maine, near the village of Dettingen, whilst the French, to the number of 16,000, watched them from the hither bank.

George II. inspected the 12th Regiment on Blackheath on April 27th, 1742, before embarking at Deptford for the war. A young officer, named James Wolfe, carried the King's colours; a year later he was Adjutant, and his letter to his father well describes the battle :—

"Hôchst,

"July 4th, N.S., 1743.

"Dear Sir,—This is the first time I have been able, or have had the least time to write, otherwise I should have done it when my brother did. The fatigue I had the day we fought, and the day after, made me very much out of order, and I was obliged to keep my tent for two days. Bleeding was of great service to me, and I am now as well as ever.

"The army was drawn out this day se'n-night between a wood and the river Maine, near a little village called Dettingen, in five lines—two of Foot and three of Horse. The cannon on both sides began to play about till nine o'clock in the morning, and we were exposed to the fire of theirs (said to be above fifty pieces) for near three hours, a great part of which flanked us terribly from the other side of the water. The French were all the while drawn up in sight of us on this side. About twelve o'clock we marched towards them; they advanced likewise, and, as near as I can guess, the fight began about one. The Gens d'Armes, or Mosquetaires Gris, attacked the first line (commanded by Lord Albemarle), composed of nine regiments of English Foot, and four or five of Austrians, and some Hanoverians. They broke through the Scotch Fusiliers, who they began to attack upon; but before they got to the second line, out of two hundred there were not forty living, so they wheeled and came between the first and second lines (except an officer with a standard and four or five men, who broke through the second line and were taken by some of Hawley's Regiment of Dragoons), and about twenty of them escaped to their army, riding through an interval that was made for our Horse to advance. These unhappy men were of the first families in France. Nothing, I believe, could be more rash than their undertaking.

"The second attack was made on the left by their Horse against ours, which advanced for the first time. Neither side did much, for they both retreated; and our Horse had like to have broke our first line in the confusion. The Horse fired their pistols, which, if they had let alone, and attacked the French with their swords, being so much stronger and heavier, they would certainly have beat them. Their excuse for retreating: they could not make their horses stand the fire.

"The third and last attack was made by the Foot on both sides. We advanced towards one another, our men in high spirits, and very impatient for fighting, being elated with beating the French Horse, part of which advanced towards us, while the rest attacked our Horse, but were soon driven back by the great fire we gave them. The Major and I (for we had neither Colonel nor Lieutenant-Colonel), before they came near, were employed in begging and ordering the men not to fire at too great a distance, but to keep it till the enemy should come near us; but to little purpose. The whole fired when they thought they could reach them, which had like to have ruined us. We did very little execution with it. So soon as the French saw we presented they all fell down, and when we had fired they got up and marched close to us in tolerable good order, and gave us a brisk fire, which put us into some disorder and made us give way a little, particularly ours and two or three more regiments, who were in the hottest of it. However, we soon rallied again, and attacked them with great fury, which gained us a complete victory, and forced the enemy to retire in great haste. 'Twas luck that we did give way a little, for our men were loading all the while, and it gave room for an Austrian regiment to move into an interval, rather too little before, who charged the enemy with great bravery and resolution. So soon as the French retreated, the line halted, and we got the sad news of the death of as good and as brave a man as any amongst us—General Clayton, who was killed by a musket ball in the last attack. His death gave us all sorrow (so great was the opinion we had of him), and was the hindrance of anything further being done that day. He had, 'tis said, orders* for pursuing the enemy, and if we had followed them, as was expected, it is the opinion of most people that of the 27,000 men they brought over the Maine, they would not have repassed with half that number. When they retreated, several pieces of our artillery played upon them, and made terrible havoc. At last we followed them, but too late; they had almost all passed the river. One of the bridges broke, and in the hurry abundance were drowned. A great number of their officers and men were taken prisoners. Their loss is computed to be between six and seven thousand men, and ours three thousand.

---

* In all probability the French would have sustained a total overthrow, had the proposal of Lord Stair for cavalry to pursue been carried out. His advice was not heeded; the King seemed ruled by his Hanoverian officers, which so disgusted Lord Stair, that "finding himself reduced to the condition of a statue with a truncheon in his hand," he took the first opportunity to resign. He died in 1747 aged seventy-four years.

"His Majesty* was in the midst of the fight; and the Duke behaved as bravely as a man could do. He had a musket shot through the calf of his leg. I had several times the honour of speaking with him just as the battle began, and was often afraid of his being dashed to pieces by the cannon balls. He gave his orders with a great deal of calmness, and seemed quite unconcerned. The soldiers were in high delight to have him so near them. Captain Rainsford behaved with the greatest conduct and bravery in the world. I sometimes thought I had lost poor Neb when I saw arms, legs and heads beat off close by him. He is called 'the Old Soldier,' and very deservedly. A horse I rid of the Colonel's at the first attack was shot in one of his hind legs, and threw me; so I was obliged to do the duty of an Adjutant all that and the next day on foot, in a pair of heavy boots.† I lost with the horse furniture and pistol, which cost me ten ducats; but three days after the battle got the horse again, with the ball in him, and he is now almost well again, but without furniture and pistols.

"A brigade of English and another of Hanoverians are in garrison in this town, which we are fortifying daily. We are detached from the grand army, which is encamped between Frankfort and Hanau, about twelve miles off.

"They talk of a second battle soon. Count Khevenhuller and Marshal Broglie are expected to join the two armies in a few days. We are very well situated at present, and in a plentiful country. Had we stayed a few days longer at Aschaffenburg we had been all starved, for the French would have cut off our communication with Frankfort.

---

* King George II., who died in 1760.

† To be marching about in a pair of heavy boots on the day of battle greatly enhances the value of a horse—

"He's a tried and trusted soldier,
So is my horse."—SHAKESPEARE.

In later years (1864) the same old corps—the 12th—possessed a tried and trusted soldier named Thomas; he was handed over to the regiment in payment of a debt of cards, and for fifteen years and more (their term of foreign service) did that good old horse do duty with the spurs.

"Poor Captain Merrydon is killed. Pray, mine and my brother's duty to my mother. We hope you are both perfectly well.

"I am, dear Sir,

"Your dutiful and affectionate son,

"J. WOLFE."

Wolfe was promoted Captain for this battle, and transferred to the 4th Regiment.

Small as was the victory of Dettingen, it produced amazing results. The French evacuated Germany, and the English and Austrian armies appeared on the Rhine; and the excitement at home in consequence was excessive. In one of Walpole's letters we read: "I expect to be drunk with hogsheads of Maine water, and with odes to His Majesty and the Duke, and *Te Deums*. We are all mad—drums, trumpets, bumpers, bonfires. The mob are wild and cry: 'Long live King George, the Duke of Cumberland, and Lord Stair, and Lord Carteret, and General Clayton that's dead.'"

Soon after Marshal Saxe* entered Flanders, laid siege to Tournay, and defeated the English at Fontenoy, to the tune of old Pierre the Drummer:—

"God save the beloved† King Louis
(For so he was nicknamed by some);
And now came my father to do his
King's orders and beat on the drum;
My grandsire was dead, but his bones
Must have shaken, I'm certain, for joy,
To see Daddy drumming the English
From the meadows of famed Fontenoy."

---

* Marshal Saxe, born 1696, died 1750, was one of the greatest captains of his day; a martyr to gout (at Fontenoy he did all his work *dans une voiture d'osier*)—he was intensely human, and always in touch with his men. He was a profound military writer; "Ses Réveries," "Adrienne le Couvreur"—the latter work being the life of one, a genius as an actress, and with a personality equally bewitching. She died, it is said, by the poison of a rival, in the arms of Voltaire her trust in and admiration of Marshal Saxe was a feature in her life.

† *Bien aimé.*

A letter from Captain Field, who took command of the 12th Regiment (then in Marshal Wade's Division) when Colonel Duroure was mortally wounded, is here given :—

"We attacked a numerous army, entrenched with a multitude of batteries, well placed both in front and flank. The action began at five o'clock in the morning and lasted till two o'clock in the afternoon. There has been a great deal of slaughter, particularly amongst the infantry—officers more in proportion than soldiers. The enemy's army was supposed to be 70,000 and ours about 50,000. The soldiers behaved with the utmost bravery and courage during the whole affair, but were rather rash and impetuous.

"Notwithstanding the bravest attempts made to conquer, it was not possible for us to surmount the difficulties we met with. The 12th suffered very much—eighteen officers and 300 men killed and wounded; amongst the latter is Major Rainsford. The number of subalterns killed was astonishing, so much so that the French King, looking upon the English that were killed, said: 'Ma foy, ces gens meritoient de vivre,' and Marshal Saxe said: 'Cette poignee de gens m'a fait plus de peine que tout le reste."

Does not history repeat itself, when we read of the gallant fight of the same regiment at Colesberg in South Africa? (Jan. 6th, 1900.)

> "How can I see the brave and young
> Fall in the cloud of war, and fall unsung."—ADDISON.

And in more glowing terms does the Irish poet Davis sing to the heroism of the day, and show how the field was fought and won :—

> "'Push on my household cavalry,' King Louis madly cried:
> To death they rush, but rude their shock—not unavenged they died.
> On through the camp the column trod—King Louis turns his rein:
> 'Not yet, my liege,' Saxe interposed, 'the Irish troops remain;'
> And Fontenoy, famed Fontenoy, had been a Waterloo,
> Were not these exiles ready then, fresh, vehement and true."

"At a crisis in the battle, a colonel took occasion to remind his men that they were not to swim the Scheldt, as they did the Mayn at Dettingen, and then made them all huzzah! An officer of the

French Guard seeing this, also came from the ranks, addressed the men, and tried to make them huzzah! but not with the same result; from this has grown the fable, *Messieurs les Gardes Francaises, tirez*, when the answer came: *Messieurs les Gardes Anglaises, tirez nous les premiers, nous riposterons.*"*

The battle of Fontenoy was fought on April 30th, 1745, when English, French, Scotch and Irish became intertwined like a mingled skein, to unravel which sends into history some grand events, to be enacted by the self-same men in other ways and in other climes.

It was at Fontenoy that Marshal Saxe stated to his Sovereign that England could ne'er be beaten except in London; and events soon shaped themselves accordingly, for Charles Edward, encouraged by the result of Fontenoy, was prompted by the French King to try his luck in Scotland. Marshal Saxe went with him in command, but a violent storm at sea caused them to return. This was but a temporary check, for the Pretender again set sail and landed at Eriska on the 23rd July, 1745, with a staff of seven and the troops that Louis lent him. Lally, who had received General's rank at Fontenoy, was his military adviser, and Sheridan (ancestor of our Brinsley) his aide-de-camp.

Young Howe (Earl Howe of Nile) here first sets foot in the vestibule to the Temple of Fame, for the "Baltimore," in company with another armed vessel, fell in with two French frigates of thirty guns, crowded with troops and ammunition for the Pretender. Howe immediately ran the "Baltimore" between them, when a desperate and bloody action commenced. Howe was soon severely wounded, and carried off the deck to all appearance dead. Scarcely had his wound been dressed than he flew again to his post, and was received with shouts of joy by his sailors. The action was continued with redoubled spirit, when the French ships sheered off. Howe was made post-captain, April 10th, 1746.

In this same year, with Howe, we must welcome into the Temple of Fame two grand Centurions—Lord Anson and his famous

* Letter from Lord George Hay to his brother.

ship, the "Centurion," of fifty guns. The latter had carried Anson's flag for seven years and more; it was one of a squadron of six, sent in 1740 to harass the Spanish settlements in the Pacific, and to destroy their trade. The voyage took Anson round Cape Horn, when tempestuous weather scattered some and lost the rest of his squadron; so he alone, with his good "Centurion," took many a rich merchantman and one or two towns in Peru, and on June 30th, 1743, captured the Manilla galleon, bound for Spain with immense treasure (over a million and a half of dollars). Hence all English silver coins from 1745 to 1746 have beneath the bust of George II. the word "Lima," the chief town of the Pacific coast. Later, in 1747, Howe captured, off Cape Finisterre, a French fleet of six men-of-war—"L'Invincible" et "La Gloire." The captain of the former, on handing his sword to Anson, politely and pithily remarked: "Monsieur vous avez vainçu l'Invincible and La Gloire vous suit."

Anson was always beloved by his men. The poet Cowper, in his "Castaway," speaks of his grief over a lost sailor:—

"No poet wept him; but the page
Of narrative sincere,
That tells his name, his rank, his age,
Is wet with Anson's tear;
And tears by bards or heroes shed,
Alike immortalize the dead."

England was not prepared for this French invasion. The King was absent in Flanders, and the Ministry refused to act; thus the Pretender soon gained successes, first at Preston Pans, where the Royalists under John Cope soon took to flight.

"Yet in that craven, dread-struck host,
One val'rous heart beat keen and high
In that dark hour of shameful flight,
One staid behind to die.
Deep gashed by many a felon blow,
He sleeps where fought the vanquished van:
Of silvered locks and furrow'd brow,
A venerable man."*

* Colonel Gardiner, slain at Preston Pans, 1745.

Gardiner had played his part in the venture of the first Pretender in 1715, for he was on "special mission" in Paris with Lord Stair, when he left with important despatches bearing on the discovery of the design which the French King was then forming for the invasion of Great Britain. On reaching England, he ventured to predict that the French King's health was such that he would not live six weeks. Spies at St. James's reported this to the Court at Versailles, which made the return of Gardiner a little difficult, for he might be offered a lodging in the Bastille. However, Gardiner faced the situation, and returned to his master, Lord Stair.

At the first banquet after this incident, when the King of France and Lord Stair and Gardiner met, the monarch laboured hard to be extra vivacious, and in an erect posture ate much more heartily than was good for him. He repeated many times to the Duke of Bourbon, so that his envoys might hear: "Il me semble que je mange pas mal pour un homme qui devoit mourir si tot." The dinner disagreed, and Louis XIV. was dead within the week; when humorous people said that Gardiner had killed the King.*

Prince Charles loitered for six weeks at Edinburgh before proceeding into England, where at Newcastle was Marshal Wade,† lately back with the Army from Flanders. Prince Charles decided to give him the slip, and to march straight to London. He reached Carlisle, which capitulated on November 17th. At Manchester, a gentleman of family and fortune, Colonel Townley by name, turned out and joined him at the head of 1000 volunteers.

---

* Louis XIV. was born in 1638, and succeeded his father, Louis XIII., in 1643. He was surnamed Dieudonné, for he was born after twenty-three years of a sterile marriage.

†Marshal Wade had been Commander-in-Chief of Scotland, and was renowned as a road maker. He employed 600 soldiers daily at this task.

"Had you seen these roads before they were made,
You would lift up your hands and bless General Wade,"

was a refrain at the time. He died in 1781, aged eighty. His aide-de-camp at Newcastle was one Captain Stringer Lawrence, 14th Regiment, of later Indian renown.

The next halt was at Derby, only 120 miles from London, which latter place went into panic, for the city could only muster one Regiment of Guards (which Hogarth has shown us in their march to Finchley), three Regiments of Line, and one squadron of Cavalry, with a few Militia.

So far, no further—a halt was called, for discretion now said "Retreat." Many French and Irish, straight from Fontenoy, awaited him; this enabled Charles to lay siege to Stirling Castle and to defeat General Hawley at Falkirk on January 17th, 1746.

The fruits of this victory were very transient, for the victors required rest and food, which were not forthcoming. A cry had risen: "It is better to be killed in England than to be starved in Scotland." Thus, starving and dispirited, Prince Charles marshalled his Highlanders for the last time on Culloden Moor, and all was over in a short half hour.

Lord Albemarle, with his son, Lord Bury, fought together as General and Aide-de-Camp at Fontenoy and Culloden. At the latter place, before the battle, Lord Bury, in showy uniform, was stealthily waylaid by a sturdy Highlander in mistake for the Commander-in-Chief, Cumberland. The shot took no effect. A Royalist soldier, who witnessed this, soon turned the tables, for his musket shot took full effect on this sad Highlander. It was when the two armies confronted each other that Lord Bury went forward to verify if there was a battery or not. The solution came, for it shortly opened fire, and thus the battle began. He next was sent to carry the news of victory to the King, who immediately ordered him a thousand pounds.

Cnmberland soon went to London, and Lord Albemarle succeeded him as Commander-in-Chief in Scotland.

A Highland lady, Flora Macdonald, known for her devotedness to the cause, secretly conveyed Prince Charlie, in the guise of an Irish female servant, to her brother's house in Skye, until opportunity came for his return to France, on September 20th, 1746. Flora was afterwards arrested, and sent to London for trial, but was pardoned.

Thus the son and the grandson of James II. had played their stake and lost—the former in 1716, at Sherifmuir; the latter at Culloden. The closing days of Charles Edward were sad; intemperance hastened his death, which came in 1788. His younger brother, Cardinal York, died in 1807, having previously restored to George IV. the Crown Jewels and other relics and documents, which his grandfather, James II., had carried away in 1688.

John Dorman (a private in the 12th) mentions in his diary that he was one of the escort party, under Captain Eyre, to accompany the Highland chieftains Tullibardine, Balmerino, Kilmarnock, Cromarty and Lovat to London. These unfortunate peers, on pleading guilty, were beheaded on Tower Hill (Tullibardine died before his trial). The crafty Lovat, who had not appeared in arms, but was really at the bottom of the rising, was the last to suffer; as he confronted the block, sentenced to die in that brutal form disgusting and degrading to England, he repeated in a calm defiant tone the line from Horace: "Dulce et decorum est, pro patria mori."

After Culloden, the Corporation of London wished to offer the Duke of Cumberland some recognition of his services on that eventful day, and proposed the Freedom of a City Company, when up jumped an alderman and shouted: "Then let it be of the 'Butchers,' for after his success mercy was nowhere and revenge was everywhere."

At Carlisle the unfortunate Manchester Volunteers under Colonel Townley were brought to surrender:—

"My Lord,—The judges came back here last Monday; the tryals are begun and will be very tedious. The Scotch lawyers who are come here as Rebell Council are playing all the game already, even so far as to try to suborn the King's evidence. We have erected a fine new gallows, which will hold fifteen at a time; God send it may be made a proper use of."

Captain J. Dawson and eight officers of the Manchester Regiment were hung, drawn and quartered on Kennington Common

(1746), and with the following sad lament, we must draw a veil over the horrors of '45 :—

" he follow'd him, prepar'd to view
The terrible behests of law ;
And the last scene of Jemmy's woes
With calm and steadfast eye she saw.

Distorted was that blooming face
Which she had fondly lov'd so long ;
And stifled was that tuneful breath
Which in her praise had sweetly sung.

And severed was that beauteous neck
Round which her arms had fondly closed,
And mangled was the beauteous breast
On which her love-sick head repos'd.

And ravished was that constant heart,
She did to every heart prefer ;
For though it could its King forget,
'Twas true and loyal still to her.

Amid those unrelenting flames
She bore this constant heart to see ;
But when 'twas smouldered into dust—
'Now, now,' she cried, 'I follow thee.' "

The Duke of Cumberland, with the prefix "Butcher," returned to Flanders in 1747, to renew the operations so briefly stayed after Fontenoy. He met his old antagonist, Marshal Saxe, again, at Lanfeld, where he was thoroughly out-generalled and defeated. His recall again placed Albemarle in chief command, until peace was signed at Aix-la-Chapelle.

Three years later, in 1757, Lord Cumberland was again entrusted with an expedition to the Netherlands, which came to a disastrous ending on September 10th at Closterseven.

Events at sea had not gone exactly well, for the French had placed 19,000 men in Minorca, with a powerful fleet at their back, and were holding up an English garrison under Blakeney at Port St. Philip.

Admiral Byng, with ten ill-equipped ships, was sent to open up communication ; he did his best, but failed. He next engaged the

French fleet, and after a drawn battle, Fort St. Philip had nought to do but capitulate.*

The Ministry determined to sacrifice Byng, in hopes of diverting public attention from their own imbecility and neglect.

He was tried by court martial and sentenced to be shot, although unanimously recommended to mercy on the ground that he had failed solely from an error of judgment.

Strong representations were made to stay this execution, for even legally the evidence was insufficient for the verdict. Voltaire, the French Marshal Richelieu, Pitt: all intervened, but "No!" —Byng was shot at Portsmouth on March 14th, 1757.

"He fell a martyr to political persecution at a time when bravery and loyalty were insufficient securities for the life and honour of a naval officer;" or, as Voltaire puts it:

"Talking thus, we approached Portsmouth; a multitude of people covered the shore, looking attentively at a stout gentleman, who was on his knees with his eyes bandaged, on the quarter-deck of one of the vessels of the fleet.

"Four sailors placed in front of him, placed three balls in his head in the most peaceable manner, and all the assembly then dispersed quite satisfied.

"'What is all this?' quoth Candide, 'and what devil reigns here?' he asked; 'Who was the stout gentleman who came to die in this ceremonious manner?'

"'It is an admiral,' they answered.

"'Why kill the admiral?'

"'It is because he has not killed enough of other people; he had to give battle to the French admiral, and they find that he did not go near enough to him.'

"'But,' said Candide, 'the French admiral was as far from him as he was from the French admiral.'

* It is well known that the Horse Guards, or War Office of those days, found it impossible to make officers remain with their regiments in the Mediterranean garrisons; and lack of officers was one of the principal causes for the loss of Minorca.

"'That is very true,' replied they; 'but in this country it is useful to kill an admiral now and then, just to encourage the rest' —'pour encourager les autres.'"

"You have taught me," said George II. when Pitt sought to save Byng by appealing to Parliament, "to look for the voice of my people in other places than within the House of Commons."

CHAPTER III.

THE SEVEN YEARS' WAR.

THE Peace of Aix-la-Chapelle, 1748, conceded Silesia to Prussia, much to the annoyance of Maria Theresa, Queen of Austria—and in 1756 she assembled large forces in Bohemia and Moravia; for England being at war with France in North America, meant that if she could find an ally she would fight France also on the Continent. This ally she found in Frederick, making the sides England and Prussia against France, Austria and Russia.

France passed two armies across the Rhine in order to attack Hesse and Hanover. Marshal d'Estrees commanded the latter, when he easily defeated Cumberland at Hastenbeck.

Cumberland followed up his defeat by signing away his troops, and surrendering Hanover, Hesse, Brunswick, and the whole country between the Weser and the Rhine, to the French.

When the Duke of Cumberland had to meet the King his august father, after his disastrous Convention of Closterseven, Sept. 10th, 1757—by which 38,000 Hanoverians laid down their arms—not a word was said direct from father to son, but to others the King announced him: "Here is my son, who has ruined me and disgraced himself"—a censure, perhaps, too severe, for the King had beseeched his son to extricate him and his country from the misery and difficulties they were now groaning under—"coûte qui coûte." The remedy he adopted certainly seemed disastrous, and the cry arose: "We are no longer a nation." "Nous sommes trahis" rent the air, when up rose William Pitt; he stayed the cry, he took the helm, and put the good old ship again on a right and steady course.

Pitt recognised in Frederick the Great the genius he wanted, and Frederick saw in Pitt the man—or, as he tersely put it: "England has been a long time in labour, and has at last brought forth a man." So whilst England gave to Frederick men and money for his Continental task in fighting Austria and France, it also gave to Pitt the opportunity he sought of driving France from out of North America, and engaging her at sea whenever favoured by wind and weather.

On this arrangement commenced the Seven Years' War. Frederick soon announced himself victorious over the French at Rosbach and the Austrians at Leuthen in 1757; in fact, honours came so quickly that Walpole said, "We are forced to ask each morning what victory there is, for fear of missing one." Next came Quiberon and Quebec, with Hawke who conquered and Wolfe who bled, and then the spoils from Minden, etc., whilst Clive was creating for us an Empire in the East.

The cradling of our East and West, *i.e.*, Hindostan and Canada, took place about 300 years ago, when James I. was king. If James was weak, his ambassadors were strong, for no monarch was ever served by abler men. Take one—"Honest Tom" by nickname, Sir Thomas Roe by name, who in 1615 travelled as far as the Court of the Emperor Jehanghir at Agra, to further the East India Company, whose charter as a trading company to India had been signed in 1600.

Likewise in 1608, at Jamestown (so called after his royal master), a British Colony first struck root; alongside of whom was Champlain, in his "abitation de Quebec." "If a system existed under which people could not live at home, destroy it, root and branch," said Cromwell—for 20,000 in Charles' reign had made their exodus from England to join the Pilgrim Band.

The rival nations in both places at times agreed to differ, but at intervals broke out in unexampled temper, to wit, this year of 1757, when the same cry that "England was no longer a nation" reached far-off Hindostan. It was then that Clive, assisted by Stringer

Lawrence (14th Regiment), saw well that the French colours flying at Pondicherry, Trichinopoly, and Fort St. George, Madras, under men like Bourdonnais, Dupleix and Lally, meant India for France.

"This must not be," said Clive, who with 200 British soldiers and 300 Sepoys, raided and captured Arcot, and defended himself there for over fifty-one days, when the besiegers raised the siege and fled.

Help from home shortly reached him, and amongst the reinforcements sent out to the East Indies on this occasion was a draft from the 12th Regiment of Foot, and with it John Dorman. The vessel in which he sailed stopped for water at the Cape of Good Hope, and arrived at her destination in eleven months, which was then reckoned a good passage. Eight ships sailed under the same convoy, and in each of them about 100 soldiers. He was at the taking of Madras and Pondicherry, and received no wound at either place. He remained with the British army for three years. The climate agreed with him, but neither he nor any of the privates could acquire any share of the wealth which was then accumulated by the officers of this army. Their food was chiefly rice, and they drank arrack with their water. Those who perspired profusely, as Dorman did, enjoyed good health, but those who did not were sickly, and many of them died. The uniform consisted of nankeen coats and trousers. The intemperate use of fresh arrack caused much sickness, and several deaths in the European armies.

Other victories soon came to Clive; for on the 23rd June, 1757, he defeated Surajah Dowlah, with his army of 15,000 horse and 14,000 foot, on the plains of Plassy. This demon was fresh from the slaughter of 146 men, women and children—for, jealous of the prosperity of the English traders, he had by a trick thrust them by night into a dungeon, since known as the "Black Hole of Calcutta."

Governor Howell (a sufferer and one of the twenty-five survivors found alive the following morning) speaks of how he supported nature by catching the drops, occasioned by the heat, which fell from his head and face, and adds these words: "You cannot imagine how unhappy I was when any one of them escaped my tongue!"

What a scene! the happiness, the existence of a fellow-creature, dependent upon being able to catch a drop of his own sweat!

Clive for the next ten years was omnipotent. Europeans and natives were at his feet, for his rule was just. In spite of all his brilliant service, Parliament thought fit, sixteen years later, to enquire into what Clive had done to Omichund—an agent employed by him to instal Meer Jaffier as the next ruling nabob. Clive was subjected to the most searching examination, at the close of which he burst forth: "Good God, Mr. Chairman! I stand astonished at my own moderation."

The House of Commons, without pressing for a division, had nought to do but simply to record and fully recognise his grand and meritorious services. But to the vulgar remains always a prejudice. The pangs of wounded honour killed him, for Clive, broken hearted, died by his own hand on 22nd November, 1774.

What irony! For the man to prosecute this charge against Clive was one General Burgoyne, so soon himself to be arraigned on a charge of surrendering the King's army at Saratoga, whereby we lost America.

The treatment in France of La Bourdonnais, Dupleix and Lally—statesmen who had served their country well—was indeed far worse. La Bourdonnais, the able Governor of the Mauritius, who in 1746 took Madras, died in the Bastille. Dupleix, who by skilful campaigning and diplomacy, had acquired great power and influence for his country in the East—until held up by Clive—was only spared to die of a broken heart in 1768. Whilst poor Lally—one who had known only peril and toil, and for threescore years had buffeted against Fate's obstruction and men's perfidy—was dragged through the streets of Paris to his gibbet on the 9th May, 1766.

The greatest honour, perhaps, for Clive, came after the close of this memorable enquiry. George III. had been for some time meddling with his American Colonies, and, in consequence, was at variance with his great commoner Pitt. This interference soon became serious, and an appeal to the sword seemed inevitable. It was then the Ministers were desirous to avail themselves of the services of

Clive, for the country, alas! too late, had learnt to know his worth —for Clive, since the death of Wolfe, was England's only General.

Frederick the Great selected Ferdinand of Brunswick, a cheerful, modest man, and one of his ablest Generals, to conduct the war around Westphalia, the commencement of which had been such a fiasco under Cumberland. Ferdinand took with him, as aide-de-camps, Major Humboldt (father of the great Humboldt) and Captain George Grey (father of the great Earl Grey). The latter belonged to the 20th Regiment, recently commanded by James Wolfe, who had left a few months earlier for America, to command a brigade at the siege of Louisberg.*

During May and June, 1759, Marshal Contades had outmanœuvred Ferdinand by taking up a strong position at Minden, hoping thus to detain his foe, whilst he marauded with detachments elsewhere. Ferdinand determined to spring upon this position by a sudden wild attack, as a letter from Captain Montgomery and the diary of Private Gorman, both of the 12th Regiment, will show:—

Dear Madam.

The pursuit of the Enemy, who have retired with the greatest precipitation, prevents me from giving you so exact an

---

* Wolfe's first success was at Louisberg, where, from the ships of Boscawen, he landed his division and invested the place; it stood a siege of six months and was then carried by assault. Admiral Boscawen had an interesting career. Born in 1711, he commanded the "Shoreham"—one of the six ships with Vernon at Portobello. In his many fights, strange to say, he took three French men-of-war, and on each occasion the same officer, Captain Hoquart, became his prisoner. We next hear of him conveying troops to Madras, where he was given a military command, and laid siege to Pondicherry. During the siege of Louisberg he kept off all marauders in the shape of French men-of-war, who employed every artifice to victual the town when environed by Wolfe. Gardiner was one of his captains, who, with the "Monmouth" (470 men and 64 guns), espied the "Foudroyant" (1000 men and 84 guns), and engaged her, saying to his men: "She looks more than a match for us; but I will not quit her while this ship can swim, or I have a soul left alive." The fight was long and furious, and Gardiner fell mortally wounded; but ere he died, he begged of his commander to nail the colours to the mast, and this was done. The "Foudroyant" shortly after flew the flag of Nelson.

account of the late most glorious victory over the French Army, as I would, had I almost any leisure, however, here goes as much as I can.

We marched from our Camp between 4 and 5 o'clock in the morning, about 7 drew up in a valley, from thence marched about 300 yards, when an 18 pound Ball came rowling gently up to us, now began the most disagreeable march, that I ever had in my life, for we advanced more than a quarter of a mile thro' a most furious fire from a most infernal Battery of 18·18 pounders, which was at first upon our front, but as we proceeded, bore upon our flank, and at last upon our rear, it might be imagined, that this cannonade would render the Regiments incapable of bearing the shock of unhurt Troops drawn up long before on ground of their own choosing, but firmness and resolution will surmount almost any difficulty, when we got within about 100 yards of the Enemy, a large body of French Cavalry galloped boldly down upon us, these our men by reserving their fire, until they came within 30 yards, immediately ruined, but not without receiving some injury from them, for they rode down two Companies on the right of our Regiment, wounded 3 Officers, took one of them prisoner with our Artillery Lieut, and whipped off the Tumbrells, this cost them dear for it forced many of them into our rear, on whom the men faced about and 5 of them did not return, these visitants being thus dismissed, without giving us a moments time to recover the inavoidable disorder down came upon us like lightning the glory of France in the persons of the Gens d' Armes, these we almost immediately dispersed without receiving hardly any mischief from the harmless creatures, we now discovered a large body of Infantry consisting of 17 Regiments moving down directly on our flank in Column a very ugly situation, but Stewarts Regiment and ours wheeled, and shewed them a Front, which is a thing not to be expected from Troops already twice attacked, but this must be placed to the credit of Genl Waldegrave and his Aide-de-Camp we engaged this Corps for about 10 minutes, killed them a good many, and as the song says, the rest they run away, the next, who made their appearance, were some Regiments of the Granadiers of France, as fine and terrible looking fellows, as I ever saw, they stood us a tug, notwithstanding we beat them off to a distance, where they galded us much they having rifled* barrels, and our Musquets would not reach them, to remedy this we advanced, they took the hint, and run away, now we were in hopes that we had done enough for one days work, and that they would not disturb us more, but soon after a very large body of fresh Infantry, the last resourse of Contades, made the final attempt on us, with them we had a long but not very brisk engagement, at last we made them retire almost out of reach, when the 3 English Regiments of the rear line came up and gave them one fire, which sent them off for good and all, but what is wonderful to tell, we ourselves after all this success at the very same time also retired, but indeed we did not then know, that the Victory was ours, however we rallied, but all that could now be mustered was about 13 files private with our Colonel and four other Officers one of which I was as fortunate as to be, with this remnant we returned again to the charge, but to our unspeakable joy no opponants could be found, it is astonishing, that this Victory was gained by six English Regiments of Foot without their Granadiers unsupported by Cavalry or Cannon not even their own Battalion Guns in the face of a dreadful Battery so near as to tear them with Grape shot against 40 Battalions, and 36 Squadrons, which is directly the quantity of the Enemy which fell to their share, it is true that two Hanoverian Regiments were engaged on the left of the English, but so inconsiderably as to lose only 50 men between them, on the left of the Army the Granadiers who now form a separate body, withstood a furious Cannonade, of the English there was only killed 1 Captain and 1 Sergeant some Prussian Dragoons were engaged, and did good service, our Artillery which was stationed in different places, also behaved well, but the grand attack, on which depended

---

* The principle of rifling was discovered at the beginning of the 16th century, but was not employed for warlike purposes until the middle of the 17th. In 1680, each troop of Life Guards was supplied with eight. In 1800, the 95th (now the R.B.) was armed with a 20 bore muzzle-loading rifle.

the fate of the day, fell to the lot of the 6 English Regiments of Foot, from this account the Prince might be accused of misconduct for trusting the issue of so great an event to so small a body, but this affair you will have soon enough explained to the disadvantage of a great man whose easy part, had it been properly acted, must have occasioned to France one of the greatest overthrows, it ever met with, the sufferings of our Regiment will give you the best notion of the smartness of the action, we actually fought that day, not more than 480 Private and 27 Officers, of the first 302 were killed and wounded, and of the latter 18, 3 Lieuts were killed on the spot, the rest are only wounded, and all of them in a good way except two, of the Officers, who escaped, there are only 4 who cannot shew some marks of the Enemys good intentions, and as perhaps you may be desirous to know any little risques that I might have run, I will mention those, of which I was sensible, at the beginning of the action I was almost knocked off my legs by my three right hand men, who were killed and drove against me by a Cannon Ball, the same Ball also killed two men close to Ward, whose post was in the rear of my Platoon, and in this place I will assure you, that he behaved with the greatest bravery, which I suppose you will make known to his Father and Friends, some time after I received from a spent Ball just such a rap on my Collar-bone, as I have frequently from that once most dreadful weapon, your crooked headed stick, it just swelled and grew red enough to convince the neighbours, that I was not fibing, when I mentioned it, I got another of these also on one of my legs, which gave me about as much pain, as would a tap of Miss Mathews Fan, the last and greatest misfortune of all fell to the share of my poor old coat, for a musquet ball entered into the right skert of it, and made three holes, I had almost forgot to tell you, that my spontoon was shot thro' a little below my hand, this disabled it, but a French one now does duty in its room, the consequences of this affair are very great, we found by the papers, that the world began to give us up and the French had swallowed us up in their imaginations, we have now pursued them about 100 miles with the advanced armies of the

Hereditary Prince, Wanganheim, and Urff in our front, of whose success in taking prisoners and baggage, and receiving deserters, Francis Joy will give you a better account, than I can at present, they are now intrenching themselves at Cassel, and you may depend upon it, that they will not show us their faces again during this Campaign.

I have the pleasure of being able to tell you, that Capt Rainey is well, he is at present in advance with the Granadiers plundering french baggage and taking Prisoners, I would venture to give him forty Ducats for his share of prize money.

I have now contrary to my expectations and in spite of many interruptions wrote you a long letter, this paper I have carried this week past in my pocket for the purpose, but could not attempt it before, we marched into this Camp yesterday evening, and shall quit it early in the morning. I wrote you a note just informing you, that I was well the day after the Battle, I hope you will receive it in due time. Be pleased to give my most affectionate duty to my Uncles and Aunts, my love to Bob Maxwell, Mathews, Nancy, Kitty, &c., and believe me to be

Dear Madam

with the greatest Affection

Your very dutiful Son

Hh. MONTGOMERY.

Camp at Paderborn

9th August 1759

As a list as long as that of a Pedlar would not contain the names of all my Friends, and as you know them, I shall not at present make it out, but beg of you to remember me to them every one.

The noise of the Battle frightened our Sutlers Wife into labour the next morning she was brought to bed of a Son, and we have had him christened by the name of Ferdinand.

Dorman was on the Continent during the whole of this war, and oftentimes engaged in different battles and skirmishes, the particulars of none of which he remembers, except those of Minden, where he was severely wounded in the left hand. In this memorable engagement—one of the most glorious in the English annals—Prince Ferdinand of Brunswick, with about 7000 English troops, defeated 80,000 of the French regular troops, under Contades and Broglie,* in fair battle. In the middle of this battle, Lord George Sackville behaved so extremely ill, that when the battle was over, Prince Ferdinand took his sword and sash from him and ordered him to retire from the army, as he had no occasion for his services. His command was given to the Marquis of Granby,† who had highly distinguished himself in this battle; and the unfortunate Lord George was afterwards tried and broke of his commission. His crime was "Not bringing up in due time a body of cavalry which he commanded." So great was the indignation against him at the moment, that six regiments nearly mutinied on the field, because he was not immediately shot.

Dorman was carried out of the field on a waggon, and brought, with other wounded men, to a military hospital at Bremen on the Weser. On his recovery, he was discharged with a pension of £7 18s. a year, which he forfeited eleven years afterwards, by refusing to move to a depot in England, from Strabane, where he was carrying on with some success the trade of a baker, and where

---

* Marshal Broglie (Duc de), born 1718, died 1804, of distinguished French family, succeeded Marshal Contade on the battlefield of Minden, and covered the retreat, for which service he was made a Marshal. His son was guillotined in 1794 for refusing to deprive Louis XVI. of life and rights. The Marshal reappeared, shortly after his son's death, at the head of a body of *émigres*, who, in alliance with the Prussians, were to place Louis XVIII. on the throne.

The Duchess de Broglie (born 1797, died 1838) with her mother (Madame de Stael), lie buried at Coppet. Both were women of genius, virtue and accomplishment. The daughter married the young Duc de Broglie (he being nine when his father was guillotined in 1794). For some time he was Prime Minister to Louis Philippe, and his first public act was to vote for the acquittal of Marshal Ney in 1816.

† Eldest son of the Duke of Rutland (born 1718); he won the battle of Warburg in 1760, and returned to England, becoming commander-in-chief in 1769.

he died, after a short illness, on January 13th, 1819, in his 110th year. His mother lived to the age of 113 years.

> "An old man, broken with the storms of life,
> Is come to lay his weary bones among you;
> Give him a little earth for charity."

The 12th and the other Minden regiments—the 20th, 23rd, 25th, 37th and 51st—all decorate on the anniversary of this eventful day, when in appearance they cease to be a regiment, and become one huge bouquet of roses.

It is hard to give an origin for this delightful show, except the fact that in the days of the early kings of France, it was considered good form to retain some tender feelings for any adverse opinion either from king or Parliament—the loser had to offer up a tribute.

> "'And what shall that tribute be?' asked the Count.
> 'A tribute of roses' replied the Regent."

Thus to these regiments, when they had won their case from the fierce arbiter the Sword, the loser had nought to offer but a coronet of flowers—"la baillée aux roses." The rose is also worn on Dettingen day, because the Sovereign then was present—also to commemorate his presence on any other day. This was last done when King Edward inspected the 12th at Malta in 1909.

In the following year—1760—the 12th, 20th, 5th, 24th and 37th were engaged at Warburg, under Granby, who quickly retrieved the reputation of the cavalry, so damaged in the Minden fight. Later, in October, the same regiments met again at Kirch Denkern and Luttemberg, which engagements practically closed the compact made by Pitt with Frederick.

France then sued for peace, whereby Austria and Prussia were left to carry on the war alone until February, 1763, when peace was signed. Prussia retained Silesia, which made her the leading Power in Europe.

The grief of the poor Queen of Austria, Maria Theresa, in having to lose Silesia, is well portrayed by Bartolozzi in a fascinating way. She is represented making this her celebrated oration :—

"In her distress she left Vienna and threw herself into the arms of the Hungarians, who had been so ill-treated by her father and by her ancestors. Having convened the four Orders of the State at Presburg, she appeared in the Assembly, holding her eldest son in her arms (almost yet in his cradle), and addressing herself to them in Latin, a language in which she expressed herself extremely well, she spoke to them almost in these words: 'Abandoned by my friends, persecuted by my enemies, attacked by my nearest relations, I have no resource left but in your fidelity, your courage, and my constancy. I commit to your hands the daughter and the son of your King, who expect of you their safety.' At this speech the Palatines were greatly moved; and drawing their sabres, they all cried out: 'Moriamur pro rege nostro, Maria Theresa' ('Let us die for our King, Maria Theresa.') They also gave the title of King to their Queen, and never was there a princess more deserving of this title. They wept when they took the oath to defend her; she alone appeared unconcerned; but as soon as she retired with her maids of honour, the tears ran plentifully down her cheeks."

Prince Charles of Brunswick was a General of Division in these latter battles. He married Augusta, daughter of George III., and lost his life at Jena. His son was killed at Waterloo. His grandson was one grand eccentric—he left by will a large fortune to the Prince Imperial of France, which he cancelled when the Third Empire fell, and made the town of Geneva his next heir.

Another rattle from the drummer :—

"And now daddy crossed the Atlantic,
To drum for Montcalm and his men;
Morbleu! but it makes a man frantic,
To think we were beaten again!
My daddy, he crossed the wide ocean,
My mother brought me on her neck,
And we came in the year fifty-seven
To guard the good town of Quebec.

"In the year fifty-nine came the Britons—
 Full well I remember the day—
They knocked at our gates for admittance,
 Their vessels were moored in our bay.
Says our General: 'Drive me yon red-coats
 Away to the sea whence they come!'
So we marched against Wolfe and his bulldogs;
 We marched at the sound of the drum.

"I think I can see my poor mammy
 With me in her hand as she waits,
And our regiment, slowly retreating,
 Pours back through the citadel gates.
Dear mammy, she looks in their faces,
 And asks if her husband is come?
He is lying all cold on the *glacis*,
 And will never more beat on the drum."

—THACKERAY.

Within a short six weeks of Minden, the same nations were hard at it, fighting for the mastery of the New World in far-off North America. Here the two Generals met—Montcalm and Wolfe, the former (born in 1712) a great favourite of the French King, who selected him, in 1756, to command the forces in North America, somewhat to the chagrin of Vaudreuil, who hoped to continue in his double *role* of Governor-General and Commander-in-Chief; and in Wolfe we recognise the late Adjutant of the 12th, then Colonel of the 20th, now the favourite of Pitt, who saw in him a soldier genius just fitted for the part.

It is said of Wolfe that, whilst on board the "Sutherland," waiting for the tide to serve before the signals could be given for the procession to move, he confided to his schoolfellow and friend, John Jervis (Earl St. Vincent, February 14th, 1797), his last will and wishes; his thoughts then turned on poetry, for he dearly loved a poet, and envied them their undying name; and at last, as the vessels slowly took the stream, he gave utterance :—

"To each his sufferings: all are men
 Condemn'd alike to groan;
The tender for another's pain,
 Th' unfeeling for his own.

Yet ah! why should they know their fate?
Since sorrow never comes too late,
And happiness too swiftly flies,
Thought would destroy their paradise.
No more—where ignorance is bliss,
'Tis folly to be wise."

The young pilot, to whom was entrusted the safety and secrecy of this midnight move, was James Cook, once a haberdasher; then apprentice on board a collier; next a volunteer for the Navy, where his skilful and trustworthy seamanship gave him at once a position. He was appointed master of the "Mercury," in which he sailed to the St. Lawrence, where he rendered most important service by his careful soundings of the river opposite the French camp.

Mathematics, astronomy, and geography were his favourite studies, and it is due to these accomplishments that the English Crown can boast of such gems as New Zealand, New Holland (now Australia), New South Wales, and New Guinea.

His death was tragic. Off the Sandwich Islands (his latest discovery, and named after the First Lord of Admiralty), the natives stole the cutter of his frigate, and on his attempt to recover it he lost his life—and his flesh, for his bones only could be found. This sad news was received with sorrow throughout Europe in February, 1779.

On Clapham Common there is a tree, carefully fenced in, which was planted by Captain Cook, and now treasured as a relic.

The expedition thus so skilfully piloted, and the spot chosen where to land, it remained for Wolfe, the British soldier, and the bayonet, to do the rest. First up went twenty men, climbing with hands and feet, closely followed by a larger body; and thus a continuous stream of red coats moved slowly upwards, and formed silently and in order on the plateau.

Malcolm exclaimed: "This is a serious business," as he despatched Johnstone (his aide-de-camp) for troops from the left and centre of his camp. He had expected a detachment, and he found an army, for full in sight before him stretched the thin red lines of Wolfe. Fight he must, for Wolfe was now in position to cut off all supplies. The end soon came—a shot shattered the wrist of Wolfe, another struck him, and the third lodged in his breast. "It is all over with me," he said; but not before he had heard the cry: "They run! they run!"

Montcalm was borne with the fugitives towards the town, and as he entered, a shot passed through his body. The excited crowd had seen it. "Oh, mon Dieu! mon Dieu! le marquis est tué."

Wolfe no more, the city of Quebec surrendered to Lord Townshend, second in command. The enemy was permitted to march out with the honours of war, and to embark for France.

Wolfe and Townshend had first met at Dettingen, where a poor drummer boy of the 12th was killed by a cannon ball of sorts. His brains were scattered, and perhaps some may have reached to Townshend, for he was heard to say: "I am not afraid; I am only astonished that a fellow with such a quantity of brains should be here." This remark is perhaps indicative of the man, for he was not to be caught remaining at Quebec, when perhaps by taking ship for home, he could turn his brains to good account, and be first in the field for honours. So it remained for General Murray, a gallant soldier and upright man, to "carry on."

The family of Murray then took the prefix "Wolfe"—Wolfe-Murray.

Quebec, the principal town of Canada, like a mighty salmon, had been securely hooked; it was, however, not exactly landed. It is between these two operations of hooking and landing, that the fish, with violent plunge and frantic effort, tears at the line and rod to get his liberty again—and so it was with Quebec.

Vaudreuil, the Governor of the Province, now became a great detractor of his predecessor, Montcalm, and argued that all had not been done to save the town. Rumour soon had it that Quebec was to be attacked again. It was whilst Generals Janvier and Fevrier were hard at work with ice and snow, and illnesses therefrom, that the survivor of a boat's crew, wrecked in the rapids off Quebec, told the tale that he was one of many thousands coming down with

General Levis from Montreal. A two days' fight ensued, and the English were forced within the walls of Quebec.

Bad and scanty fare were telling hardly upon the garrison, when an officer came to Murray's quarters to say a ship of war was beating up the harbour. The excitement grew intense to learn under which flag she sailed—the red of England or the white of France; and at last was spied the red cross of St. George, and the frigate "Lowestoffe" anchored in the Gulf. She brought the news that two more ships (the "Diana" and the "Vanguard") would reach Quebec within the week.

The three ships then passed the town to attack the enemy, lying up the river—they mustered six ships in all.

Admiral Vaquelin did not belie his reputation. It was only when his last round was spent that he struck his flag. The other ships, in the same predicament, threw their guns overboard and escaped. This raised the siege, and Quebec a second time was won.

Hawke had been appointed to replace poor Admiral Byng in the command of the Mediterranean fleet, and like a greyhound on the leash, straining upon the start, lay wait in Quiberon Bay.

> "In seventeen hundred and fifty-nine,
> When Hawke came swooping from the West,
> The French king's admiral with twenty of the line
> Was sailing forth to sack us, out of Brest.
>
> "The ports of France were crowded, the quays of France a-hum,
> With thirty thousand soldiers marching to the drum;
> For bragging time was over, and fighting time was come,
> When Hawke came swooping from the West!"
>
> —NEWBOLT.

A general idea of this action is best given in Sir Edward's own words: "In attacking a flying column, it was impossible, in the space of a short winter's day, that all our ships should be able to get into action, or all those of the enemy brought to it. The commanders and companies of such as did come up with the rear of the French on the 20th, behaved with the greatest intrepidity, and gave the strongest proof of a true British spirit. In the same manner, I am satisfied, would those have acquitted themselves, whose bad going ships, or the distance they were at in the morning, prevented from getting up. Had we had two hours' more daylight, the whole had been totally destroyed or taken, for we were almost up with their van, when night overtook us."

For this victory Hawke received a pension of £2000 a year, and was raised to the peerage in 1776. He died in 1781.

CHAPTER IV.

> "Behold the child, by nature's kindly law,
> Pleased with a rattle, tickled with a straw,
> But each with a destiny, unshunnable like death."

NOW that this war of seven years is over, it is prudent to take stock of the many victories, both by sea and land; for a query now arises as to "Who can carry on?" and so keep the prizes won. The answer cometh that the same decade, 1758-69, which gave to England her India and North America, and made her mistress of the seas and first of commercial nations, was also very rich "en naissance."

HORATIO NELSON was born on Michaelmas Day, 1758,—a day of naval note, for it was when Queen Elizabeth had sat down to dinner of which a goose formed a part, that news was brought to her of the total defeat of the Armada. Her Majesty then ordered that the dish before her might be served on every 29th September.

Horatio was the fifth son in a large family of eleven children. At the early age of nine he lost his mother—a loss severely felt. Her brother, Captain Suckling, kept Horatio as his own, until old enough to join his uncle's ship, the "Raisonnable" (64), fitting out to settle with Spain a dispute over the Falkland Isles, 1769-70.

MASSENA was born in 1758: The ablest Marshal of France, "l'enfant cheri de la victoire," as Napoleon named him. One word, "Massena," marks his tomb in Pere la Chaise.

PICTON was born in 1758: The famous leader of the Third or Fighting Division in the Peninsula, and the hero of Badajoz, Cuidad Rodrigo and Vittoria—killed at Waterloo—an old officer of the 12th Regiment, which quite belonged to the family of Pictons.

LAFAYETTE—For he was only two years old when his father was killed at Minden, 1759.

WILLIAM PITT, the second son of Chatham, was born in the Minden year 1759. At the early age of twenty-four, he was the first Minister of England; his power was irresistible, and all for good. What a contrast to another born in the same year, whose power was irresistible, and all for misery and death. I allude to that odious adventurer, Robespierre, who lived to enjoy the same fate he had so often dealt out to others—he was guillotined in July, 1794.

BURNS was born in 1759—the poet of Scotland, and a patriot, for his last letter reads :—

"12th July, 1796.

"A rascal of a haberdasher, to whom I owe a considerable bill, £7 14s., for patriotic volunteering uniform, taking it into his head that I am dying, has commenced a process against me, and will infallibly put my emaciated body into jail."

WILBERFORCE was born in 1759—our great philanthropist; an intimate friend of Pitt, Fox and Sheridan. He lived to see his life's desire accomplished, viz., the abolition of slavery :—

"The number of slaves brought from the coast of Africa in one year (1768), between Cape Blanco and the river Congo, amounted to one hundred and four thousand one hundred ;—Great Britain, the seat of freedom, 53,100; France, 23,500; Holland, 11,300; Portugal, 8,700; British America, 6,300; Denmark, 1,200; in fact, so many beasts of burden, only because they were of a different colour!"

Now for 1769—a year which gave to England her IRON DUKE, and the great diplomat LORD CASTLEREAGH; also rattled into life, from the most obscure parentage, babes since known in history by the names of :—

NAPOLEON—Wonderful in peace and war, Emperor and King, but to die a prisoner in exile.

HOCHE—A General of highest distinction, who died suddenly in 1797, some say by poisoning.

MURAT, to become a King; sentenced to death and shot.

JUNOT, to become a King; then to madness and suicide.

NEY, to become a Prince; sentenced to death and shot.

Soult—A fine General, who lived to argue out with Wellington his victories and reverses. Both were fêted at the Mansion House, for Soult was with us as French Ambassador at the Coronation of Queen Victoria. He died in the same year as Wellington, 1852.

Lannes—A grand soldier, killed at Essling, 1809.

Chateaubriand—A man of useful pen; for when Napoleon was at Elba, to create zeal towards the Bourbons, he published a book called "Bonaparte and the Bourbons," which Louis XVIII. pronounced as being worth an army. But before it had its spell, Napoleon was back again in power.

Cuvier—A great naturalist and scientist. He laboured to make reason the arbitrator and supreme guide of public opinion. His report, called for by Napoleon, on the "History of the Progress of Science since 1789," is most remarkable. The sword of the conqueror cleared the way for the Minister of Education.

Humboldt—Born in Pomerania, and a great traveller. His exploration of South America, for adventure and research, can never have a rival. Some of his most interesting letters are contained in his "Cosmos." He assisted at the christening of our late King Edward.

Charlotte Corday—who did not deem assassination a crime when directed against assassins, and felt uncertain which was the greater monster, Robespierre or Marat; but on hearing from the latter that he intended to guillotine the Girondists, she drew her knife and stabbed him in the throat.

Mehemet Ali—A Pasha of many tales, and like Napoleon, a grand man. He fought with us at Aboukir, and from that date he had an open arena for his strength. He lived to the age of eighty, and his life deserves a book.

It is hard to divorce Mars, the God of War—and the one who hurls the Thunder—from Minerva, the lady who presides over all handicrafts, inventions, arts and sciences; for the same spell of life, eighty years, as given to Mehemet Ali, was mercifully given to Brunel, our greatest engineer, who, as a youth, distracted by the horrors attending the execution of Louis XVI., openly avowed his

Royalist opinions and fled to America, and next to England, where he met a kindred spirit in H. Maudslay, the great *ouvrier* of the day.

In the Armada days, the "wooden walls of England" were considered, after God, as the principal protection of the country. To maintain these ships, it was forbidden (under a fine of 40s.) to fell any tree or part of a tree from off the banks of any river or within fourteen miles of the sea. This enactment naturally crippled the metal trade of England, for how were the furnaces to be supplied to cast her precious steel ?

It came to Queen Elizabeth's hearing that a Spaniard in authority had stated that it would not be difficult to conquer England, as her arms (steel) were so bad. Elizabeth promptly sent to Sweden for iron and wood, and forged her metal in Sussex and the Dean.

It is said that after Minden—when our spoils were grand—a chemist addressed the Duke of Brunswick, and proposed to change the captured arms for gold. "Non pas, repondit le Duc, j'ai besoin du fer pour resister a mes ennemis; quant a l'or j' en trouverai en Angleterre." And he was right: for the strength of England lies in the men and women of her race, and it will be a sorry day for them, if ever they allow the rust of "vae victis" to settle on their steel.

Now as each one of these many babes here mentioned have to emerge from a mental husk or shell in which their destiny lies hid, it behoves us to employ this waiting time by again listening to old Pierre, as he goes thumping with his sticks and his soul-stirring song :—

"And now I'll commence my own story:
Once more did we cross the salt ocean,
We came in the year eighty-one,
And the wrongs of my father the drummer
Were avenged by the drummer his son.

"In Chesapeake Bay we were landed;
In vain strove the British to pass;
Rochambeau our armies commanded,
Our ships they were led by de Grasse.

Morbleu! how I rattled the drumsticks
  The day we marched into York-town;
Ten thousand of beef-eating British,
  Their weapons we caused to lay down."

George III. was told by Pitt to remove "that one tax on tea, and all will go well," and he refused. It was then America drew up her Bill of Independence.

The blame of all this strife and struggle is placed by America on the shoulders of George III., for they sought, and the Colonies fought for their independence on the plea that :—

"A prince George III., whose character is thus marked by every act that may define a tyrant, is unfit to be a ruler of a free country."* It was Jefferson who drafted the Bill for Congress; it

---

* George III., in domestic ways, seemed equally annoying, if only to take his marriage in 1761, through no fault of his own, but still there was no form of ceremonial extant for the marriage of a king, hence the Heralds College went almost crazy. Henry VIII. had married all his wives according to the ritual of the Romish Church; the same with his daughter Mary. Edward VI. and Elizabeth were unmarried; James I. was married to Princess of Denmark in Norway; Charles I. was married in Paris; Charles II. was married in the Presence Chamber at Portsmouth simply as, "I, Charles, take thee, Catherine," the Bishop of London attending as witness; James II. (then Duke of York) was married, firstly, privately at Worcester House, September, 1660; secondly, to Princess of Modena, at Dover, on the day of her arrival at Dover, the Bishop of Oxford attending as witness. Anne was married to George of Denmark in 1684, and became Queen in 1702; George I. and George II. were both married before they came to the throne; hence George III. was first of any Royalty to be married as King.

Another surprise this quaint monarch gave. On February 23rd, 1789, Mr. Pitt and Lord Melville (Dundas) were dining with Lord Chesterfield, when a letter was brought to the former, which he read, and sitting next to Lord Melville, gave it to him under the table, and whispered, when he had looked at it it would be better for them to talk it over in Lord Chesterfield's dressing-room. This proved to be in the King's own hand, announcing his recovery (from madness) to Mr. Pitt. The letter ran as follows :—

"The King renews, with great satisfaction, his communication with Mr. Pitt after the long suspension of their intercourse, owing to his very tedious and painful illness. He is fearful that during this interval the public interests have suffered great inconvenience and difficulty. It is most desirable that immediate

was John Adams, a fine orator, who became its champion in debate. Both these men lived to hold twice the high office of President of the United States; and on the day to commemorate the fiftieth anniversary of their Republic, viz., 4th July, 1826—when a special banquet and other festivities were prepared—it was learned that Thomas Jefferson, of eighty-three, and John Adams, of ninety-one years, had both died that very morning.

Hostilities commenced on 17th June, 1775, for from Boston first was noticed a small breastwork on a hill—Bunker's Hill they called it—and this breastwork hourly grew more threatening, until at last Clinton,* Howe,† Burgoyne‡ and Grey, a Brigadier, did their best to take it. It cost them just 1000 killed and wounded, before Prescott§ and his 1500 Yankees thought fit to move away.

---

measures should be taken for restoring the functions of his Government, and Mr. Pitt will consult with the Lord Chancellor to-morrow morning, upon the most expedient means for that purpose, and the King will receive Mr. Pitt at Kew afterwards, about one o'clock."

There could be no hesitation on the part of Mr. Pitt. After having had the necessary conference with the Chancellor, he waited upon the King at the appointed time, and found him perfectly of sound mind, and in every respect as before his illness, competent to all the affairs of his public station. The King wrote the letter at a little table of the Queen's, which stood in his apartment, without the knowledge of any person; and having finished, rang his bell, and gave it to his valet-de-chambre, directing it to be carried immediately to Mr. Pitt. Then, even at his death in February, 1820, it seemed as if he did not wish to go alone, for his son, the Duke of Kent, a few days predeceased him—thus giving two Royal funerals within the week.

* Clinton succeeded to the command in 1788. It was he who entered into negotiations with the American General Arnold, who wished to become a Royalist and to surrender up West Point. Poor Major André was made the go-between, and he was caught and hung. Colonel Clinton was Colonel of the 12th Regiment from 1766 to 1779, when he left to become Commander-in-Chief of the Army in America. Colonel William Picton succeeded him in the 12th.

† The Howes.—Richard Howe, from our first notice of him on board the "Baltimore" in 1745, was in 1754 with Boscawen at Louisberg, and other naval victories in 1757 and 1758. In 1758 his brother, Lord Howe, was killed, fighting with Abercrombie against the French in Canada. Richard Howe then became

Then followed shortly after this, the surrender of Burgoyne and his whole army at Saratoga, 1777. To make this victory doubly sure America called upon France for aid, which was freely given. This new alliance now took away all chance of regaining our supremacy.

The blame of this disaster must rest on Lord George Germaine, who as Lord George Sackville had disgraced himself at Minden, and when found guilty by court martial was dismissed the Service and erased from the list of Privy Councillors—only to be reinstated by George III., and as Secretary of State for the Colonies, under Lord North, to conduct this War of Independence in such a way that General Grey plainly told him, when on Committee before the House: "The reduction of America is impossible with our present force; it is uncertain with *any* force; and at no time have we ever had a sufficient force. This inadequate supply of men has been the cause of all our troubles in wars past, present, and probably will be so in the future. War without her sinews is like starting out for cricket with only two umpires and a stump."

---

the Peer, and the two brothers continued in their grand career together, and were considered the ablest officers of their day. The former crowned his many victories by the battle of June 1st, 1794, in which he defeated the French, capturing ten of their ships. He was then nearly seventy (born 1725, died 1799). In 1797 he redressed the grievances of our seamen—by whom he was beloved and nicknamed "Black Dick." The mutiny was stayed, and harmony took its place. The soldier Howe succeeded General Gage in the American command in 1775, and fought at Bunker's Hill, Brooklyn, and Brandywine. He sought retirement in 1788, and died in 1814.

‡ General Burgoyne, the mover of the vote of censure on Clive, surrendered his 3500 fighting men, guns, etc., to General Gates at Saratoga. He stood his trial, which came to nothing. His literary attainments ranked high. He died in 1792. His son, Sir John Fox Burgoyne, R.E., became a Peninsular and Crimean hero, and died a Field-Marshal, living to lament the loss of his talented boy, who, with his 500 men, foundered in the "Captain" (six gun turret ship) off Cape Finisterre on September 7th, 1870.

§ And in Prescott, who kept these Generals at bay at Bunker's Hill, one recognises the grandfather of the blind historian—the author of those charming books, "Ferdinand and Isabella," and "Conquest of Mexico."

England now sought the power of Pitt (Chatham). His policy had always been to pacify America, but "No surrender!" This recent move of France intensified his pride in his country's greatness, and he entered the arena of politics afresh. The task, however, was too great, for broken with age and disease, whilst in his old attitude and gesture of debate, pleading hard for peace with America, and for every ship and soldier to be sent against the Bourbons, he fell into the swoon of death, in April, 1778.

Burgoyne's disaster at Saratoga was partially effaced by the successes of Lord Howe, and by Lord Cornwallis at Charlestown and elsewhere; but the end was soon to be, for the latter, being denied the promised assistance of Sir Henry Clinton, fell back on Virginia, and entrenched himself in the lines of York Town, where he was starved into surrender—for his only chance was a victory at sea, which never came.

"On July 31st, 1781, Admiral Sir George Rodney entrusted the command of his Majesty's fleet at the Leeward Islands to Rear-Admiral Sir S. Hood, and sailed from St. Eustatius on the following day, with the 'Gibraltar,' 'Triumph,' 'Panther,' 'Boreas,' two bombs, and a convoy for England.

"On July 5th, M. de Grasse had gone with his whole fleet and a large convoy for Martinico, and arrived about the middle of the month at Cape Francois, St. Domingo, where he was reinforced by five sail of the line. At the beginning of August he sailed from thence with his prodigious convoy, which having seen out of danger, he touched at the Havannah for money, and then directed his course with twenty-eight sail of the line, and several frigates, to the Chesapeake, where he arrived by the end of the month.

"Sir Samuel Hood, having received intelligence of this, lost not a moment in hastening to the coast of America. On August 25th, 1781, he arrived off Cape Henry, and from thence despatched a frigate with intelligence to Rear-Admiral Graves. Finding, however, that no enemy had appeared either in the Chesapeake or Delaware, he proceeded off Sandy Hook.

"On the very day of his arrival there, the Commanders at New York received intelligence that M. de Barras, who succeeded Ternay in the command at Rhode Island, had sailed three days before with his squadron to the southward. The intercepting of this squadron was an object of importance; and Rear-Admiral Graves, on August 31st, bringing out of New York to Sandy Hook five ships of the line, and one of fifty guns, took the command. Sir Samuel Hood getting under sail at the same time, the fleets proceeded together to the southward.

"The cruisers placed before the Delaware by Rear-Admiral Graves could give no certain information, and the cruisers off the Chesapeake had not joined. The wind being rather favourable, they approached the Chesapeake on the morning of September 5th, when the advanced ships made the signal for a fleet. A number of great ships, being twenty-four sail of the line, were soon discovered at anchor off Lynnhaven Bay, just within Cape Henry, extending across the entrance of the Chesapeake. They had a frigate cruising off the Cape, which stood in and joined them. As the British Fleet approached, the French immediately slipped their cables, turned out from their anchorage in some confusion, and formed without any particular regard to prescribed order, as they could come up. Wind N.N.E.

"The British Fleet amounted only to nineteen sail of the line, that of the enemy to twenty-four; so that the French had a superiority of no less than five line of battle ships. The action commenced soon after four o'clock amongst the headmost ships, pretty close, and then was nearly partial, being general only as far as the second ship from the centre, towards the rear. It ceased a little after sunset. Our fleet had ninety men killed, and 230 wounded, chiefly belonging to Sir Samuel Hood's squadron. The French during the battle had 1800 seamen and ninety officers on shore.

"The two fleets continued for five days in sight of each other, repairing their damages and manœuvring, until the French Admiral had gained his object by covering the arrival of M. de Barras' squadron and convoy from Rhode Island, when he returned with his fleet to the Chesapeake, and anchored across, so as to block up the passage. This decided the fate of Lord Cornwallis, and finished the war between England and her American colonies."

England now found herself conflicted against Europe, for Spain had joined in; for three years she held the rock and fortress of Gibraltar against famine and bombardment from a French and Spanish army. "The strength of a fortress is the strength of the man who defends it," and in Elliott (afterwards Lord Heathfield) Gibraltar possessed a man replete in military skill and resource. He was abstemious, never partaking of animal food or wine; an example which had a powerful effect on the men, for they submitted cheerfully to all privations.

The garrison at the time consisted of the 12th, 39th, 58th and 72nd, two Hanoverian regiments, artillery and engineers.

The grand attack was on September 13th, 1782. On the land side were stupendous batteries, mounting 200 pieces of heavy ordnance, supported by 46,000 men, under the Duc de Crillon. On the sea side were the combined fleets of France and Spain, numbering forty-seven sail of the line, besides numerous frigates and ten formidable battering ships. The brave men under Elliott threw 5000 red-hot shot on that memorable day; and the attack was beaten off at all points.

"Paul Jones," the well known fighting man, was a terror to our commerce at the time. He caught sight of some cargo boats bound for Scarborough, escorted by Captain Pearson, of the "Serapis." "Paul Jones" duly set off in chase, giving scarcely time for Pearson to dock his boats and to get to sea again. The two then met, and a deadly conflict followed, when the "Serapis" at last surrendered; but ere Jones could reach his prize, his own ship, the "Richard," sank. Now the vanquished saved the victors, and both went into Port L'Orient together—but only just in time, for so riddled and dismantled was the "Serapis" that she dipped her bow and filled. Pearson was knighted by George III. and Jones received a sword of honour from Louis XVI.

The curtain was rung up once again in 1813—the events of which, perhaps, centre round the heroism displayed in single ship contests; and history records the plucky deeds of Commodore Perry. These successes on the part of America roused the indignation and spirit of British officers, so much so that Captain Brooke, cruising off the port of Boston, formally sent in a challenge for young Lawrence, of the "Chesapeake," to come out and fight. Thousands of spectators lined the beach to witness the approaching conflict, expecting but one result, for in guns and crew the "Chesapeake" was so vastly superior; but in fifteen minutes the "Chesapeake" was in tow of her conqueror, though her brave commander knew it not—young Lawrence had been killed, and thus the war ended.

The last fortress or citadel we attacked was New Orleans, defended by General Jackson. Some days of useless bloodshed here occurred, for Peace had been signed, and the "cease fire" had not been sounded, at least loud enough for the combatants to hear.

This chapter seems incomplete without one mention of the great man Washington, the American Commander-in-Chief. In 1753, when the encroachment of the French in Canada became so very serious, he fought right royally and upheld the Flag, until General Braddock arrived from England with two British Regiments and some Highlanders.* It was then he was informed that, although a Colonel in the Virginian Army, he must now become a simple Captain, and be subordinate in that rank to all King's officers. This he would not do, so he resigned his commission, only to be approached again—and this time with a request to retain his rank.

Braddock's first brush with the French and Indians around Fort du Quesne became a rout. Braddock was slain, and Washington badly wounded. He was now made Commander-in-Chief of the Virginian Forces, only to be annoyed again by the claims and pretensions of officers classed as "King's." However, he served

---

* No better troops could be than these Highlanders, in spite of all their recent distractions when they wished Prince Charlie to remain as King.

cheerfully under General Forbes, when Fort du Quesne was captured in 1758, and it became Fort Pitt.

Washington then married and settled down at Vernon. Later, of course, he had to side with his country in her differences with England; and to his honour be it said, that during his second Presidency (in 1793), when England was at war with France and the United States wished to join in with their former ally against their former enemy, their Uncle Sam—the President—on no account would allow it.

In 1802 the "Edinburgh Review" was started, and the colours, buff and blue, are taken in admiration of the man—for that was the colour of his uniform during the War of Independence.

The last days of 1799 saw his end; when Napoleon, in his Order of the Day, requested all colours to be draped, and the army to be placed in mourning for a week. No mention was made of Lafayette in this Order, for it was Lafayette who outfitted at his own expense (in the early days of the war—1777) a large ship with a contingent of men to fight for Washington; he also outfitted Paul Jones.

CHAPTER V.

"That right in peace, which here we urge in war."

THE American War had thus ended ingloriously for the British arms. William Pitt was keeping France and England together by a commercial treaty, which gave free trade between nation and nation, and enabled subjects of both countries to reside and travel in either without let or hindrance. In fact, the blessings of peace were with us, and a message went forth to the army: "Let them have pay, and part."

But not for long—for on January 3rd, 1793, England was again at war with France. Roaring war it soon became, for every European nation (not forgetting our Empire in the East) joined in—but England's share was greatest, for both by land and sea, and for over twenty years, was she engaged in this great Napoleonic struggle.

Napoleon, we recollect, was born in 1769—the same year as Wellington and in the same decade as Nelson. Each one possessed a power sufficient to make his name a marvel and a retrospect for ever; for was it not Napoleon, his admirals and marshals, that gave to Wellington and Nelson and *their* centurions:

> "The stern joy which warriors feel
> In foemen worthy of their steel"—

men who, by voice and presence, amid the din of battle, when fear and danger came, inspired fresh courage, and gave to the dying flame of war a further fury, and an impetus which sent—and kept—the flag of England ever flying at the fore.

It may be interesting to trace the early career of Wellington and Napoleon, for their lives seem to act as mighty loadstones one

St. Nicholas Church, Brighton.

To face page 61.

towards the other—Come! Come hither! Approach! Come near!—until the final grip at Waterloo.

Wellington, as a preliminary step to Eton, was placed under the care of the Vicar of Brighton.* The parishioners have of late erected a very handsome cenotaph to record the fact that he worshipped in their church (St. Nicholas') and brought honour to his country; also in the adjoining churchyard is a monument to one Phœbe Hessel, who was born at Stepney, in the year 1713. She served for many years as a soldier in the 5th Regiment of Foot, in different parts of Europe, and in the year 1745 fought under the command of the Duke of Cumberland at the battle of Fontenoy, where she received a bayonet wound in her arm. Her long life, which commenced in the reign of Queen Anne, extended to that of George IV., by whose munificence she received comfort and support in her latter years. She died at Brighton, where she had long resided, on December 12th, 1821, aged 108 years.

The Navy can likewise boast of a long roll of female warriors. Lord Howe, after his glorious fight of June 1st, specially thanked the sailors' wives who happened to be on board the English ships; for they fought with the most determined valour at the guns, encouraging and assisting their husbands. One woman in particular was "Tom Bowling" by nickname; she had served twenty years as a boatswain's mate, and was in receipt of a good pension from the Chatham Chest.* In later years she was not so circumspect as Phœbe Hessel—she defied all convention, and slept oftener in gaol than out of it. She always dressed as a man, and lived to a very advanced age.

---

* The son of this future hero died suddenly at Brighton Railway Station on August 13th, 1884—the same day that his father's statue was being removed from Hyde Park Corner to Aldershot.

† The Chest at Chatham, which gave gratuities for wounds, and small annual pensions to the wounded in proportion to their sufferings, was established by Queen Elizabeth in 1588, and kept going by the Government, and 6d. a month from the pay of sailors (voluntary).

In 1788, being a year of peace, the latter contributions equalled £6,608.

,, 1797 ,, ,, war ,, ,, ,, £45,700.

It is easy to imagine young Arthur Wellesley, aged twelve, lost in wonderment at Phœbe Hessel, aged sixty-seven, as she attended the Sunday service, and perhaps disclosed the arm that had been so badly wounded at Fontenoy—grand Fontenoy!

Wellesley was not long at Eton, for his father, the Earl of Mornington, died in May, 1781, leaving a widow and nine children in so-called straitened circumstances, which meant a less expensive education. Accordingly Wellesley was sent to Angiers, in France, for a course of military tuition, about the same time that Napoleon entered the French Military School at Brienne as a King's Pensioner.

The latter soon showed a disposition for war and destruction; he liked to seek retirement in the playground, and to erect small fortresses, the tracings of which were somewhat original. This love of solitude and silence was not his real character; he knew he had intellect, but he hated to be considered poor, as the following letter shows:—

"Military School, Brienne,
"5th April, 1781.

"FATHER,—If I am not to be allowed the means, either by you or my protectors, to keep up a more honourable appearance at the school I am in, send for me home, and that immediately. I am quite disgusted at being looked upon as a pauper, and of seeing my insolent companions, who have only fortune to recommend them, smile at my poverty; there is not one, but who is far inferior to me in those noble sentiments which animate my soul.

"What, sir! shall your son continually be the butt of a few paltry purse-proud fellows, who sneeringly joke upon the privations I experience? No, father! I hope not. If my condition cannot be bettered, remove me from Brienne; put me to some mechanical trade if it must be so; let me but find myself among my equals, and I answer for it, I will soon be their superior. You may judge of my despair by my proposal; once more I repeat it, I would sooner be foreman in a workshop, than be sneered at in a first-rate academy.

"Do not imagine that this letter is dictated through a vain desire of indulging myself in expensive amusements; they have not the smallest temptation for me; I only experience a wish to show my fellow students that I have the means of procuring them as well as they.

"N. BUONAPARTE."

Other fulminating letters of Napoleon to his father had to be stopped by his mother, who returned them with wholesome advice. "If ever I receive any similar epistles, I have done with you for ever," writes Lætitia. Carlo Bonaparte, the father, died in 1785, leaving a family of eight.

"Argentum deficit" may thus have acted in inverse ratio on the characters of Arthur Wellesley and Napoleon.

With the former, the mere fact of his removal from Eton took him away from the atmosphere of wealth, and perhaps quieted and toned down his aspirations—whereas with Napoleon, as a King's Pensioner, to be thus thrown and kept amongst the rich (as his letter shows) was unbearable; it made him taciturn, jealous and so ambitious, that he longed to be the giant of his times, and mighty in history.

Wellington and Napoleon obtained their first commissions early in 1787; and Nelson (now a post captain of 1781), with services up-to-date both varied and brilliant, was in the same year (March, 1787) united in marriage to the widow of Dr. Nisbet, of Nevis. The wedding was grand, and H.R.H. the Duke of Clarence (William IV.), who was on cruise with Nelson, gave the bride away.

Napoleon Bonaparte, on leaving the Military College in Paris, was reported on as follows:—

"Napoleon Buonaparte, a Corsican by birth, reserved and studious, neglectful of all pleasure for study; delights in important and judicious readings; extremely attentive to methodical sciences; moderately so as to others; well versed in mathematics and geography; silent, a lover of solitude, whimsical, haughty, excessively prone to egotism, speaking but little, pithy in his answers, quick and severe in repartee; possessed of much self-love, ambitious and high in expectation. This young man is worthy of protection."

This young man, "so worthy of protection," soon had a record. His favourite expression ran: "I pant after fame and notoriety; I wish to hold the world in chains." Whilst at college, he made one great friend, with whom he corresponded freely. On entering the service their paths became divergent—his friend went to Martinico and Napoleon remained in Paris. His first letter to his chum is dated De Guise, September 3rd, 1789.

The contents of the next chapter are taken from two letters of Napoleon to his college friend, D——, and bear date "De Guise, September 3rd, 1789," and "Toulon, December 17th, 1793." The latter describes the Revolution—events of which are known to everyone—but the original manner in which Napoleon writes, giving his ideas and opinions, creates an air of novelty, hence their publication. The replies to these letters, and also other letters of Napoleon (1790 to 1793) were seized by the police from his friend's house and handed over to Fouché.

CHAPTER VI.

"I am a lieutenant; is not that something? Well, it had but little effect upon me. To tell you the truth, the more decorated my uniform becomes, the higher my wishes aspire."—NAPOLEON.

"WHAT wonderful events have transpired since your departure! Men and affairs, master and servant, subject and prince, are no longer the same, nor in the same place. What, you will ask, is the king doing? Alas! he is acting the part of a good man, and of an excellent father of a family; he speaks with mildness and good nature to men who deserve to run the gauntlet. I would have made any sacrifice to have been his minister only for a month. I am only one-and-twenty, my friend, but I would have handled the innovators rather roughly. The prince, no doubt, would have been afflicted at my severity, for his fault is that he is too indulgent; but his family, his faithful servants, and every honest man in France, would have thanked me in their hearts.

"The very dregs of the people, whom I would, for the moment, have opposed, would now be under an obligation to me; for in the manner they are proceeding, it may yet cost them the last drop of their blood by their opposition; and yet, to stop the first inroads of an inundation, which will perhaps overwhelm the whole of Europe, the plan to be pursued was but trifling. I would have commissioned six regiments of cavalry, or more if necessary, to bring before me the haughty innovators, and thus would I have addressed them: 'Gentlemen, if it be within your jurisdiction to remonstrate with your sovereign, surely it was not becoming your duty, I will even say your dignity, to imitate those ballad singers, and place yourselves at the corner of every street to carol forth couplets to the populace, which the king alone ought to have been acquainted with. You have not only committed an error, but even a crime. Be it as it may, the king my master, as just as he is indulgent, will take your complaints into consideration, and forgive you, on condition that you will no longer secretly tamper with the people. If you counteract this order, you may expect to be sent to Toulon or Marseilles to ply the oar as galley slaves.'

"A threat, by the bye, which I would quickly and strenuously have put into execution, if the delinquents had swerved in the least from their duty. This remonstrance given in private, which I would have backed by that which causes every undertaking to succeed—an imposing force, ready for action—would, without fail, have repressed the indiscreet, or rather interested zeal of these remonstrance makers; the consequences would have been that the people would not have even thought of complaining.

"This is, my dear friend, the first spark of a conflagration which has kindled, and is gradually increasing; a conflagration, smouldering in the embers, from which I may hope to realise part of my expectations.

"Alas! why is our king so virtuous, so humane, and so mild? Why does he always show himself a good father, while his ungrateful and rebellious children are constantly plotting his ruin?

"He who first advised the king to assemble the States General, would well deserve the name and fate of Prometheus. What I predicted respecting the States General was realised much sooner than I had imagined—that impolitic and dangerous measure of convoking a large assembly of men in times of public effervescence puts every passion on the alert.......

"8th September.—I resume my pen, D——, and cannot help expressing my regret at your not being on the spot. The country which you have left, and the countrymen you have abandoned, are no longer morally in existence; there are absolutely new laws, new people, new manners.

"However dark the political clouds may appear, as they are settling over the French horizon, you would scarcely credit, when informed, my friend, that it is only the foreboding of the most dreadful storm. Yes, D——, the vessel of the State is threatened with hurricanes and multiplied destruction; I do not think it in the power of mortals to stem the raging winds that roar around it; a divinity alone can safely guide it into harbour.

"This penetration into the convulsions our country must necessarily undergo, makes me six feet higher than I was; I fancy to myself that fortune wishes to realise the truth of the note placed by my name on the list of candidates in the Military College. Yes, I aspire at an eminence, and circumstances will favour me; we can no longer doubt it. You will, then, D———, recollect the conversations of our youth, my first desires, my expectations, and the sentiments I confide to you in this despatch. You will say to yourself: 'He cherished every future hope by present anticipation, and my friend was not indulging in idle fancies.' May my first sheets be a convincing proof to you that I am insensibly approaching the goal....

"Toulon, December 17th, 1793.

"From Besancon, I arrived at Paris on the 10th of September, 1792. This was a new modelled city, a nursery for mean shufflers, seekers after preferment, cut-throats, and weak petty citizens, driven headlong from crime to crime, and who, after having proceeded so far, find it difficult to retreat; such a gang, collected in this gulf of perdition, was perfectly conformable to my aim. On the 22nd of the same month, royalty was abolished and the republic ordained; it was then that I chuckled with expectation, and not without a motive: brutes alone have no foresight.

"In the month of December I was still at Paris to solicit promotion. During all the time not given up to my solicitations, I was employed in sifting into men and motives; nor could I occupy myself more profitably. I frequented assiduously the sittings of the Convention; there I learned that the philosopher and the observer are one and the same person, and that at times, each may acquire more knowledge in fathoming a volcano than in investigating a bed of plants. Half a century devoted to study, and all the books that have been printed, could not have instilled the hundredth part of those valuable notions which I acquired in the bosom of this representative chaos. All the elements of moral evil were, I believe, there in fermentation. If fear still kept within it a few honest individuals (a circumstance probable enough), I am certain they must have been upon thorns.

"The 9th of December, the pleader Target informed me that the virtuous Louis XVI. would be brought to the Bar of the Convention on the 11th. At this information one half of my frame sprang with hope, and the other half shuddered with horror.

"A just and benevolent king, who has nothing wherewith to reproach himself but his unbounded clemency, brought on a sudden before his subjects, who have constituted themselves his judges, to be hereafter his executioners—such a man, I say, compelled to lay aside his name, his title, and his rank, is one of those terrible spectacles that rend the heart of the good man and faithful subject; that are awful and painful to the philosopher; that are rapturous and sublime to the ambitious. How blunted soever may be the feelings of a man, he would purchase, even at the price of his blood, the right of assisting at such a meeting.

"I arrived at the Convention; I seated myself; the monarch enters. 'Oh, Nature! Who ordered thee to imprint on the foreheads of legitimate sovereigns so much grandeur and dignity? Why am I not gifted with the same majestic appearance? How well would it suit my projects! But, no; indiscreetly liberal to this thy favourite class, thou art a rigid step-mother to others.' Such were the jealous reflections that assailed me, D——, I must confess it, at the sight of the king of France; reflections which, on an attentive consideration, were succeeded by milder sentiments. Nature, said I, is in the right, in having distinguished these privileged classes. Who will captivate by favour and benevolence, if it be not the man who governs his equals? This title alone, whether acquired justly or not, is sufficient to conciliate to us the protection of supernatural and divine beings.

"My looks were again turned towards the accused monarch. The serenity of his soul could be traced on his calm and dignified countenance; his innocence was manifest in his least motions; and the bench of criminals, on which he had been placed by crime and ambition, appeared to me a splendid throne of glory and majesty. Excuse me, my friend, if to display these noble scenes to you, I make use of an heroic style; but you know that everything that strikes my mind, assumes the height of a cedar. My thoughts are lofty, therefore my expressions are so; and that which would be ridiculous in another, in me is only the effect of nature.

"The king spoke; he was at once concise and sublime. I expected, at one time, that his judges would fall down at his feet; but I was deceived; they possessed the audacity and energy of wickedness. Silent, they looked upon the king and shuddered not. Neither the rank of the accused; his innocence, which no person present doubted; the serenity of his countenance, nor the august appearance of his whole person, had any effect on the iron hearts of these men, determined to brave both celestial anger and human revenge. Such audacity on their part raised them in my estimation. At least, said I, they act their part admirably; and their criminal insensibility is some recommendation: 'For if virtue has its heroes, so has guilt.'

"Shall I, D——, entrust you with a secret? Well, then, these guilty wretches made so deep an impression on me, that had it depended on myself to have been either a mere witness of the solemn condemnation they were about to pronounce, or one of those who condemned him, I should not have hesitated in my choice; as a subject of the prince, I would have taken my place among his judges. But there my ambition would have known where to stop; I would have boldly spoken in favour of the monarch, and probably have saved him, If I am mistaken, it is an idea that has often possessed me. Learn, D——, how I would have steered my course in this impetuous ocean. I would, in the first place, have secured a method for escaping; for he is only half a hero who falls a victim to his laurels. This important precaution being first taken, I would have rushed forward into the assembly, not to refute the accusation brought against the king of France—whose death was already secretly planned; and who was merely led forth for form's sake—but to cut short all deceitful decorum and proclaim the secret intentions of the guilty directors. Thus would I have addressed them:—

"Representatives, what has the King of France to do here? Why act a prelude to his death by judiciary forms? Why impose on yourselves the unpleasant task of supporting the majesty of his looks, the serenity of his countenance, the sublime self-consciousness of his innocence? Why torment yourselves to find out charges against him? Why require him to justify himself, when the irrevocable decree of his condemnation is written on your hearts? What! you, who have trampled under foot all decorum; you, who have broken through every tie, and destroyed every privilege; do you lower your tone now in order to feign, to ape justice? Gentlemen, you are out in your parts. On the footing you are now, your sovereign can no longer exist; you know it yourselves, but dare not proclaim it, as if afraid of withstanding the indignation of other nations. In the name of your formidable renown, be not mollified; be great and sincere in this illustrious murder; let one of you, like a sublime regicide, thus address Louis XVI.: 'Successor of Henry IV., henceforth France will no longer obey one king alone; we impose on it seven hundred sovereigns, consequently you can no longer live; prepare for death.'

"Do not you think, my friend, that such a speech would have changed the appearance of things? Would the Assembly have listened to it with indifference? Would it not, in short, have roused the energy of those who, in the Convention, were secretly inclined towards the prince? I am convinced that the numerous sensations produced would have brought on a better order of things. As to myself, wisely making my escape after the meeting, I would have taken refuge with the French princes, and trusted to them for my safety and advancement.

"These were, my friend, the aerial castles my imagination formed in the midst of this popular vortex. The monarch is at last directed to provide himself with a counsel, and his guards re-conduct him to the Temple.

"Next day I was informed that the pleader Target had refused his assistance to his sovereign. This was, in the full force of the term, blotting his own name from the annals of immortality. But this coward, out of prudence, replied to this: 'By risking my own life, I should not save his.' In that respect he was right, and had I been in his place, I should, perhaps, have acted in the same manner; for to tell you the truth, I love my own life. Malesherbes, Tronchet and Deseze,* devoted subjects whom I could imitate, but whom I would place on my right hand if I were a king, united themselves to defend the grandson of Saint Louis. Should they survive this courageous act, I will never pass by them without a respectful bow.

"20th December.—Having gone to Versailles on some private business, I did not return to Paris before the 16th of January. I lost, consequently, three or four scenes of this ambitious tragedy; but I was present on the 18th at the Convention. Ah! D———, in spite of all that may have been advanced by the furies of the revolution, a king is not a mere man. His head falls, it is true, like that of a peasant; but he who commits the murder, inwardly shudders at the deed; and if superior motives which direct him did not benumb his senses, he would not dare to utter the fatal condemnation.

"I eagerly contemplated the determined wretches who were about to pronounce sentence on their virtuous sovereign; I watched every motion of their countenances; I probed into their very hearts. The magnitude and importance of the crime alone supported them; the name and rank of their victim secretly alarmed them; and had they hesitated, the prince would have been saved. Unfortunately, they had said to each other: 'If this head do not fall to-day, ours must hereafter drop under the axe of the executioner.' This thought,

* This trio were devoted servants of the king. Malesherbes was a savant, and a man of heroic character, and has left a name endeared to France. He was guillotined the same day as his daughter and grand-children.

more than anything else, influenced the votes. What pen would be sufficiently skilful to depict the situation of the Assembly. Pensive, mute, hardly daring to breathe, their eyes were fixed alternately on the accused, the judge and the counsel. Unparalleled circumstance! D'Orleans exclaims: 'I vote for death......!'* Electricity itself could not produce a more instantaneous effect. Directors and judges rose dismayed, and the court re-echoed with a murmur of horror. One man alone, immovable as a rock, remained seated. I was that man! The reason I gave myself for this insensibility was found in my ambition; and as an ambitious man, the Duke of Orleans' action appeared to me perfectly natural. He aimed at a throne which belonged not to him, and certainly such an acquisition could not be made by a possessor of virtue and general esteem.

"Now, my friend, I must be brief; dismal subjects are not to my taste. The king was condemned to death; and on the 21st of the same month, if the French name was stained with an odious epithet, the martyrology was augmented with an illustrious name.

"What a city, D——, Paris presented on this conspicuous day! The populace was no longer furious, but horror-struck; some looked pensive, others as if struck dumb. The streets were deserted, and the whole population remained inactive. Houses and palaces were changed into sepulchres; the very air smelt rank with slaughter. At last, the grandson of Saint Louis was led to the scaffold between two files of mournful automata, lately his subjects.

"If any person should be present, were it even your father, when you are reading this dispatch, conceal carefully from him what follows, my dear friend; it is a stain on my character. Napoleon Buonaparte, deeply affected at the destruction of a mere man, and compelled to take to his bed in consequence of this impression, is an incredible fact, although a true one; so extraordinary an event, that I cannot own it without a blush and a despicable opinion of myself.

---

* Dame Du Barry, Josephine Beauharnais and Philippe d'Orleans were prisoners at the same time in the Bastille. At the trial Josephine was acquitted; d'Orleans was sentenced and guillotined on November 6th, and Dame Du Barry on December 8th, 1793.

Yes, D———, I had that weakness, glorious to any other, but disgraceful to me, who already wished to have nothing in common with the tenderness of the human heart.

"The night before the 21st of January, it was impossible for me to close my eyes, nor could I account for it. I rose early and visited all the quarters where the soldiers were drawn out; I wondered at, or rather I despised the silly complaisance of forty thousand National Guards, of whom nine-tenths were mechanically acting the parts of under-executioners. I met Santerre at the Gate St. Denis; a numerous retinue attended him. I would willingly have cut both his ears off; but as that was out of my power, I contented myself with casting a disdainful look on him. Such a wretch appeared to me unworthy of his mission. The Duke d'Orleans would have suited me in this station; he at least would have dishonoured himself; his aim was a crown, and we are all aware that such a motive causes many things to be overlooked.

"I crossed the boulevards and arrived at the Place de la Revolution. I had not till then seen the fatal invention of Doctor Guillotin; a cold perspiration seized me the moment I fixed my eyes on it. A stranger, who was next to me, attributing my paleness to the interest I took in the fate of the king of France, said to me: 'Don't be alarmed, he will not fall; the Convention only wish to prove the power they possess; his pardon awaits him at the foot of the scaffold.' 'If that be the case,' replied I, 'the Conventionalists are little aware that the gibbet awaits them; for never did rogues deserve it more. He who attacks a lion and does not wish to be torn into pieces, ought not to wound him, but kill him on the spot.' A mournful and sepulchral sound was heard: it was the august victim. I pressed forward; I pushed and was pushed; at last I approached as near as it was possible, but in vain—the scaffold was concealed by the armed force. A beating of drums suddenly disturbed the mournful silence of an immense crowd. 'It is the signal of his pardon,' said the stranger to me. 'So, then,' rejoined I, 'his condemnation was only intended as an insult by the Convention. If so, committing a crime by halves is a triple offence.' A momentary silence ensued; something suddenly fell heavy on the scaffold. This noise struck to my heart; I asked a gendarme what it was. ''Tis the fall of the knife,' answered he. 'Then the king is not saved? He is dead!......he is dead!' A dozen times I pronounced these words: 'He is dead!' For a few moments I was bereft of my senses, and not knowing who had extricated me from the crowd, I found myself on the Quai des Theatins. There I recovered, in some degree, my senses; but could articulate nothing but 'He is dead!' I returned home in a melancholy state, and an hour elapsed before I completely recovered. So much weakness and pusillanimity appeared ominous to my future projects. I gave myself such a lecture as I would not have easily borne from any other person.

"In order to dissipate my melancholy thoughts, I went to pay a visit to the Deputy Barbaroux, on whom I frequently called. But, alas! what an unexpected spectacle: his family were in tears, and I found that unhappy man completely distracted in consequence of the remorse he felt at having given his fatal vote against his sovereign. In the same manner as a Roman Emperor once cried out for the restoration of his legions, so did the Deputy cry out for the restoration of Louis XVI. Such imbecility on his part provoked me against him; I had forgotten that I, who had not the same cause to reproach myself, had been nearly in the same situation as he. What, said I to myself, is it such feeble-minded beings as these who rush into the gulf of heinous atrocities? Why did not the unhappy wretch remain peaceably in the bosom of his family concerns? I was on the point of returning home, when Fouché and Carnot entered. These were at least men, determined regicides! No impulse foreign to their ambitious projects had induced them to vote for death; they had reasoned well on their crime. In their actions as well as in their words could be constantly traced the tenacity of their opinions. If these men, thought I, have sinned, it was certainly upon mature reflection and for some purpose.

"Both of them shrugged up their shoulders at the sight of their delirious colleague. It was the first time I had seen Fouché, but by his appearance and discourse I could easily see he would go great lengths. I left Barbaroux's house, determined in my own mind never again to see a man ambitious enough to commit a heinous crime, and cowardly enough to repent of it. Whoever in a similar case cannot free himself from an accusing recollection, ought to kill himself, and not lament like a woman. Barbaroux, however, did not think proper to take this resolution, for he soon re-appeared on the political theatre, from which the executioner, with the stroke of an axe, drove him some time after.

"Excuse, my friend, the rapid view I have taken of this terrible catastrophe. It is not my intention to give you a detailed history of our revolution; I merely wish to particularise the principal facts, to enable you to judge of the events, and, what flatters me the most, to express to you with sincerity what I myself am, and my opinion of these political convulsions. You may easily collect the scattered outlines in these despatches, and compose a total, which will give you the measure and proportion of my ideas. If some of my propositions appear bold paradoxes, I forewarn you that you may expect many similar ones; I will, however, add that these apparent paradoxes cease to be such as soon as we see things as they really are. Examine thoroughly what follows—you will be convinced it is so.

"If the ambition of power and the love of riches, in certain individuals, have brought the king to the scaffold, both these passions have stirred up avengers in his cause, in his very executioners. Thousands of factions sprang from the grand national faction, and successively butchered each other. I mean not to fix your attention on this slaughter; my object is to fix it on something nobler and greater, and that is on myself. How great, indeed, I appeared in my own eyes! France streaming in blood, and groaning under her wounds, while I was secretly intoxicated with joy; a delight so much the less deserving of censure, as it was involuntary. Yes, D——, your friend felt happy in spite of himself. He did not reason on the cause of his joy, but welcomed it. Each faction, as it fell under the axe of another factious body, appeared to him to remove another obstacle to the high expectations raised in his breast by ambition. I hurl contempt at him who would condemn sentiments which could not be replaced by others.

"Had I not been ambitious, my want of fortune and the disorders of my country would have impelled me to take advantage of circumstances; and yet, when the royal cause was completely ruined, I was undecided as to the choice of the faction to which I should devote myself. Here, D——, I must place one of the most painful circumstances of my life; a circumstance which I should certainly have related before the death of the king, but which I have only kept back through the shame of owning it.

"What could, in fact, afflict me more sensibly at that epoch than the indifference or the contempt of any man? What! for me to condescend to write, and not to receive a reply!...... You know me, D——; you are aware what an exalted opinion I have of myself. Need I, after that, tell you how I resented the treatment of M. Montmorin? But this circumstance must be more amply detailed; history will necessarily gain by the knowledge.*

"If I needed any more to convince me how very wrong it was not to try the plan which I had conceived and traced, I will add that the axe of the regicides, delivered into the hands of the executioner, has brought the king, the queen, and her beloved sister to an untimely end. Their son, their young son, the presumptive heir, languishes in a prison, and is put to torture each successive day by a wretch who ought to die on the scaffold. Madame, like an angel of sweetness, sufferance and compassion, groans under pangs and painful recollections on a pallet in a dungeon. The unfortunate and elegant Princess de Lamballe, the valuable and virtuous friend of a queen yet more unfortunate, has been torn piece-meal, and her limbs, dripping with blood, were borne on the heads of her assassins,

* The plan which Napoleon had conceived and traced was to save his sovereign. The means he wished to adopt for preserving throne and king were grand in every detail, and were forwarded to M. Montmorin, Minister of State, on April 2nd, 1792; and naturally Napoleon was greatly incensed at no acknowledgment.

who paraded them through the capital. At Versailles, Brissac, Lessart, and thousands of others have fallen slashed and bleeding into the tomb. I shall proceed no further; the noise of the revolutionary hatchets, falling on the surface of this fair country, loudly condemns the rejection of my memorial, and compels me to repeat: What worse could have happened?

"A few months ago I took a journey to Corsica, to take advantage of any circumstances that might offer. I found people's minds there as well disposed as I could wish; the effervescence was general; meetings were forming, the members of which were composed of ambitious villains and petty citizens. On all sides were bandied about the words 'Liberty and Equality.' The opportunity was favourable; I entered into the sport as others did, although I cordially despised my companions. My fellow citizens were ensnared by my fair pretences, and named me Commander of the National Guard.

"Paoli, in the meantime, was forming very different plans. His fortune and reputation being made, he wished for a peaceable and steady revolution. Two parties were formed; we became enemies. I kept an equal balance with him for some time; but his name, his exploits, his intrigues gained an ascendancy, and my family, with myself, were banished from Corsica. Never did I so much regret knowing the value of existence; I could have blown my brains out; but convinced that of all acts of cowardice suicide was the greatest, I determined to live. Madame Buonaparte said to me one day, 'Why do you give yourself up to such transports of rage? It is so noble, so great to show oneself superior to a reverse of fortune! and what is it after all? a contrariety, a mere petty vexation. What would you think if I were to tell you that it is perhaps a fortunate circumstance; that the decree which proscribes you may become the diploma of the most brilliant dignities, and the brevet of a high renown. Napoleon, Corsica is but an uncultivated rock, an imperceptible and miserable speck of land; France on the contrary is extensive, rich, populous, and in flames. This, my son, is a noble conflagration, worth risking being scorched at.'

"Never did a mother receive so cordial an embrace from a son as Madame Buonaparte received, in return for the observations she had just made. How did her words vibrate in my ears! 'Corsica is but an uncultivated rock, a speck of land; France on the contrary is extensive, rich, populous,' etc. This was enough to bring me back to milder sentiments, and render my exile much less painful to support. This, D———, is what happened to me in Corsica; this is what I wished to conceal, and which I now briefly relate to you without comments and reflections.

"This first slight received in Corsica, and my memorial at Court unanswered, irrevocably threw me into the revolutionary party. To what cause could I then devote myself, since the minister actually removed me from that of the sovereign?

"I received orders to proceed to Toulon.* There I was compelled, as well as several others, to undergo many vexations, by manœuvring as directed by men entirely ignorant of the military art—I allude to those representatives sent by the Convention to head the armies, by which they were universally despised.

"I was directed to cannonade the forts of La Malgue and Malboquet, to drive the enemy from their position. Barras and Fréron, commissioned to direct the siege, assumed the men of

---

* A group of three, with scores of other young English officers, received their baptism of fire at Toulon (November and December, 1793). The three were Graham, Beresford and Hill, the only Peers made for the Peninsular War.

Graham, aged 45, in search of adventure for the cure of grief, became a volunteer under Hood and Lord Mulgrave; he died in 1843, after earning his Peerage for Peninsular work (Lord Lynedoch).

Beresford—Ensign 6th Foot—a grand General—worked up the Portuguese Army into fighting trim, and gained laurels in every battle, and a Peerage for the Peninsular War (born 1768, died 1854). He would have succeeded Wellington in the field had any untoward event happened.

Hill, as Ensign 36th Foot, was aide-de-camp to Mulgrave and O'Hara, and was quickly noticed for his coolness and bravery. He died in 1842, with a Peerage for Peninsular work; Commander-in-Chief of English Army. When Wellington was Prime Minister, he was frequently offered the Governor-Generalship of India. Born 1772, died 1842.

Junot was a sergeant.

experience, and regulated my batteries; I left them to do as they chose, being well aware they were acting wrong. Ten days, in short, passed, and I had merely wasted powder and shot. It was then time, I thought, for me to prove to the representatives that if they knew how to make fine speeches and gibbet antagonists, they knew nothing about destroying a fortress. I said not a word to them of my plan, for fear of engaging in a dispute with them. In the night I fixed, to the eastward of my other redoubts, a plain battery of eight pieces of twelve pounders and two howizers; the work was carrying on when they appeared. 'What is this battery for?' said they to the Captain employed. 'A new one ordered by the Commander.' I was within three yards of them, and heard everything, but said not a word. Fréron viewed with his telescope, and desired them to leave off, that the battery was useless. I immediately came up. 'This battery shall remain,' said I; 'I answer with my life it will succeed; follow your employment, and leave me to follow mine.' Astonished at such boldness, the Commissaries were silent for a moment. Barras placed himself in the angle designed for the howizers, examined for some time their supposed direction, and retired at a little distance, taking Fréron by the arm. I know not what he said to him, but returning to me, 'Well, go on, we shall see whether you do not presume too much.'

"No, I did not presume too much; I had discovered the weak side of the two forts. I took them at half-flank; by this means three-fourths of the defenders were exposed to my fire. The day after, I made my first assault; the success trebled my expectations. My balls completely swept away the ramparts, and the forts were a few days after taken. What will, no doubt, surprise you, my friend, is that I had the idea of this battery before the representatives had ordered those which only served to lengthen the attack and waste ammunition. I might certainly have made the observation to the Commissaries, and pointed out its utility; but that would only have been rendering a service to the State, and what I most feared, to the Deputies, who probably would have arrogated to themselves all the honour. How much wiser did I act! None but a blockhead can deny me this consequence. The representatives, with whom I had acted bluntly, and whom, perhaps, I had inwardly humbled, were prudent enough not to resent it. Pleased at having gained possession of so fine a harbour, they recompensed me for the part I had taken in the acquisition, by naming me Brigadier-General; I am waiting for the commission to sign in that quality. Not that I am prevented from so doing at present through modesty, but that I fear the Convention may not ratify the determination of the representatives, as their power is subject to so many variations. Be it as it may, I trust that this despatch, which has been taken up at different times, will furnish you matter for reflection, not only on the great catastrophes of which your country is the witness and the victim, but also on the expectations I form for myself. Your approbation, I am well aware, will not extend to all the circumstances of my life, but I owe you no ill-will for that; you shall, nevertheless, be my confidant and friend, as much as one man can be so towards another. I am, moreover, persuaded that I alone can appreciate and approve myself. This conviction is one of my greatest satisfactions. Why have I anything in common with other mortals? I would wish to be completely a man apart. I possess, however, the sole approbation that I aspire after—that is my own.

"I conclude for the present, my dear D———, happy in the expectation that these sheets will convince you that I am making hasty strides towards the object of my wishes, which you may rest assured, should I ever reach, the friend of my youth and the confidant of my most secret thoughts shall not be forgotten. Then will be the time, my friend, when I shall be able to give you proofs of the inviolable attachment with which I subscribe myself yours,

"Napoleon Buonaparte."

CHAPTER VII.

"**Ships are but boards, sailors but men.**"

THE National Convention declared war against England early in 1793; and with flying colours, on June 21st, young Edward Pellew (Lord Exmouth) entered Portsmouth Harbour with "La Cléopatre" in tow—the first prize of the war—proving that he had the gifts of initiative and resource. "For a captain deserves more praise for any particular detached action with the enemy, than one serving in a fleet who only obeys the orders and signals of his superior officer."

Pellew reports his capture as under:—

"Here we are, thank God! safe after a glorious action with 'La Cléopatre,' the crack ship of France (40 guns and 320 men). We dished her up in fifty minutes, boarded, and struck her colours. We have suffered much, but I was long determined to make a short affair of it. We conversed before we fired a shot, and then, God knows, hot enough it was, as you will see by the enclosed list of killed and wounded."

On Sunday, the 23rd, the French captain, who fell soon after the action began, was buried by Captain Pellew in Portsmouth Churchyard. The body was followed only by his own officers; the inscription on the coffin was "Citoyen Mullen, slain in battle with 'La Nymphé,' June 18th, 1793, aged 42."

Young Pellew, when placed in command of his frigate (36) the "Nymphe," owing to the scarcity of able seamen, was unable to make up her complement. In this dilemma, he wrote to his brother at Falmouth to assist him in procuring a crew. "Sailors," he said, "*if possible*, and if not sailors, Cornish miners, for they understood discipline and could wrestle, therefore board a vessel." Eighty miners entered; and his brother, Captain Israel Pellew, likewise came to work the guns, for he was an excellent artillerist.

At this same time the Convention, or Republican Government of France under Robespierre, was engaged in subduing the Royalists of the South. One army had taken Marseilles; but Toulon had sought protection from the English, and was holding out under the flag of Lord Hood, Commander-in-Chief in the Mediterranean. Hood had already the Spaniards on his side, but requiring further help, he sent Nelson with despatches to Sir Wm. Hamilton, our Minister at Naples, acquainting him with the situation, and petitioning for more troops, either Neapolitan or Sicilian. General Acton sent a contingent of both. At dinner on the night of Nelson's arrival, Sir William told his lady that he was about to introduce a little man to her acquaintance, who could not boast of being very handsome; "but," added Sir William, "this man, who is an English naval officer—Captain Nelson—will become the greatest man that ever England produced. I know it from the few words of conversation I have already had with him. I pronounce that one day he will astonish the world. I have never entertained any officer at my house, but I am determined to bring him here. Let him be put in the room prepared for Prince Augustus."

The troops Nelson went to requisition duly reached Toulon in October, 1793, making a total of 16,000 men.

Robespierre and his agents soon plotted with the Spaniards that on being attacked they were to make off at once in flight. The fight occurred, and upon General O'Hara perceiving the *emeute* on the part of the Spaniards, he hastened to rally them, but quickly fell severely wounded. With him was Lieut. Duncan, R.A., who was bayonetted in the breast, but escaped to the beach a good half mile distant; here the bayonet was withdrawn, and he was carried by the sailors to the nearest ship. O'Hara was taken prisoner.

The French despatch reads thus:—

"Cette sortie enlevé a leur armée plus de douze cents hommes; tant tués que blessés et faits prisonniers; parmi ces derniers, plusieurs officiers d'un grade superieur; et enfin leur General en Chef, M. O'Hara, blessé d'un coup de feu au bras droit. . .

"Parmi ceux qui se sont le plus distingués, et qui m'ont le plus aidè a rallier et pousser en avant, ce sont les citoyens Buonaparte, Commandant l'Artillerie; Arena and Gervoni, Adj. Generals.

"Dugommier,
"General-en-Chef."

Five years later, when Napoleon writes to Kléber from Cairo, September 10th, 1798, he contrasts the conduct of his Contre-Amiral Duchaila with O'Hara:—

"Quant a la conduite du Contre-Amiral Duchaila il êût été beau pour lui de mourir sur son banc de quart, comme du Petit Thouars.

"Mais ce qui lui ôte toute espècé de retour a mon estime, c'est la lâche conduite avec les Anglais depuis qu'il a été prisonnier. Il y a des hommes qui n'ont pas de sang dans les veines.

"Il entendra donc tous les soir les Anglais en se soûlant de punch, boire a la honte de la marine française. How different with O'Hara, l'orsqúil ful fait prisonnier à Toulon sur ce que je lui demandais de la part du General Dugommier, ce qu'il desirait, répondit, 'Etre seul et ne rien devoir a la pitié.'

(Signe) "Buonaparte."

Lord Hood, now finding the position no longer tenable, prepared for sea, but not before Sir Sydney Smith,* Admiral Trogoffe (a French Royalist), Captain Elphinstone (Lord Keith), Beresford, Hill, Graham,

---

* Napoleon, at St. Helena, states: "Sydney Smith landed at Havre, for some sottise of a bet he had made, according to some to go to a theatre; others said it was for espionage; however that may be, he was arrested and confined in the Temple as a spy; and at one time it was intended to try and execute him. He is active, intelligent, intriguing and indefatigable; but I believe he is *mezzo pazzo*." Sydney Smith escaped from the Temple on the eve of Napoleon's departure for Egypt, May, 1798.

Iremonger, etc., had burnt the ships, stores, arsenals, etc., and had embarked all non-combatants and Royalists.

Lord Hood's next move was to drive the French from Corsica. Accordingly Colonel Moore (Sir John Moore) and Major Koelher were sent ahead to reconnoitre and report; upon which report, Lord Hood proposed that the 11th, 25th, 30th and 69th Regiments, with a full complement of sailors and their guns, should land and capture Bastia. However, in this scheme General Dundas would not co-operate; he preferred waiting for a further reinforcement of 2000 men he expected from Gibraltar. Lord Hood could not brook this delay; he therefore determined to attempt the reduction of Bastia with the naval force entrusted to his command, and ordered the details of the 11th, 25th, 30th and 69th Regiments to return on board the respective ships on whose roll they were borne as marines, when Lord Hood (as Admiral) disembarked them again as troops under his own command, and commenced the siege of Bastia.

Captain Nelson had rejoined from Naples, and with Hunt, Sericold and Bullen, commanded the brigade of bluejackets ashore.

Bastia surrendered on May 21st, 1794, to "a detachment of British seamen and marines or soldiers acting as such, and who had no tents but such as were made of sails, and no other battering train than the lower deck guns of line-of-battle ships."

Lord Hood rendered to those serving under him unstinted praise:—Lieutenant-Colonel Villettes, Major Brereton, Captains Nelson, Hunt, Bullen, Sericold, Wolseley, Hallewell, Inglefield and Knight, Lieutenants Gore, Hotham, Stiles, Andrews and Brisbane—and concluded his despatch thus:—

"I cannot but express in the strongest terms the meritorious conduct of Captain Duncan and Lieutenant Alex. Duncan of the Royal Artillery, and Lieutenant de Butts of the Royal Engineers; but my obligation is particularly great to Captain Duncan, as more zeal, ability, and judgment was never shewn by any officer than was displayed by him; and I take the liberty of mentioning him as an officer highly entitled to his Majesty's notice."

Nelson lost an eye at this siege by the up-kick of some sand and gravel, caused by a passing shot. His health suffered in consequence, for the wound was most painful. He was heard to say when the "Gazette" was published—not *apropos* of the Duncan mentioned, for he dearly loved a soldier—"They have not done me justice; but never mind, I'll have a 'Gazette' of my own."

Lord Hood was warmly thanked by both Houses of Parliament. His age was then seventy-four, and surely a less strenuous life was needed. So he accepted his well-earned retirement. He died at Bath in 1816, aged ninety-two. Lord Hood and Lord Bridport were brothers.

A question may now be asked—"Why had this attack on Bastia been deemed impracticable and visionary by General Dundas?" The readiest answer seems that the army at the commencement of this war (1793-4-5) was, in conjunction with the Austrians and Russians, on service against the French in Flanders, and was commanded by H.R.H. the Duke of York, of no military aptitude whatsoever. So in November, 1794, when Lord Hood was thanked by Parliament for doing soldier's work at Bastia, we have before us a letter from the Head Quarters of our army at Arnheim, November 11th, 1794:—

"We are really come to such a critical situation, that unless some decided, determined and immediate steps are taken, God knows what may happen. Despised by our enemies, without discipline, confidence or exertion among ourselves, hated and more dreaded than the enemy even by the well-disposed inhabitants of the country, every disgrace and misfortune is to be expected. You must thoroughly feel how painful it must be to acknowledge this, even to your lordship, but no honest man who has any regard for his country can avoid seeing it.

"Whatever measures are adopted at home, either for removing us from the Continent or for our remaining, something must be done to restore discipline and the confidence that always attends it.

"The sortie from Minequen on the 4th was made entirely by the British, and executed with their usual spirit; they ran into the French without firing a single shot, and consequently lost very few men—their loss was when they were afterwards ordered to retire."

The 12th Regiment had been sent from Ireland to reinforce the troops under the Duke of York for this campaign in Flanders, together with the 38th and 55th Regiments, and the 8th Light Dragoons. The regiment, commanded by Major Bewes, took part in numerous operations, and was complimented twice in General Orders for its gallant behaviour, and for the readiness and good-will with which it had borne so many and great fatigues.

The battle of Boxtel, fought by the 12th, 33rd, 42nd and 44th Regiments, and some squadrons of cavalry, together with a brigade of guns, against the French, is interesting to the British as being the maiden battle of our illustrious Duke of Wellington. The Duke was then Lieutenant-Colonel Wellesley, commanding the 33rd Regiment.

A detachment of the 12th Regiment, in this battle, defended a post with great gallantry against an overwhelming force; their loss was Lieutenant Eustace, three sergeants, one drummer and forty-four rank and file.

It is too tedious to wade through the particulars of this Continental campaign. It is sufficient to say that after the battle of Boxtel, the Duke of York, with his army, was driven across the Meuse; they fought again at Bommel, captured Venloo and Minequen, and then retired behind the Rhine—all of which movements put to a severe test the strength of the officers and soldiers; for besides a deficiency of provisions, their marches lay along roads choked with ice and snow. The 12th Regiment was reduced in this campaign from 815 to 429 rank and file.

Upon receipt of this grave military intelligence, Pitt insisted upon the immediate recall of the Duke of York from service. On his return home, however, the king placed him at the War Office as Commander-in-Chief, with the rank of Field Marshal; and the General Dundas aforementioned was of the same school of tactics of whom the King and Court had the highest opinion, so tightly had he dressed, and so accurately had he drilled the Guards. General Dundas became Commander-in-Chief, 1809.

How this state of things must have worried a promising soldier like Arthur Wellesley, who, we remember, entered the army in the same year as Napoleon (1787)—first into the 73rd Regiment; then lieutenant into the 76th; in 1789 promoted to the 12th Light Dragoons; in 1791 promoted captain of the 58th; in 1792 exchanged into the 18th Light Dragoons; in 1793 obtained majority in the 33rd Regiment; and in 1794 succeeded to the lieut.-colonelcy. He was on service throughout the above wretched campaign in Flanders, being present at Boxtel, Venloo and Minequen. His next move was in 1795 to embark with the 33rd Regiment for the West Indies,* for concurrent with the fighting under York in Holland, just narrated, was an expedition to the West Indies under Sir Charles Grey and Sir John Jervis. Grey we recognise as the young aide-de-camp at Minden, and the colonel of the 20th—next brigadier. His nickname was "No Flint." So much did he value infantry fire, that he loved to decide himself the when and where for his men to fire, and he took away the flints to prevent all heedless shooting.

With this expedition were the two flank companies of the 12th. The islands, then so wealthy, must be wrested from the French. All went well at first, and England flew her flag over each for a while; but then came sickness, followed by a French attack, when alone from Guadaloupe, 125 men crept out like skeletons to surrender.

Thus had England three expeditions on her hands at once—Toulon, Holland and the West Indies—all three dismal failures, due to being weak everywhere and strong nowhere. When the flank companies rejoined, their strength was one officer, one sergeant, one man.

Wellesley never reached the islands, for after five weeks' ineffectual attempts to get out to sea, the squadron was obliged to

---

* Fort Royal was the birthplace of Josephine (June 23rd, 1763). A master sculptor has given great charm to this Creole face—now reposing as a statue in the Market Square.

return to Portsmouth to refit. Then the destination of the 33rd was changed, and in April, 1796, Colonel Wellesley sailed for the East Indies.

"There's a Divinity that shapes our ends,
Rough hew them how we will."

France, now so proud of her feats on land, expected the same at sea, so only gave the command of her ships to men of a determined spirit—staunch Republicans, and such enthusiasts as to take an oath never to strike the national colours. Such were the men Lord Howe had to contend against in his glorious actions of May 29th and June 1st, 1794, when he dismasted thirteen out of twenty-seven line-of-battle ships, and took seven. This defeat came to the French as a great wound to their national vanity.

When the King, Queen Charlotte, and three Princesses went to Portsmouth to welcome Howe on his return, he said: "'Tis not I; 'tis those brave fellows," pointing to the seamen, "who have gained the victory."

The "Queen Charlotte," Lord Howe's flag ship, was blessed with a wonderful Commander—Bowen by name. He addressed Lord Howe frequently during the action by his title, and the following grateful and animated reply was overheard by the officers on board: "Mr. Bowen, you may call me 'My Lord' and 'My Lord!' You yourself deserve to be a prince!" This officer was at once deservedly raised to the rank of Post-Captain.

The "Naval Chronicle," of this same good officer Captain Bowen, gives the following extract of a letter from an officer on board his Majesty's ship 'Argo,' dated Port Mahon, May 29th, 1799:—

"We are just arrived from Algiers, where we have been to negotiate with the Dey for a supply of fresh provisions for the Army and Navy at Minorca. Captain Bowen was so much in favour with the Dey and Regency there, that he procured the freedom of six poor fellows—British subjects—that had been confined in slavery upwards of fourteen years. They are now on board, enjoying the blessings of true liberty, serving their king and country, and have requested me to get their names inserted in some of the English newspapers, that their friends, if any living, may know that they are now happy and comfortably situated. Before we came away, the Dey, as a further mark of his friendship for Captain Bowen, requested him to accept of a rich Turkish sabre and two fine Arab horses."

The Reverend Bowen Armstrong, here in Brighton, is a great-grandson of the same.

Admiral Hotham, in 1795, gained further triumph, for in a stiff fight in the Leghorn Roads he captured the "Ca-Ira" (of 80 guns) and the "Censeur" (of 74); and in the same year (June 23rd, 1795), Alexander Hood, younger brother to Lord Hood, after a well-fought and glorious action off Port L'Orient, captured "Le Formidable" (74), "Le Tigre" (80), and "Alexander" (74). Hood in this action took charge of his own ship, for the pilot refused to move nearer to the coast. The vessel, perhaps, was dangerously near, for five of the French Captains were broken for not taking the "Queen Charlotte." Alexander Hood was created Baron Bridport for this victory. He died in 1797.

In proportion as the success of the English at sea continued, so did the rivalry of other Powers increase. Spain threw in her lot with France, and ordered her fleet (twenty-seven sail of the line) to effect a junction with the French at Brest, and to enter the English Channel.

To prevent this junction became the question of the moment. At first it seemed impossible, for Jervis had but fifteen ships of the line; but remembering

> "Our doubts are traitors, and make us lose
> The good we oft might win, by fearing to attempt,"

was off at once, and had the glory of capturing four of the largest Spanish ships; the others that could, sought shelter in Cadiz.

Lady Hamilton had warned Jervis also of another likely coalition, for during the siesta of the King of Naples a letter had fallen from his pocket, in which the King of Spain begged of his brother to join him in this fight. Nelson's share in the fight is given in his own words:—

"At one p.m., the 'Captain' having passed the sternmost of the enemy's ships which formed their van and part of their centre, consisting of seventeen sail of the line—they on the larboard, we on the starboard tack—the Admiral made the signal to tack in succession, but perceiving all the Spanish ships to bear up before the wind, evidently with an intention of forming their line, going large, joining their separated divisions, at that time engaged with some of our centre ships, or flying from us—to prevent either of their schemes from taking effect, I ordered the ship to be wore, and passing between the 'Diadem' and 'Excellent' at a quarter past one o'clock, was engaged with the headmost, and, of course, leewardmost, of the Spanish division. The ships, which I knew, were the 'Santissima Trinidada' (126), 'San Josef' (112), 'Salvador del Mundo' (112), 'San Nicholas' (80), another first-rate, and a 74, names unknown.

"I was immediately joined and most nobly supported by the 'Culloden' (Captain Troubridge). The Spanish fleet, not wishing, I suppose, to have a decisive battle, hauled to the wind on the larboard tack, which brought the ships above mentioned to be the leewardmost and sternmost ships in their fleet. For near an hour, I believe (but do not pretend to be correct as to time), did the 'Culloden' and 'Captain' support this apparently but not really unequal contest; when the 'Blenheim' passing between us and the enemy gave us a respite, and sickened the Dons.

"At this time the 'Salvador del Mundo' and 'San Isidro' dropped astern, and were fired into, in a masterly style, by the 'Excellent' (Captain Collingwood), who compelled the 'San Isidro' to hoist English colours; and I thought the large ship, 'Salvador del Mundo,' had also struck; but Captain Collingwood, disdaining the parade of taking possession of a vanquished enemy, most gallantly pushed up, with every sail set, to save his old friend and messmate, who was to appearance in a critical state; the 'Blenheim' being ahead, the 'Culloden' crippled and astern. The 'Excellent' ranged up within two feet of the 'San Nicholas,' giving a most tremendous fire. The 'San Nicholas' luffing up, the 'San Josef' fell on board her; and the 'Excellent,' passing on for the 'Santa

Trinidada,' the 'Captain' resumed her station abreast of them, and close alongside. At this time, the 'Captain' having lost her fore-top-mast, not a sail, shroud nor rope left, her wheel away, and incapable of further service in the line, or in chase, I directed Captain Miller to put the helm astarboard, and calling for the boarders, ordered them to board.

"The soldiers of the 69th, with an alacrity which will ever do them credit, and Lieutenant Pearson of the same regiment, were almost the foremost on this service. The first man who jumped into the enemy's mizzen chains was Captain Berry, late my first lieutenant (Captain Miller was in the very act of going also, but I directed him to remain); he was supported from our sprit-sail yard, which hooked in the mizzen rigging. A soldier of the 69th Regiment having broke the upper quarter-gallery window, I jumped in myself, and was followed by others as fast as possible. I found the cabin doors fastened, and some Spanish officers fired their pistols; but having broke open the doors, the soldiers fired, and the Spanish Brigadier (commodore with a distinguishing pendant) fell, as retreating to the quarter-deck. I pushed immediately onwards for the quarter-deck, where I found Captain Berry in possession of the poop, and the Spanish ensign hauling down. I passed with my people, and Lieutenant Pearson, on the larboard gangway to the forecastle, where I met two or three Spanish officers prisoners to my seamen; they delivered me their swords. A fire of pistols, or muskets, opening from the Admiral's stern gallery of the 'San Josef,' I directed the soldiers to fire into her stern, and calling to Captain Miller, ordered him to send more men into the 'San Nicholas,' and directed my people to board the first-rate, which was done in an instant, Captain Berry assisting me into the main chains. At this moment a Spanish officer looked over the quarter-deck rail, and said they surrendered. From this most welcome intelligence, it was not long before I was on the quarter-deck, where the Spanish captain, with a bow, presented me his sword, and said the Admiral was dying of his wounds. I asked him on his honour if the ship was surrendered? He declared she was; on which I gave him my hand, and desired him to call on his officers and ship's company and tell them of it, which he did; *and on the quarter-deck of a Spanish first-rate, extravagant as the story may seem, did I receive the swords of vanquished Spaniards, which, as I received, I gave to William Fearney, one of my bargemen,* who put them, with the greatest *sang froid,* under *his arm.* I was surrounded by Captain Berry, Lieutenant Pearson of the 69th Regiment, John Sykes, John Thomson, Francis Cooke, all *old 'Agamemnons,'* and several other brave men, seamen and soldiers. . . Thus fell these ships!

"N.B.—In boarding the 'San Nicholas,' I believe we lost about seven killed and ten wounded, and about twenty Spaniards lost their lives by a foolish resistance. None we lost, I believe, in boarding the 'San Josef.'

(A COPY) (Signed) "HORATIO NELSON.
"RALPH WILLETT MILLER.
"T. BERRY."

After St. Vincent, Nelson wrote the following:—

"'Irresistible,' off Lisbon,
"February 26th, 1797.

"SIR,—Having the good fortune, on the most glorious 14th February, to become possessed of the sword of the Spanish Rear Admiral Don Xavier Francesco Wintheysen, in the way set forth in the paper transmitted herewith; and being born in the county of Norfolk; I beg leave to present the sword to the city of Norwich, in order to its being preserved as a memento of this event, and of my affection for my native country.

"I have the honour to be, Sir,
"Your most obedient Servant.
"HORATIO NELSON.

"To the Mayor of Norwich."

Nelson received for this action a pension of £1000 per annum for his many wounds and scars. Before the issue of this grant, custom required that he should distinctly state his services. The following is a copy of his letter:—

"To the King's Most Excellent Majesty, the Memorial of Sir Horatio Nelson, K.B., and a Rear-Admiral in your Majesty's Fleet.

"That during the present war your Memorialist has been in four actions with the fleets of the enemy, viz., on March 13th and 14th, 1795, on July 13th, 1795, and February 14th, 1797; in three actions with frigates; in six engagements against batteries; in ten actions in boats employed in cutting out of harbours; in destroying vessels; and in taking three towns. Your Memorialist has also served on shore with the Army four months, and commanded the batteries at the sieges of Bastia and Calvi.

"That during the war he has assisted at the capture of seven sail of the line, six frigates, four corvettes, and eleven privateers of different sizes; and taken and destroyed near fifty sail of merchant vessels; and your Memorialist has actually been engaged against the enemy upwards of one hundred and twenty times.

"In which service your Memorialist has lost his right eye and arm, and been severely wounded and bruised in his body. All of which services and wounds your Memoralist most humbly submits to your Majesty's most gracious consideration.

(s) "NELSON."

The gentlemen of England who lived at home at ease could hardly have fathomed all this; for on the Mayor and Corporation of Yarmouth presenting Nelson with the Freedom of the City, and when the usual oath was tendered to his Lordship, who placed his left and only hand on the Bible, the Town Clerk officiously exclaimed: "Your right hand, my Lord." His Lordship, with a good-humoured smile, replied that he only wished he had.

At 3 p.m. February 15th, Sir John anchored with the fleet, including his four prizes, in Lagos Bay, where he put on shore about 3000 Spanish seamen and soldiers. Scarcely, however, had the *Te Deums* and rejoicings for this great victory died away at home, than the country learnt with great sorrow of the discontent of her Navy. The men were not happy, so much so that a message from his Majesty to the Commons had been handed in by Lord Grenville, in extract thus :—

"That amongst that class of men to whom the country had at all times looked up as its greatest and best bulwark of defence, it was, however, certain that a very large part of that body of men were in a state of the greatest insubordination and mutiny; and that such a disgrace to the character of the seamen had never before happened in the annals of our history, and every principle of regard to the honour and interest, nay, the dearest interests of the British Empire, required that such a mutinous and rebellious spirit should be immediately counteracted and crushed."

Mr. Pitt then rose, and in the course of his speech, said: "But, sir, when we recollect the conduct and character of the English sailors, a race of men who established the glory and renown of the British Navy, and have raised the country to such power and security; I say, sir, when we recollect that conduct, I say, sir, that it cannot be in their hearts such principles have originated. There must be domestic enemies who have been endeavouring to pervert the principles of the sailors. We must harbour against them a greater degree of indignation than against the misguided men who have been the object of their seduction."

Mr. W. Smith, of Norwich (grandfather to Florence Nightingale); Mr. Sheridan (great-grandfather to the late Lord Dufferin)*; Mr. Dundas (afterwards Lord Melville); Mr. Percevel (assassinated May 11th, 1812), all followed on in debate, when the Bill was read a second time, and Mr. Pitt gave notice that the House would proceed to business to-morrow at four o'clock.

---

* "Good God! sir! let us recollect that one remedy was the system of barracks. We were told that the measure carried this good effect with it, that it would keep the soldiers out of the reach and influence of sedition; that if the people could not be made dumb, the soldiers should be made deaf. What is the consequence?—the evil has increased."

The Stock quotations for that day were 3 per cent. Consols, 47½.

In the local news of the day we read that "The 'Sans Pareil,' 'Russell,' and 'Cæsar' were ordered yesterday for the Downs. The ships got up one anchor this morning, but the 'Cæsar' has refused to go to sea this evening until there is a fresh regulation about prize money."

Sir John Jervis continued off Cadiz, and a great jealousy had taken place between the line-of-battle ships and the frigates, as the latter made rich prizes, while the former were employed in the unprofitable employment of blocking up the port of Cadiz—consequently had no chance of prize money.

Notwithstanding these dissensions and sad misunderstandings, the British Navy was at heart as true as steel, and never more terrible to its enemies than in this eventful year, when Duncan, to prevent the Dutch from quitting the Texel to effect a junction with the French at Brest, set out with the "Venerable" (74) and "Adamant" (50), and there kept watch over the Dutch De Winter with fifteen sail, until reinforcements could reach him—for Duncan's other ships had joined the mutineers at the Nore. "In war take all 'vantages"—and Duncan caused frequent signals to be made, as if the main body of the fleet was in the offing. This stratagem succeeded. In the middle of June his penitent ships returned and watched until October, when a portion of the fleet had occasion to return to England to re-victual. The Dutch then broke the blockade. Duncan soon heard the news, was quickly on their heels with his eleven re-victualled ships, and without even forming line he broke through the enemy's array in only nine fathoms of water—and captured many prizes. This hard-fought "stand up" fight raised great enthusiasm at home. Duncan became Lord Camperdown and Onslow a baronet. Alas! one captain—of the "Agincourt" (64)—brought his vessel out with not one man killed or wounded, for he had avoided the peril. He was tried by Court Martial, and declared to be thenceforward incapable of serving in the British Navy. Duncan's prizes all reached the Nore on October 17th, 1797.

This victory came as a blessing in disguise to rebellious Ireland, for had the Dutch fleet escaped and become a fleet in being, it surely would have hardened their resistance.

It was during this fight that De Winter sent up a broom to his masthead, evidently as a signal "to sweep the English off the seas." A prompt reply came from Duncan, for he plaited some bunting into the shape of a whip—now called the pennant—and De Winter was whipped off the scene.

The great merit of Admiral Duncan consisted in running his fleet between the enemy and a lee shore—a precursor of the tactics of the Nile, for Nelson was always proud and pleased to say that Duncan had taught him a lesson, or a precept, which he put into practice the following year.

CHAPTER VIII.

"To be governed by our equals is well; to be governed by our superiors is better; but to be governed by our inferiors is pitiful."

"How hard it is to climb the steep, where
Fame's proud temple beams afar."

BUT not so with Napoleon, for in less time than it takes many an officer to obtain one step in rank, this wonderful man had shot off like a meteor through the sky; and from 1787 to 1804, a short seventeen years, he had graduated as under:—

Born	August 15th, 1769
Entered Military School at Brienne	1781
Transferred to Military College, Paris	1785
Lieutenant, 1st Artillery Regiment,	September 17th, 1787
Captain	February, 1792
Chief de Battalion	October, 1793
General of Brigade	February, 1794
General-in-Chief, Army of Interior ...	October, 1795
General-in-Chief, Army in Italy ...	February, 1796
Married	March 9th, 1796
First Consul	December, 1799
Consul for life	August 2nd, 1802
Emperor	May 18th, 1804
Crowned	December 2nd, 1804

"En 1810 alors l'empire eut 130 departements; il s'etendit de l'océan Breton aux mers de la Grèce, du Tage jusqu'a l'Elbe, et 120 millions d'hommes, obéissant à une seule volonté, soumis à un pouvoir unique et conduits dans une même voie crièrent, vive Napoléon en huit langues différentes."

He had seen his country pass from the meeting of the States-General in 1789 to the dethronement of the King on August 10th, 1792—"this was the reign of constitutional royalty." Second, from the abolition of royalty, 1792, till October 26th, 1795—this was the period of "the revolutionary government and reign of terror, ill-disguised under name of Republic." Third, from the establishment of the Constitution of the third year, October 26th, 1795, till the re-establishment of the Monarchy, May 18th, 1804—this was the reign of Republicanism. Next the Empire. And how was all this accomplished?

From Toulon Napoleon returned to Paris, where he lived in great privation. In this condition he came across, by chance, his old friend, D——. "The villains have cashiered me . . . robbed me of my rank . . . my labours . . . my bread . . . my fortune . . . my future hopes . . . my immortality. They said I was a revolutionist;" and taking the arm of his friend, the two went to meet an Englishman named B—— at an adjoining coffee-house, where projects for quitting the country were discussed. Spain, England, America, all offered advantages, but Constantinople seemed to please Napoleon best. The friends met in frequent consultations, and plans were fairly settled, when Napoleon, on returning to his lodgings one evening, found the following little note:—

"Mr. Buonaparte is requested to call upon me this evening about ten o'clock. I have something important to communicate to him.

"Yours,

"Sept. 27th, 1795. "BARRAS."

"Mr. Buonaparte, we are threatened with a counter revolution; the National Convention is threatened*; the troops are excellent,

---

* Barras first offered this *role* to Menou, who refused it. Then he took pains to find out the whereabouts of "le petit faquin, le petit drole de corse," who had done so well at Toulon.

but they want a chief. I immediately thought of you. You must head them instantly," said Barras.

In a few days all was over—for from the corner of Dauphin Lane to St. Roch Steps, a few volleys....some broken windows, and certain splashings—and the counter revolution gave up their arms.

Napoleon was not pleased with Barras, for the latter took all the honour of the day (September 13th) to himself, and was warmly congratulated on the dispositions he had made—and no one deigned to address a single word to the real actor. "To-morrow," said Napoleon, "I will call upon Barras." From the interview Buonaparte came out not quite so dissatisfied with his patron as the day before, for Barras gave full appreciation of his services, saying that his conduct on the 13th had marked out his qualifications, and asked to provide for him. "But before anything is done, you must enter into some matrimonial alliance—somebody of rank, some person of distinction; this will give you some stability." Napoleon felt humbled and surprised at this comical proposal, but not sufficiently so to prevent him accepting Barras' invitation to a theatre or supper* the following evening. It is needless to say Miss F—— was not the one chosen by Napoleon; but to the widow he was all attention. Events did not admit of a long engagement, for on March 9th Napoleon and Josephine were married; and as a wedding gift he received from Barras the command of the army in Italy.

On March 11th he left Paris to assume command, reporting his arrival thus:—

"Napoléon Buonaparte, General en Chef de l'Armée Italie.

"Au Directoire Executif.

"Je crois utile citoyens directeurs, de vous donner mon opinion sur les généraux employés a cette armée. Vous verrez qu'il en est fort peu qui peuvent me servir.

"BERTHIER—Talens, activité, courage, caractere tout pour lire.

* The supper party consisted of Barras, Veuve Beauharnais, Miss F—— and Napoleon.

"AUGEREAU—Beaucoup de caractère, de courage, de fermeté, d'activite; a l'habitude de la guerre, est aimé du soldat, heureux dans ses opérations.

"MASSENA—Actif, infatigable, a de l'audace, du coup d'œil, et de la promptitude à se décider.

"SERRURIER—Se bat en soldat, ne prend rien sur lui, ferme, n'a pas assez bonne opinion de ses troupes, est malade.

"SAURET—Bon, tres bon soldat, pas assez eclairé pour être général, peu heureux.

"ABATTUCI—Pas bon a commander cinquante hommes.

"MACQUART—Pas de talens, brave.

"GAUTHIER—Bon pour un bureau," etc., etc., etc.

Surely envy, hatred, malice, must have rankled in the breasts of the above, who had grown grey in the service and in danger, and were now to be subordinate to a mere upstart, who up to date had only swept away a few fortifications at Toulon, and shot a few Parisians on the steps of St. Roch.

On April 1st, 1796, Napoleon convened his first War Council, and asked all his generals to discuss how hostilities should commence.

Napoleon listened in silence to all that was advanced, and awaited an opportunity to speak. General Rampon soon afforded him one by saying: "We cannot act with too much circumspection; we have to oppose experienced generals, who are defending their native soil. Banhew, Kray, Merfeld, Laudon and Bellegarde, are men that will take every advantage offered them. Inured to battle, they will neglect no means of resistance."

Napoleon then rose and said: "Allow me, gentlemen, to lay before you some new ideas. War hitherto has been carried on in a theatrical and effeminate manner. We are not going to give each other rendezvous, and with our hats under our arm, to say to the enemy, 'Gentlemen, please to fire first, we never do' [referring to the English Guards at Fontenoy, when the Colonel cried out: 'Gentlemen of the French Guards, fire! We never fire first.' 'Please to fire yourselves,' was the answer.] Such silly gentlemen as these must be no patterns for us, entrusted as we are with the high interests that at present set us in motion. To slash our enemy, rush like a torrent on their battalions, and lay them in the dust: this is restoring warfare to its primitive state. Thus did Alexander and Cæsar rush forward. Experienced generals, you say, are to oppose me. So much the better, gentlemen, so much the better. Their experience will not gain battles with me," etc., etc.

This ardour and impetuosity amazed his audience. Even Massena was thunderstruck at what he had heard, and General Rampon seemed to think there was something in this Corsican fellow.

Napoleon, as we can see, had not called this Council to demand advice, but to prove that he knew how to conceive a grand plan of operations. He had no prestige to work upon, like his predecessors, Condé, Turenne and Villars. He must make it by the sweat of his brow, which accordingly he did; for in the marvellously short space of time from April, 1796, to December, 1797 (when Napoleon was again in Paris), he had defeated the Austrians at Montenotte, Fombro, Pizzighitone; had forced the bridge at Lodi; defeated Beaulieu at Valleggio, and laid siege to Mantua; had won the battles of Lonato, Castiglione, Medola, Roveredo, Calliano, Bassano, Caldiero, Arcola and Rivoli, and taken Mantua—surely sufficient to create an awe and a fascination for all time.

Pitt hated Napoleon: "His name will be recorded with the horrors committed in Italy in the memorable campaign of 1796 and 1797—in the Milanese, in Genoa, in Modena, in Tuscany, in Rome, and in Venice."

His entrance into Lombardy, April 27th, 1796, was heralded thus: "Nations of Italy! The French Army is come to break your chains; the French are the friends of the people; in every country your religion, your property, your customs, shall be respected."

And again, from Milan on May 20th: "The French, victorious, consider the nations of Lombardy as their brothers."

In testimony of the above, contributions were levied of six million sterling. The churches were given up to indiscriminate plunder. Every religious and charitable fund, every public treasure, was confiscated, and the country was made the scene of every species of disorder and rapine. Napoleon's concluding remarks in extracting peace ran thus: "Gentlemen! all discussion is at present extraneous—I have traced out this ultimatum with the point of my sword, and my horse is saddled in case of objection."

He sent Berthier with this treaty to the Directory, adding "that one enemy was humbled, that the war with Austria was terminated, and that now was the time to prosecute their operations against this country;" concluding his despatch, "the kingdom of Great Britain and the French Republic cannot exist together."

And soon was put *en train* that wonderful expedition comprising artists, scientific men, linguists, and all sorts of instruments, implements and machinery—for new countries were to be seized and colonised. It sooned leaked out that the expedition was to make for Egypt, as a step to stamp out England's prowess in the East.

In the meantime, it is excusable to linger on for a while to see how this Wonderful Man was received in Paris on his return from Italy.

On December 10th, 1797, the Directory gave a triumphant fête to welcome Buonaparte. It was held in the great Court of the Luxembourg Palace; at the further end of this Court was an altar and a statue of Liberty, at the foot of which the five Directors sat, dressed in Roman costume.

The ministers, ambassadors, etc., were seated on benches arrayed in the form of an amphitheatre, and behind them were the reserved seats for invited guests; the gardens and adjacent streets were crowded with people. Mme. Récamier sat with her mother on one of the reserved seats. See had never seen General Buonaparte, but shared the general enthusiasm for the young hero. Born December 4th, 1777, she grew into a woman of remarkable beauty and influence; marvellous tact, sweetness of temper, kindness of heart, and forgetfulness of self were her leading characteristics. She was beloved by all.

Where she sat, Mme. Récamier could not distinguish his features; so she took advantage of a moment when Barras was replying to the General, to rise and look at him. By this movement she displayed her whole person; the eyes of the crowd were attracted to her, and she was greeted by a long murmur of admiration. This sound did not escape Buonaparte. He turned his head quickly to see what it was that could divert attention from him—the hero of the assembly. He perceived a young woman dressed in white, and he gave her so harsh a look that she hastily sat down.

Thus was the power of the Directory strengthened by these exploits of Napoleon. Envy probably entered into the hearts of the other Directors; and their words may have concealed their thoughts, as each one in turn had to deliver speeches in his praise, for the thoroughness of Napoleon had left nothing for the others to do beyond falling in with Napoleon's next idea—Egypt.

It was on May 20th, 1798,* that Napoleon, with Admiral Brueys, and a fleet of 194, amidst the wildest enthusiasm, quitted the harbour of Toulon, sailing eastwards.

* Barras, the Managing Director, a man of rapid, courageous, decisive action, and who had fought against us at the siege of Pondicherry, now retired into private life; he died in Paris, 1829.

CHAPTER IX.

NELSON ON THE TRACK.

ON May 20th, 1798, when Napoleon left for Egypt, Nelson at once received orders to sail with the "Vanguard," "Orion," "Alexander" (all 74's), and three frigates, to watch the movements of this vast armament. On May 17th, he captured a small corvette, but could learn nothing of their destination; and on May 22nd, which proved to be the day on which the French armament left Toulon, Nelson lost all the yards from off the "Vanguard," and his three frigates disappeared from sight in a most violent squall of wind. Assistance for a time was refused him from St. Peters, Sardinia, whither he was driven; and it was not until June 4th that succour came to him—ten sail of the line and a fifty gun ship, under Captain Troubridge.

Napoleon possessed himself of Malta on June 15th, at the time when Nelson was off the coast of Tuscany; Naples, June 20th; and Messina June 22nd.

Nelson, at the latter place, learnt for certain that Malta had fallen on the 15th, and that the French armada were off Crete *en route* to Egypt. So he left Messina at once for Alexandria, arriving on June 29th. There the game of "hide and seek" became interesting; for Napoleon, when off Crete, learned from a passing frigate, "La Justice," that Nelson had been seen at Naples. Napoleon, therefore, instead of making direct for Alexandria, manœuvred to attack Africa from the Cap d'Azé, twenty-five miles from Alexandria.

Thus Nelson and Napoleon were, on June 29th, only these twenty-five miles apart—so far and yet so near. Had Nelson possessed his frigates just to scout, and had not the atmosphere been so hazy, there might have been another tale to tell.

Nelson, hearing and seeing nothing of the French, set off for Alexandrietta, Rhodes, Syracuse, from which latter place he wrote, dated July 18th :—

"It is an old saying, 'The devil's children have the devil's luck.' I cannot find, or to this moment learn, where the French fleet have gone to. My ill-fortune has proceeded from my want of frigates."

And further, to his friends the Hamiltons :—

"My Dear Friends,—Thanks to your exertions, we have victualled and watered; and surely, watering at the fountain of Arethusa, we must have victory! We shall sail with the first breeze, and be assured, I will return either crowned with laurel or covered with cypress."

Information again pointed to Alexandria, so Nelson, with all possible sail, left Syracuse, and was rewarded on August 1st by seeing the port of Alexandria filled with ships, all with the French flag flying at the fore—for Napoleon had been in Egypt exactly one month, had won the battle of the Pyramids,* and his headquarters were at Cairo, from whence he wrote to Brueys :—

"Au Caire, le 12 Thermidor Au XI.,

"30 Juillet, 1798.

"Lettres du Général Buonaparte a l'Admiral Brueys, 1798.

"Je reçois à l'instant et tout à la fois vos lettres depuis le 25 messidor jusqu'au 8 thermidor. Les nouvelles que je reçois d'Alexandrie sur le succés des sondes me font espérer qu'à l'heure qu'il est, vous serez entré dans le port. Je pense aussi que le Causse et le Dubois sont armés en guerre de manière à pouvoir se trouver en ligne, si vous etiez attaqué; car enfin deux vaisseaux de plus ne sont point à négliger.

* Du haut de ces Pyramides quarante siecles vous contemplent.

"Le Contre-Amiral Perrée sera pour longtemps nécessaire sur le Nil, qu'il commence à connaître. Je ne vois pas d'inconvénient à ce que vous donniez le commandement de son vaisseau au citoyen . . . Faites là dessus ce qu'il convient.

"Je vous ai écrit le 9. Je vous ai envoyé copie de tous les ordres que jai donnés pour l'approvisionnement de l'escadre; j'imagine qu'à l'heure qu'il est, les cinquante vaisseaux chargés de vivres sont arrivés. Nous avons ici une besogne immense, cést un chaos à débrouiller et à organiser qui n'eut jamais d'égal. Nous avons du blé, du riz, des legumes en abondance. Nous cherchons et nous commençons à trouver de l'argent; mais tout cela est environné de travail, de peines et de difficultés.

"Vous trouverez ci-joint un ordre pour Damiette, envoyez-le par un aviso, qui, avant d'entrer, s'informera si nos troupes y sont. Elles sont parties pour s'y rendre, il y a trois jours en barques sur le Nil: ainsi elles seront arrivées lorsque vous recevrez cette lettre; envoyez-y un des sous-commissaires de l'escadre pour surveiller l'execution de l'ordre. Je vais encore faire partir une trentaine de batiments chargés de blé pour votre escadre.

Toute la conduite des Anglais porte à croire qu'ils sont inferieurs en nombre, et qu'ils se contentent de bloquer Malte et d'empêcher les subsistances d'y arriver. Quoi qu'il en soit, il faut bien vite entrer dans le port d'Alexandrie, ou vous approvisionner promptement de riz, de blé, que je vous envoie, et vous transporter dans le port de Corfou; car il est indispensable que, jusqu'a ce que tout ceci se décide, vous vous trouviez dans une position à portée d'en imposer à la Porte. Dans le second cas, vous aurez soin que tous les vaisseaux frégates vénitiennes et francaises qui peuvent nous servir, restent à Alexandrie.

(Signé) "BUONAPARTE."

However, from the moment Nelson, on August 1st, 1798, perceived the position of the French, his intuitive genius displayed itself; and the idea struck him that where there was room for a

French ship to swing, there must be space for a British ship to anchor between it and the shore. Accordingly, he determined that a part of his fleet should engage them on that side, whilst the rest should bear down on the other, and thus enclose them between two fires. And so, soon, the following:—

"Lettre du Général Buonaparte a la Citoyenne Brueys.

"Au Caire (1798).

"Votre mari été tué d'un coup de canon, en combattant à son bord. Il est mort sans souffrir, et de la mort la plus donce, la plus enviée par les militaires. Je sens vivement notre douleur. Le moment qui nous sépare de l'objet que nous aimons est terrible; il nous isole de la terre; il fait éprouver au corps les convulsions de l'agonie. Les facultés de l'ame sont anéanties, elle ne conserve de relations, avec l'univers, qu'au travers d'un cauchemar qui altère tout. Les hommes, paraissent plus froids, plus égoistes qu'ils ne le sont réellement. L'on sent dans cette situation qui si rien ne nous obligeait à la vie, il vaudrait beaucoup mieux mourir; mais lorsque après cette premiere pensée, l'on presse ses enfants sur son cœur, des larmes, des sentiments tendres raniment la nature, et l'on vit pour ses enfants; oui, madame, voyez dès ce premier moment qu'ils ouvrent votre cœur à la mélancolie: vous pleurerez avec eux, vous éleverez leur enfance, cultiverez leur jeunesse; vous leur parlerez de leur père, de votre douleur, de la perte qu'eux et la république ont faite. Après avoir rattaché votre ame au monde par l'amour filial et l'amour maternel, appréciez pour quelque chose l'amitie et le vif interêt que je prendrai toujours à la femme de mon ami. Persuadez vous qu'il est des hommes, en petit nombre qui méritent d'etre l'espoir de la douleur, parce qu'ils sentent avec chaleur les peines de l'ame.

(Signé) "BUONAPARTE."

Thus, hardly had Admiral Brueys time to read Napoleon's letter to himself (dated July 30th), before Nelson had won his Peerage; for between sunset and sunrise on August 1st and 2nd, Nelson had found sufficient time to prove that he was still most formidable at sea, and able to destroy what proved to be by no means a phantom fleet.

Nelson shortly after—bereft of arm and eye, and with a fractured skull—proceeded to Naples to recruit; but not until he had written the following to Earl Spencer, the First Lord:—

"My Lord,—Were I to die this moment, 'Want of Frigates' would be found stamped on my heart. No words of mine can express what I have suffered, and am suffering, for want of them. Having only the 'La Mutine' brig, I cannot yet send off Captain Capel, which I am very anxious to do; for as an accident may happen to Captain Berry, it is of some importance, I think, for your Lordship to be informed of our success as speedily as possible. If the King of Naples had joined us, nothing at this moment could prevent the destruction of the store ships, and all the transports, in the port of Alexandria; four bomb vessels would burn the whole in a few hours; but as I have not the means I can only regret the circumstance. I send you a pacquet of intercepted letters, some of them of great importance; in particular one from Buonaparte to his brother. He writes such a scrawl as no one not used to it can read; but, luckily, we have got a man who wrote in his office, to decypher it. Buonaparte has differed with his generals here; and he did want—and if I understand his meaning, does want—and will strive to be, the Washington of France. 'Ma mere' is evidently meant for 'my country.' But I beg pardon; all this is, I have no doubt, well known to Administration. I believe our victory will, in its consequences, destroy this army; at least, my endeavours shall not be wanting. I shall remain here for some time. I have thought it right to send an officer (by Alexandretta, Aleppo, and Bussorah)

December 14th.—The Navy Board purchased the hulls of the following prizes taken by Lord Nelson at the Nile for the sum of £117,000, viz., "Franklin" and "Tonnant" (80 guns), "Le Spartiate," "Aguilon," "Conquerant" and "Souveran Peuple" (74 guns). The two last were only valued at £5000, the "Franklin" £30,000. "La Guillaume Tell," the only ship that escaped from the battle, was taken by Captain Blackwood in 1800.

overland, to the Board of Control, that they may give the necessary directions for paying the officers' bills. If it should have gone to the East India Company, I hope their Board will forward it.

"Ever believe me,

"Your Lordship's most obliged and obedient servant,

"HORATIO NELSON."

Napoleon, after quitting Salahie, took eight days to cross the desert. In this march, he lost many men and horses, by bad provisions and for want of water, as well as by the swords of the Arabs, who never ceased to harass him.

On leaving the desert, he took Darissa Gaza, Lidda and Ramleh, which was ill defended by the troops of Ghezzar (Butcher) Pacha.

On March 3rd he was close to Jaffa, which he took by assault on the 7th; the Turks made a vigorous resistance, and the assault cost Napoleon over 1200 men. On March 17th Buonaparte entered Caiffe, and on the 18th he invested St. Jean d'Acre.

On March 3rd, Sir Sydney Smith arrived at Alexandria, and took over from Troubridge the chief command of the naval forces. On the 7th he left for the coast of Syria, and on the 11th reached Caiffe.

On the 15th he steered for St. Jean d'Acre to consult with Ghezzar Pacha. Thus, having got the start of Napoleon by two days, they prepared for a vigorous defence. On the 16th, Sydney Smith, after a chase of three hours, took, off Cape Carmel, the whole French flotilla—under the command of Eydoun, Chef de Division—laden with heavy cannon, ammunition platforms, and other articles necessary for Napoleon to undertake the siege.

The artillery, consisting of forty-four pieces, was immediately mounted on the ramparts of Acre against the lines and batteries of the enemy, as well as on the gun vessels, which latter were employed against the enemy's flank. The nature of the ground permitted the French to carry their trenches close up to the ditch of the place.

Thus the siege commenced, and went on without intermission for seventy days.

Buonaparte, at the last assault, was distinguishable in the centre of a semicircle; he was violent and excited, dejected and down-hearted. He raised the siege on May 22nd, 1799, leaving his remaining guns and stores behind him. His principal park of guns, we note, were captured at sea on the 16th by Sydney Smith, and had been used against him.

This, and other disquieting events, convinced Buonaparte that England was mistress of the seas, and that his past year of warfare had yielded nothing; so, hurriedly penning a proclamation to his army—"Intelligence from Europe has decided my departure for France; I leave the command of the Army to General Kleber"—he left for Paris in October, 1799.

With England mistress of the seas, how had Napoleon, with a fleet of four from Egypt, managed to land in France? England and Austria groaned at this, and the former showed her teeth in the caricatures of the day: "Dans l'une ou representait Nelson s'amusent a draper Lady Hamilton pendant que la frégate la Meuron passait entre les jambes de l'amiral.'

One poor Turkish admiral had his head taken off for not stopping him at Alexandria.

Naples, in early 1799, was in revolution; their Majesties had fled, and Carraccioli, head of the Marine, had followed suit, for what was he to do? It was on reaching Naples that Lord Nelson saw a flag of truce flying on board the "Seahorse" (Captain Foote), and also on the castles of Ovo and Nuovo. His lordship at once made the signal to annul.

General Mack, with Austrian troops, was supposed to protect the frontier; but for want of pay his army had melted away, and thus the French had entered Naples. Royalists and republicans soon made a feud, and in this disorder Nelson might have acted as the timely mediator. But no! Captain Foote is ignored, and scenes such as these occurred:—

It was on June 29th, 1799, as eighteen ships lay in battle order in the beautiful bay of Naples, that a country boat, guarded by ragged ruffians, brought on board Lord Nelson's flagship, the

"Foudroyant," the Prince Carraccioli, Admiral of the Neapolitan fleet, haggard with misery and want. "I am accused," said the Prince, "of deserting my King in distress and leaguing with his enemies. The accusation is so far false that the King deserted me and all his faithful subjects." Such was his defence when tried by Court Martial the same afternoon. The sentence of the Court ran: "You shall be hanged by the neck at the yardarm of your own flagship in two hours from this time, and may God have mercy on your soul."

The King of Naples, Sir John Acton, and other foreign Ambassadors, a few days after, joined the fleet of Nelson. The King was addicted to early rising, and next morning there was his Majesty on deck, gazing with intense anxiety on some distant object. At once he turned pale, and letting his spy glass fall on deck, uttered an exclamation of horror, for there, full upon him, with his face much swollen and discoloured by the water, and his orbs of sight started from their sockets by strangulation, floated the ill-fated Prince, Admiral Carraccioli.

The old man's gray hair streamed in the light breeze that rippled the placid waters of the lovely bay; the King and Court were alarmed, and looked very pale; for all the superstition of the Italian school was called into play by this extraordinary and fearful apparition.

The priesthood at last were summoned, for there were many on board, when one more adroit than his brethren told the King that the spirit of his unfortunate admiral could not rest without his forgiveness, which he had risen to implore.

At a Court of Directors, held on Wednesday, April 24th, 1799, it was resolved unanimously "That the thanks of this Court be given to the Right Honourable Rear-Admiral Lord Nelson for the very great and important services he has rendered to the East India Company, and by the ever-memorable victory over the French fleet near the mouth of the Nile on August 1st, 2nd and 3rd, 1798; and for his magnanimous conduct on that occasion to request his lordship's acceptance of 10,000 guineas."

On September 24th, 1799, his Majesty opened Parliament with a long and energetic speech, complimentary to the late exertions of Captain Sir Sydney Smith, at St. Jean d'Acre. Then Lord Spencer, Lord Hood, Lord Grenville, etc., etc., spoke in highest praise of this gallant defence, which lasted without intermission for seventy days, and moved the following: "The thanks of the House to Captain Sir W. Sydney Smith, and the British seamen under his command, for their gallant and successful defence of St. Jean d'Acre against the desperate attack of the French army, under the command of General Buonaparte." Motion agreed to *nem dis.* And to Lord Spencer, the First Lord, the country struck a medal to congratulate him on his wise counsels.

CHAPTER X.

"Storms that might veil his fame's ascending star."

NAPOLEON had purposely said "No!" to Josephine when she asked to accompany him to Egypt; for in case Fortune failed to smile, she was to observe and acquaint him how affairs went on in France. He also asked Lucien and Joseph to watch—"to protect my retreat," as he expressed it.

Fortune frowned and forced Napoleon to raise the siege of St. Jean d'Acre, and to retreat to Egypt in a manner so wanton and cruel, that the poor sick and wounded were even given poisoned drugs.

The "Conquest of the East" was to Napoleon now but a shattered dream. "Ce't homme m'a fait manquer ma fortune," he spitefully said of Sir Sydney Smith; so after reaching Cairo, and slaughtering some more Turkish janissaries, Napoleon handed over the command of the Army to Kléber, and embarked on August 22nd for France, with Lannes, Murat, Berthier, and one thousand men.

The news from France of late had been distinctly bad. The Jacobins and Royalists were at issue, and the Directory now meant Anarchy—hence France in difficulty became once again Napoleon's opportunity, for on landing at Fréjus on October 9th, 1799, he was received with shouts of welcome. *Le Journal de Paris* proclaimed: "Vive la Republique! Bonaparte est debarqué à Fréjus!" And from that hour Napoleon became "L'homme necessaire," ready to take his part as Cromwell in the following play*, which was acted on the 18th Brumaire (November 9th) at La Salle de L'Orangerie, Paris.

"A la vue du général le plus violent tumulte éclate—on n'entend plus que le cris de "A bas le Dictateur! A bas le Tyran!"

The firmness and coolness of Lucien as President of the Five Hundred, when Buonaparte appeared in their midst to clear the Hall, came as oil to troubled waters. He explained to them that his brother was a staunch Republican; he came as a reactionary, as a restorer of order—not as a Revolutionist—and had only brought soldiers to act on emergency; and thus the crisis passed. A Directory of five became a Consulate of three, and Buonaparte was First Consul. Paris that night was placarded as under:—

"PROCLAMATION

"Issued by General Buonaparte at 11 o'clock, 19th Brumaire (November 10th), 1799.

"On my return to Paris I found discord reigning among all the constituted authorities," etc., etc., etc., and concluded thus: "Frenchmen! You will doubtless recognise in this conduct

---

* Cromwell started up from the Council with marks of violent indignation in his countenance, and hastened to the House with a detachment of 300 soldiers, whom he posted at the door and in the lobby.

He then entered, and addressing himself to his friend St. John, told him he was come to that which, to his great grief of soul, the Lord had imposed upon him. After having sat some time to hear the debates, when the Speaker was about to put the question, he suddenly rose up, and in the most opprobrious terms, reviled them for their ambition, tyranny, extortion and robbery of the public. After this torrent of general obloquy, he stamped upon the floor, and the soldiers entered the House; then, addressing himself to the Members, "Get you gone," said he, "give place to honest men; you are no longer a Parliament; I tell you, you are no longer a Parliament, the Lord has done with you." Sir Henry Vane rising up to remonstrate this outrage, Cromwell exclaimed: "O, Sir Harry Vane! Sir Harry Vane! the Lord deliver me from Sir Harry Vane!" He took hold of Martin by the cloak, saying: "Thou art a miscreant!" Another he reproached as a thief, a third as a drunkard, and a fourth as an extortioner. "It is you," added he, "that have forced me upon this. I have sought the Lord night and day, that He would slay me rather than put me on this work." Pointing to the mace he bade a soldier "take away this bauble." Then he turned out all the Members, ordered the door to be locked, and putting the key in his pocket, retired to his lodging in Whitehall.

the zeal of a soldier of liberty, and of a citizen devoted to the Republic. Views directed to conservation, to protection, to liberal purposes, will resume their influence by the dispersion of incendiaries who oppressed the Council, and who, though they rendered themselves the most odious, never ceased to be the most contemptible of mankind.

(Signed) "BUONAPARTE."

Fouché, the Minister of Police, had also his little say, for at the same time he wrote to—

"Manager, Opera Comique National, Favort Street.

"A piece entitled 'The Mariniers of St. Cloud' has been performed in your theatre. The intention is laudable, no doubt. But too much detail calls up old recollections, which it is necessary to efface. When all the passions subside before the law; when we wish to suppress, for the sake of internal peace, all our resentments; when this wish is strongly expressed by the people and by the magistrates, and that they give a striking example, no one can be permitted to pursue an opposite conduct.

"You will obey this sentiment, citizens, and I confide sufficiently in your patriotism to believe that you will make, without my sending an order, a sacrifice of your piece, since public tranquillity requires it.

"Health and fraternity,

(Signed) "FOUCHÉ,

"Minister of Police."

A youth, aged sixteen, was present at this performance; and when he saw Napoleon act the part of Cromwell, the warrior blood he had received from his fathers boiled in his veins. His vocation was decided; he became a soldier from that hour, and dreamt of nothing but battles. Every other career he held in contempt. Ségur was his name. Two years, however, passed before Ségur gave full adhesion to the First Consul. More than once during that time he was nearly siding with the rivals of the Chief. His mind at last was fixed—Napoleon and the Grand Army became his one and all. Napoleon and Ségur served together in twenty-one campaigns.

One of the First Consul's projects was the levying of a Volunteer Regiment, to be exclusively composed of young men armed, equipped, and mounted at their own expense. Ségur joined. His father approved, but how would his grandfather, the old Marshal and Minister of War under Louis XV., receive him?

The old Marshal received him much as an old Roman would have received a son who had broken the military oath, or returned without his shield, "parma non bene relictâ."

"I arrived early and approached his bed in the most submissive attitude. 'You have proved wanting,' he sternly began, 'to all the traditions of your ancestors. But it is done; think well of it; you are voluntarily enrolled in the Republican Army. Serve in it frankly and loyally, for your course is taken, and it is no longer the time to turn back from it.'

"Then seeing me bathed in tears, he melted, and with his only remaining hand taking mine, he drew me towards him; then giving me twenty louis—it was almost all he possessed—he added, 'Come, there is something to help you in completing your equipment; go, and at least sustain with bravery and fidelity, under the flag you have thought fit to select, the name you bear and the honour of your family.'

"Fifty years have passed, and I never think of this noble and painful counsel, of this manly and touching benediction, without being moved to the bottom of my heart."

The winter of 1799-1800, after the Revolution (18th Brumaire), was very brilliant in Paris, and fêtes, etc., had to be given in honour of the First Consul. Lucien then filled the post of Home Minister, and it was at his house that the first banquet was held. Madame Récamier attended. Soon she noticed a gentleman standing before the fireplace in the salon. In the dim light, she took him for Joseph Buonaparte, whom she had met frequently at Madame de Stäel's, and she bowed pleasantly. The salutation was promptly returned, but with an expression of surprise. She saw instantly her mistake, and recognised the First Consul. Her impression of him was very different from that which she had received at the Luxembourg, and she was struck with his mild expression. While he was talking with persons about him, he held the hand of Lucien's little daughter —a child of four years—whom he at last forgot. The child, tired of her captivity, began to cry. "Ah! pauvre petite!" he exclaimed in a tone of regret, "I had forgotten thee!" More than once in after years Madame Récamier recalled this excess of apparent kind-heartedness, and contrasted it with the harshness of his proceedings towards others and herself.

Napoleon now changed his address. From his small house in the rue de Victoire, the First Consul and La Citoyenne Josephine moved into the Palais du Luxembourg, thereby realising a good portion of his ambitious scheme, viz., to sleep in the bedchamber of the Kings of France. The *Journal de Paris* commented that: "Buonaparte finds time for everything: he assisted for three hours at the Institute;" and that as he was mounting a spirited horse, a citizen came up and assisted him. "I ought to mount easily," observed Napoleon, as he thanked him, "for I am not weighty." "Pardon me," replied the citizen, "you are a counterpoise to all the Powers of Europe." He thus passed those early days in developing to the full his magnificent organising powers, and giving to France her civil code and laws which have lasted ever since.

Ugly and disquieting tales regarding the Egyptian campaign had followed Napoleon to France; and he had learnt that it was Kléber's intention to disclose all, and that Tallien was to arrange the *dossier*.

Kléber, we know, was murdered on June 14th, 1800; but was it through the rage of the Corsican or the patriotism of the Arab? Tallien, after Kléber's death, was put under arrest and sent to France for trial, for endeavouring to excite mutiny in the Army of Italy. Fortunately for him, he was taken on the voyage and brought as a prisoner to England, which probably saved his life. General Desaix was not so fortunate, for on his return to France, Carnot, who was then War Minister, appointed him to command the Reserve in the Army now leaving for Italy. This appointment was not agreeable to Buonaparte, for he knew that Desaix had been in the plot (to disclose all) with Kléber, Tallien, Regnier, etc.

It is also well known that Napoleon and Desaix were closeted together for hours on the eve of Marengo, and that on the morrow Desaix received a shot from behind and was stabbed in the back. Rapp and Savary were his two aide-de-camps. After the murder or death of Desaix, Rapp and Savary were appointed by Napoleon to be his aide-de-camps. Desaix and Kléber were killed on the same day, June 14th, 1800.

Menou, an incapable general, by right of seniority took command of the Army. After his severe defeat by Abercrombie at Canope and Alexandria, in March, 1801, he capitulated; and on August 1st, 1801, the French Army embarked on English ships, and Egypt knew them no more.

Contrary to diplomatic usage, Napoleon wrote in haste to the King of England, to acquaint his Majesty that he had been made Premier Consul, and to demand peace, saying: "La guerre qui depuis huit ans ravage les quatre parties du monde doit-elle être eternelle? N'est il donc aucun moyen de s'entendre." William Pitt saw through this advance of the First Consul as only a skilful trick to conciliate public opinion, so sent a polite refusal.

England, with any other statesman but Pitt at the helm, would have yielded—at the time the country was emerging from a most disastrous campaign; for she had allowed H.R.H. the Duke of York to try his hand once again at active campaigning, and, in conjunction with Russia, to assume command of an expedition to expel the French from Holland.

After several battles, in which the Duke had been defeated by General Brune, he was forced to capitulate at Alkmaar and surrender 41,000 men. After some *pourparlers* the French agreed to the re-embarkation of the Allies—the English and the Russians—provided they departed before November 1st, 1799, and left behind them the artillery and 8000 French and Dutch prisoners, who had been captured on some former occasion.

These victories of Brune saved France from invasion in the north; whilst Massena, with his victories over Souwaroff at Zurich and elsewhere, saved her in the east. She was, however, menaced by the Austrians on the Rhine and from the side of Italy.

And this was how the English troops came tumbling home from Holland :—

"Dover, November 19th, 1799.

"Several ships of war and transports passed by this day with troops from the Texel, bound to the westward. 'L'Espion,' I hear, is quite lost, but the crew are all saved."

"Deal, November 19th.

"The 'Fury' bomb arrived in the Downs this morning, and brought over Colonel Heath and Colonel Clipstone of the Guards, and Major Cuyler of the — Regiment. They bring an account of the loss of the Dutch twenty gun ship, the 'Valk,' on the island of Ameland. Lieutenant Hill and nineteen men of the 23rd Regiment and five Dutchmen were the only persons saved. Four officers and 200 men of the 23rd Regiment were lost. The names of the officers are Lieutenants Hanson, Vischer, Maclean and Hoggard."

"Hull, November 18th.

"The 'Janus,' of Whitby, arrived off that port on Thursday, with 295 Russian troops from the Helder, from whence they have been fifteen days bound for Guernsey. They were separated from the rest of the fleet by stormy weather. The same afternoon arrived in Shields harbour, the 'Latona,' 'Wheldon' and 'Janus,' all with Russian Cossacks, etc., on board, but bound for Guernsey or Jersey. The above troops are bold looking athletic men, amounting to upwards of 1000 rank and file. The Cossacks all wear long beards. The Duke of York had actually proposed to introduce this Russian force to coerce and civilise Ireland, and would have done so had not the better sense and feeling of Cornwallis prevailed."

Thanks to our fleet, the chagrin caused by this shore defeat became just bearable by the taking from the French of the Ionian Islands, Minorca and Malta.

Napoleon now turned to good political account the Russian prisoners taken by the French in this war. He used them as an instrument to sever Russia from England; for to each man he gave a new gun, and an entire fresh uniform of the corps to which he belonged, and sent them all back home without ransom, exchange or condition of any sort. This ingenious munificence was not thrown away, for Czar Paul from an ally of England became her declared enemy, and he joined the league now created by the First Consul for her destruction.

England could still boast of the following Powers as faithful to her, viz., Turkey, Portugal, Austria, Bavaria, Wurtemburg, Naples and Sardinia; and as the war rolled on, Austria planned her campaign against France to be thus:—To have two armies of 120,000 men, one in Italy under Baron de Melas, the other commanded by Marshal de Kray, to besiege Genoa, pass the Var, invade Provence, then to rally 20,000 English assembled in the Isle of Minorca.

The First Consul, anxious to retake Italy, placed Massena, with 36,000 men, on the defensive, to watch Melas around Genoa and along the littoral as far as the Var; and to Moreau he gave an army of 120,000 to assume the offensive and cross the Rhine at Bale and Constance, to debouch on the left flank of Marshal Kray. Napoleon at the head of a third army, called the "Reserve," was to assemble at Lausanne, Vevey, and cross the Alps to give the decisive blow in the rear of Melas—or in the words of Napoleon, as he left Paris after looking at his map:—

"Ce pauvre M. de Melas se repliera vers Alexandrié. Je le joindrai dans les plaines de la Scrivia, et je battrai, là, là,—il indiquait San Giulano près de Marengo."

Fleets at the same time were kept in readiness at Havre, Boulogne and Antwerp, for on Napoleon's return, the Czar Paul and the First Consul were to invade England. Man proposes; God disposes—for Paul was assassinated on March 25th, 1801,* and his

* On March 23rd, Paul wrote to the King of Prussia, that if he did not at once declare war against England, he would march 80,000 men across the frontier.

Count Pahlen, Governor of St. Petersburg, saw this despatch ere it went, and wrote in pencil: "His Imperial Majesty is to-day in bad health; serious consequences may ensue."

The despatch on reaching Berlin was held over for a day, and next day came the news that the unhappy monarch was no more.

successor, Alexander, refused further alliance with Napoleon. Hence the latter now turned his thoughts to peace, and drew up preliminaries, which were signed in October, 1801.

"Peace! peace! when there is no peace!" How could it be otherwise with Napoleon, the child and champion of Jacobinism? However, the satirists were pleased to call it "The First Kiss, or Napoleon embracing Britannia."

How often *Punch*, in his happy way, gives the political situation at a glance; so in the cartoons of that day we have the following:—

"Madame, permettez moi de vous assurer de ma profonde estime pour votre seduisante personne et de sceller sur vos levres divine mon eternal attachement."

"Monsieur, vous êtes certainement un gentlemen fort bien èléve et quoique vous me faissiez rougir, votre baiser cependant est si doux que je ne puis pas vous refuser; mais malgré cela, je suis persuadée que vous me tromperez à nouveau!"

CHAPTER XI.

"The Kingdom of Great Britain and the French Republic cannot exist together."

CONCURRENT with the Continental events narrated in the last chapter, the hand of Napoleon was equally visible in Ireland, India, the Cape, and elsewhere—his emissaries had gone forth to preach that the world could not permit of France and England; there was not room for both—the latter must go!

England was fortunate in possessing at this time a good statesman and soldier in Lord Cornwallis. It was only for England to be in trouble, when "Send for Cornwallis!" became a national cry. As a soldier he had commanded the 12th Regiment, and served with distinction in Germany and America and at home as Master-General of Ordnance; while he had received many times from Pitt the offer of high Cabinet rank. He was three times Governor-General and Commander-in-Chief in India, where he died (at Ghazeepore) in 1805.

It was on his first appointment as Governor-General in India that the same meddlesome tactics of George III., which had cost to England her American Colonies, were repeated by the Regent (George IV.) the son; for in a letter to his brother, the Archbishop of Canterbury, Lord Cornwallis wrote:—

"You tell me that I am accused of being remiss in my correspondence with a certain great personage. Nothing can be more false, for I have answered every letter from him by the first ship that sailed from hence after I received it. The style of them, although personally kind to excess, has not been very agreeable to me, as they have always pressed upon me some infamous and unjustifiable job*—which I have uniformly been obliged to refuse—and contained much gross and false abuse of Mr. Pitt, and improper charges against other and greater personages, about whom, to me at least, he ought to be silent."

Other letters followed, which all go to prove that the First Gentleman of the Day was an unscrupulous jobber. They met with no response beyond—

> "I am armed so strong in honesty
> That they pass by me as the idle wind
> Which I respect not."—SHAKESPEARE.

Cornwallis, on his return to the Mother Country, found—as he had often predicted—that the French fully meant the invasion of our shores. Consequently, he was on the spot to take command, for it was early in 1796 that a French fleet, commanded by de Galle, with 15,000 men under Hoche†, sailed from Brest for Ireland. A fierce storm dispersed the greater part, leaving but a few to reach Bantry Bay. However, next year General Humbert was more successful; he landed with 1200 men and marched inwards to Killala, where he turned the Bishop's (Dr. Stock) Palace into his headquarters. After scoring other small successes, and creating many panics, General Humbert was cleverly caught by Cornwallis, Lake and Crawford at Ballynamuck, where, thus surrounded, he surrendered.

Further trouble came in the Irish Rebellion of 1798 (the French Jacobins again), the settlement of which was given to three kind and

---

* He proposed to the Governor-General to displace a black named Alii Khan, who was Chief Judge of Benares, in order that a youth named Pelligrini Treves, the son of a notorious London money-lender, might be appointed to that office.

† Hoche (born 1769, died 1797) crammed more into his short life than Napoleon into his. Son of an innkeeper, he joined as simple *soldat* at the age of sixteen, and rose to highest rank. He held Dunkirk against the Duke of York; defeated the Austrians at Weissenberg (December 26th and 27th, 1793); and suppressed the Royalists in La Vendee and Brittany. In 1795 he was successful against the Quiberon Expedition; in 1796 in Ireland; and in 1797 gained many successes in the Austro-Italian Campaign. He died in September, 1797—some say POISONED.

worthy men—Cornwallis, Lake (Lord) and Moore (Sir J.); and this is how the event was voiced in the English House of Commons:—

"London, June 25th, 1798.

"After the rejection of Lord George Cavendish's proposed resolutions in the House of Commons, on Friday, recommending lenient measures towards Ireland, Mr. Fox moved the following proposition:—

"'Resolved—That this House, understanding it to be a matter of public notoriety that the system of coercion has been enforced in Ireland with a rigour shocking to humanity, and particularly that scourges and other tortures have been employed for the purpose of extorting confessions—a practice justly held in abhorrence in every civilised part of the world—is of opinion that an immediate stop should be put to practices so disgraceful to the British name; and that our best hopes of restoring permanent tranquillity to Ireland must arise from a change of system, as far as depends on the Executive Government, together with a removal from their stations of those persons by whose advice those atrocities have been perpetrated, and with regard to whom the afflicted people of Ireland can feel no sentiments but those of resentment and terror.'

"The discussion of this resolution kept the House sitting till about half past four o'clock on Saturday morning, when it was rejected by a majority of 142—62 voting for, 204 against it.

"Another meeting of the noblemen and gentlemen who disapprove of the system pursued by Government towards Ireland took place on Saturday evening at Norfolk House, when it was determined to make a fourth and final Parliamentary effort in favour of that country in the House of Lords on Wednesday next."

The obituary of that day (June 30th, 1798) was sad and interesting—"On Saturday, at St. Thomas' Hotel, Berkeley Square, her Grace the Duchess of Leinster." The recent calamities she had witnessed in Ireland, and particularly the melancholy fate of her brother-in-law, Lord Edward Fitzgerald, are supposed to have so preyed on her mind as to have occasioned her premature death. On

June 4th, Lord Edward Fitzgerald died in delirium in the gaol of Newgate.

Lord Edward had served in the 19th Regiment during the greater part of the American War—a capable officer, a man of honour and humanity, esteemed for his frankness, courage and good nature. After the war he was placed on half pay, until brought in as Major into the 54th Regiment, then quartered at St. John's, Brunswick (May, 1788.)

In 1791 the Regiment landed at Portsmouth, where Lord Edward received a letter from Lieutenant-Colonel Bruce (of the same corps) from Naples, telling him he was in a rapid consumption, and advising him to take proper measures for succeeding him.

Lord Edward fully expected that the Lieutenant-Colonelcy would have been given to him without solicitation. On hearing that Colonel Sturt succeeded to the commission—soured with disappointment and fired with indignation—he left for Paris at the end of the year, and there he fell a prey to the extravagant political theories of the French. His name was struck out of the Army List, but he was allowed to sell his commission. He married a bright little French woman—conjecture says Pamela, the daughter of the Duc d'Orleans ("Egalité"), who, we remember, was guillotined. Poor Lord Edward was only secured after a most desperate resistance whilst leading on the Rebels.

Another actor in this Irish *émeute* was a Mr. Bagenal Harvey. He, unfortunately, was Commander-in-Chief of the rebels when the dreadful massacre at Scullabogue happened. To his credit, he at once issued orders proclaiming the penalty of death against such persons as had murdered their prisoners—for which humane effort he was deposed.

After this he hid in one of the Saltee Islands. On his trial, he said: "He became a member of the Irish Union imagining that its only object was to reform the Constitution; only recently he discovered the Popish priests were deeply concerned in it, and that the extermination of Protestants was their main design." When asked about the bloody business of Scullabogue, he much shocked, replied

"That it was brought about by an infamous sanguinary Popish faction." He was executed.

The King's troops at Arklow, under Colonel Walpole, were foolishly ambushed, and nearly all slain. The rebel General, Father Michael Murphy, was in command at the time. It was while heading his column at the Charter School, where they showed a little reluctance to advance, that he took out of his pocket some musket balls, which he said were fired by the enemy, and some of which had hit him without wounding him; and others he had caught in his hands. He assured them at the same time that the balls of heretics could not injure them, as they were under the protection of the Almighty, in whose cause they were fighting. By this strategem he prevailed on many of the deluded wretches to follow him—until he was shot, when they thought his tale was blarney, and made their own retreat.

Other distressing incidents happened around Kildare. One night the Limerick mail coach was stopped and plundered, and Lieutenant Giffard, of the 82nd Regiment, was cruelly massacred. Being asked by the rebels who he was, Giffard replied he was an officer proceeding on his way to Chatham. They next demanded whether he was a Protestant, and being answered in the affirmative, they informed him he must die unless he joined and helped them in their attack on Monastreven, and denied his God and King. On his firm reply, "No! Never!" the rebels assaulted him; he drew his revolver and stood at bay, until wounded to the death.

But, now! A sudden thought strikes me—"Let England and Ireland swear an eternal friendship," and byegones be forgot; so we will change the scene, and join the 12th, so soon to leave for India.

CHAPTER XII.

REGIMENTAL LIFE IN THE OLDEN DAYS, 1796-1830.

"In 1796, I apprized my Government of the departure of the Dutch Squadron from the Texel, bound with troops to the Cape of Good Hope; and I had the good fortune to transmit this intelligence to Admiral Elphinstone (Lord Keith) in India, in time to enable the squadron to return to the Cape, and to capture the Dutch Squadron, and so save that valuable conquest to our country."

"In 1796, I counteracted the agents of a French Mission, sent to Egypt to inveigle the Beys of Egypt into the designs of the French for passing an army through Egypt by the Red Sea, to strengthen Tippoo Saib, and to annihilate the British dominion in the East Indies."

SUCH was the political situation on England's road to India, when a young officer, Lieutenant Elers, from the 90th—a regiment raised by Graham (afterwards Lord Lynedoch, whom Napoleon had at Toulon described as a daring old man)—joined the 12th Regiment on exchange. The full Colonel then was W. Picton, successor to Sir Henry Clinton, late Commander-in-Chief in America. It was when Picton went to Court to thank the King for the honour thus conferred, that His Majesty replied: "You must thank Captain Picton who commanded the 12th Grenadiers at Minden."

The regiment was at Newport; so from Cowes he hired a chaise, and was put down at "The Bugle." There he heard that the regiment was on parade a short way out of town. It was a punishment parade, which the present writer will not describe, for although he has witnessed many he wished they had all ended thus. The sister service tells of a bluejacket, who was stript and tied ready for punishment, when his ready and rhyming wit came to his rescue, for knowing that his commander had an aversion to a cat, he spoke to him as follows:—

"By your honor's command, an example I stand
Of your justice to all the ship's crew;
I am hampered and stript, and if I am whipt,
'Tis no more than I own is my due.
In this scurvy condition, I humbly petition
To offer some lines to your eye:
Merry Tom, by such trash, once avoided the lash,
And, if Fate and you please, so may I.
There is nothing you hate, I'm informed, like a cat;
Why, your honor's aversion is mine:
If puss, then, with one tail, can make your heart fail,
Oh, save me from that which has nine!"

N.B.—He was pardoned.

The punishment parade being over, a few magic words brought the regiment into column; and on the word "Quick March," the band struck up, when the last joined, with his friend Lieutenant Hon. John Meade, followed them into town.

The battalion dismissed, Lieutenant Elers was introduced first to the Major, Jack Picton, the very image of his brother, Sir Thomas, killed at Waterloo; then next to Captain Allen and Captain Craigie of the Grenadiers.* All three had been at Gibraltar in the days of its great siege.

The same regiment, in the same barracks, on August 1st, 1909, commemorated the 150th anniversary of Minden by an "at home" or grand ball.

The day at length (June, 1796) arrived for the regiment to embark on four East Indiamen of 800 tons. The separation of wives from husbands not on the married roll, always renders this a scene truly most heartrending.

1796.

TWELFTH (OR THE EAST SUFFOLK) REGIMENT OF FOOT.

COLONEL ... ...William Picton.
LIEUT.-COLONEL...Henry Harvey Aston (k. in duel, Dec. 23rd, 1798).
Thomas Grey (d. at Cape, 1796).

---

* The Grenadiers and Light Infantry were Companies élite—they often preceded the regiment on service; in fact they were *en evidence* everywhere.

MAJOR ...	...John Picton.
	James Bellairs.
CAPTAIN ...	...James Allen (d. at Seringapatam, 1799).
	Thomas Craigie.
	Joseph Moore.
	Walter Ruding.
	T. Hayward Winstone (d. at sea, Madras, 1797).
	Thomas Woodhall (d. at Madras, 1802).
	Thomas James.
	John Wilton (d. Pondicherry, 1800).
	William O'Brien.
LIEUTENANT	...Robert Swyer (d. Pondicherry, June, 1798).
	William Omly.
	John Crawford.
	Mark D. Buckeridge (k. at Seringapatam, 1799).
	Robert Nixon.
	Nicholas Eustace.
	Patrick Moyna (d. at sea).
	John Parker (d. at sea, 1801).
	George Nixon (k. Seringapatam, 1799).
	William Eslam (d. Trichinopily, 1804).
	John Cassidy (d. April, 1798).
	Andrew Kirkwood.
	John Willock.
	Hon. John Meade.
	Samuel Percival (k. Seringapatam, 1799).
	William Morris.
	Richard Ashton.
	James Cavendish (d. Seringapatam, 1805).
	Mathew Price.
	George Eld Darly.
	George Elers.
	William Gahan (d. June, 1798).
	Charles Morgan.
	Thomas Falla (k. Seringapatam, 1799).

LIEUTENANT	...James Seaton (k. at Toulouse, 1814).
	T. Dundas Campbell.
	R. Sale (k. at Moodkee, 1846).
	— Shaw (k. at Seringapatam).
	Thomas W. Edwards (k. at Seringapatam, 1799).
	Alexander Sinclair.
	John Jordan (d. at sea, Batavia).
	William Langford (k. at Wallajahbad, 1800).
	W. Gahan (k. Seringapatam, 1799).
	Henry M'Keady.
	Richard Bayley.
SURGEON ...	...J. Bagot (k. Seringapatam, 1799).
	Thomas Gray.
CHAPLAIN...	...John Edwards.
ADJUTANT	...Joseph Moore.
QUARTERMASTER...	Alexander Stuart (d. Seringapatam, 1805).

And so this casualty list continued until the last joined, Richard Bayley, came out at the top about 1828.

The ships lay off St. Helens until the 27th, when the last farewells were said, and they soon became like specks in an expanse of sea and sky.

In the old days "St. Helens" was the great roadstead for the Navy and other marine; the Dover or shore of St. Helens being too often selected as the place to finish off some foolish quarrel. It is now, however, turned to better account, viz., the links of the Royal Isle of Wight Golf Club. The tower of the derelict church on the shore is still of use as a landmark. Formerly it was the custom for the bluejackets to break from the altar, arches or graves, sufficient stones to scrub their decks—which from their sanctity gave the name of "holystoning."

A doggerel rhyme, somewhat as under, was used as an accompaniment to the scrubbing process :—

> "Old Father Neptune, with glass to his eye,
> On the beach at St. Helens a church he did spy;

There was no tower or steeple,
And hardly any people
Ever went inside the walls
Of this derelict St. Paul's.

"'But,' said Father Neptune, 'from this very land,
I'll get what I want of holystone and sand;'
So he holystoned his father, he holystoned his mother,
He holystoned his sister, and likewise his brother;
And then after all, not content with that,
He squeegeed the cook, and dumbscraped the cat."

The 2nd 12th Regiment made a similar voyage just seventy years later; and as all voyages are alike, the following stanzas, written on the Line, will equally do for both :—

1864.

Indiaman "Aliquis."

"From Queenstown sheets, hats, handkerchiefs and caps moved with a will,
The band blew out their sharps and flats till they were nearly ill;
'St. Patrick's Day' and 'Auld Lang Syne' and 'Home, Sweet Home,' I ween
Were heard on shore by Cove's good folks, and then 'God Save the Queen.'

"In this way left the Second Twelfth their country's shores, 'midst hope
That with treason, war and all that's dire it might successful cope;
That in India, where in bygone days the Twelfth had earned a name,
It might flourish as in those old times, and yet enhance its fame.

"On, on we sail'd, the Em'rald Isle receding from our view;
But now, that awful *mal-de-mer* made some feel—rather blue;
This feeling did not last, however, for most, till next day,
And ere another week had passed we sighted Biscay's Bay.

"Had Dibdin, now (when writing that fine nautical old air
That begins, 'Loud roared the dreadful thunder!') seen Biscay, he'd stare—
Have stared, I should have said, for Dibdin is alas! no more—
When we passed Biscay, though, there were no signs of thunder's roar.

"The weather since we left the shores of Erin's isle so dear
To all her sons (and others, too), had been most calm and clear.
We'd passed Madeira, 'most in view, then having settled down,
All minds prepared to patient wait until sighting the Maidoun.

"Now Mr. Gardiner, with great kindness, said he'd—twice a week—
Lecture on all that appeared—of places passed to speak;
The kindness was not thrown away, for many did attend,
And, when the lecture was concluded, sorrow'd for the end.

"'A library also there was—a hundred books or more,
And from these books our trav'llers gained no small amount of lore;
Draughts, dom'noes, cards and every game to pass away the time;
Pens, ink and paper, too, for all who wished to write and rhyme.

"'The officers, too, had their games, and played upon the poop;
They'd quoits and games, and also guns, with which sea-birds to shoot;
But though this fun was carried on in style of great beauty,
Yet neither men or officers omitted any duty.

"'The line at last we safely reached, in spite of weather calm,
To keep a celebration of th' event was thought no harm.
'Slasher and Crasher,' preceded by a well delivered prologue,
And then the niggers of the Twelfth all going, yes, the whole hog!

"'But as every medal (as we know) has always a reverse,
I will not keep the better side unto, but show the worse.
The duty on the ship at last began to get so hard,
That swabbers had to leave their work and do both watch and guard.

"'We sail'd three hundred miles on the ninth day of September
(Less fourteen)—that's worth noting down—and if I well remember,
It was the greatest distance made in one day noted down,
And at the time we weren't far off a straight line with Cape Town.

"'October 3rd—was thrown o'erboard, at even on that day,
'The Dead Horse'—which was done to celebrate the claim to pay,
The sea-kit debt now being clear'd; in India lots of money
When once on land all thought to get; the 'vags' were very funny.

"'Preceded by a fancy band of whistle and tin pot,
Drumm'd *a la* tom-tom the cortege in solemn order got;
And all the 'vags' and 'skylarkers' that were within the ship,
Procession went in round the deck; then did the 'Dead Horse' dip.

"'On that e'en we'd another sight—a sunset most sublime,
And one so full of brilliancy we'd not seen for some time;
And in the e'en the moon appear'd with pale and silv'ry light—
For full three weeks we'd not had such a cheerful lovely sight."

This was the calm before the storm, for in the Roaring Forties the waves went mountain high, rendering it impossible for our Poet Laureate to continue—hence a break. It is, however, sad to say that this bright and cheerful writer fell a victim to disease within a year of his arrival—Arthur Leroux Whipple, a name well known in Plymouth.

Now we return to 1796. On arrival at the Cape, the Bay was full of ships of war, for there lay Lord Keith's squadron and the Dutch fleet he had taken only six days before. The regiments on shore were the 33rd, 78th, 80th and 86th. In the former was Colonel Arthur Wellesley, just turned twenty-seven, all life and spirits; in height he was five feet seven inches, with a long pale face, large aquiline nose, clear blue eyes, and a very black beard. He spoke quickly, with a slight lisp. He had one other peculiarity—the lobe of the ear united to the cheek. (So did Byron's.) Dancing, duelling and cricket kept them all employed.

On November 10th, 1796, the 12th, 86th and 94th Regiments—in all eleven large Indiamen, with the "Fox" frigate as escort—sailed from Table Bay. It was all excitement, for at any time they might encounter some French frigates known to be cruising in the Bay of Bengal; so hours were spent at drill, and also at the guns; but no encounter came, and Madras was reached in safety in the new year, 1797.

The 74th Regiment, with whom the 12th were so often to fight shoulder to shoulder in India, Africa and elsewhere, warmly welcomed them on arrival; so did the 25th Light Dragoons (then commanded by Lieut.-Colonel Cotton, a pretty faced active little man of twenty-five, so soon to become the war-worn Lord Combermere). There was also awaiting them a young officer, Lieutenant Robert Sale. He had been posted to the 12th on promotion, and with them he remained until given the command of the 13th Light Infantry, with which regiment he made such grand history during the first Afghan War.

From January to August, drill went on without intermission. Colonel Baird (afterwards Sir David Baird, Bart.) frequently attended parade, and expressed his high approval of the battalion as regards their drill and discipline. Lieut.-Colonel Wellesley, 33rd Regiment, was in the same brigade. The name of Lally again crops up, for it looked at one time as though one young subaltern contemplated matrimony with the beautiful Harriet Lally, natural daughter of the famous French general of Fontenoy, Culloden, Pondicherry, etc.

This attachment, together with the amusement of gambling, had, however, so much crippled his finances, that matrimony became a venture on which he could not well embark.

For the loss of Pondicherry, poor Lally had to suffer. He was sentenced to death on a vague and frivolous accusation. This judicial murder was exposed by Voltaire, and the sentence was reversed; but who could remit the punishment? His son, the Comte de Lally-Tolendal, retaliated, and in a most powerfully written play, "Le Comte de Strafford, tragedie en cinq actes, et en vers, Londres, 1795," he wrote:—

"Quoique la comparison ne pût pas s'établir sous tous les rapports, cependant le Comte de Strafford decapité à Londres au mois de Mai, 1641, et le Comte de Lally, decapité à Paris au mois de Mai, 1760, offraient mille traits de ressemblance dans leur caractère, leur conduite, leur infortune, leur mort.

"Tous deux avaient aimé passionément leur roi, l'un en ministre et en favori, l'autre en serviteur et en soldat.

"Ce que la perfidiè puritaine avait fait contre l'un, la perfidie jesuitique l'avait fait contre l'autre. Strafford livrant Newcastle aux Ecossais, n'avait rien de plus absurde que Lally livrant Pondicherry aux Anglais. . . ."

In early August, 1797, rumour had it that an expedition to the Luconian Islands was in contemplation, and Manilla was mentioned as the objective.

At Clifton is a monument to the gallant officers and men who had fallen victims to a previous capture of Manilla, under the command of Sir William Draper, a soldier and scholar. He saved the place from plunder under promise of a ransom of £1,000,000. The money was never paid. "Junius," however, alleged it was, and brought charges against Draper. He entered with his pen into every controversy, and, need I say, swept the field. The venom of his shaft and the vigour of his bow became a most appalling dread to all, for they had in him an enemy more unscrupulous than themselves. Sir William Draper, the Guards, the King—all were attacked. To the latter he wrote: "Lay aside the wretched formality of a king, and speak to your subjects with the spirit of a man, and in the language of a gentleman."

Lord Beaconsfield described a bore as one who asked the question: "Who was 'Junius?'"

"I am the sole depositary of my secret,* and it shall die with me," said "Junius."

A very strong case of authorship was made in favour of Lord George Sackville, who, our readers will remember, lost his opportunity at Minden, 1759.

The strongest clue to authorship, however, rests with Sir Philip Francis, for he was at the Court of France in 1761, with Hans Stanley, on a diplomatic mission, when he saw "lacérés et brûlés en la Cour du Palais," the Jesuitical books of disputes between the King of France and this order.

This incident "Junius" narrated fully in his earlier writings; neither Lyttleton, Temple, Burke, Sackville, etc., could have been in Paris at the time.

However, it is fair for "Junius" to be allowed his secret, and right for us to return to our soldiers now starting for Manilla. Unfortunately, when the day of embarkation came (August, 1797), the surf on the Madras coast was at its very worst, causing the regiment, by the upsetting of two Massula boats, to lose many men.

This was a sad commencement. Further sorrows and disappointments dogged the voyage, and the greatest confusion, discontent and almost mutiny prevailed. Europeans and Rajput sepoys were all crowded into the same vessels, where each caste, sect or religion had to lead a life apart, for fear of one class contaminating the other, more especially in their water or food arrangements.

The expedition had no sooner landed than it was recalled; so after a repetition of the above discomforts, they were back again in

* It is well known the handwriting of papers sent to Press was always that o a woman.

Madras by December 15th, 1797. It was during their short absence that the French squadron (the same seven frigates whose attack was expected when the regiment was traversing the Bay of Bengal on their passage to India) had succeeded in driving an Indiaman ashore, and completely disabling three of our seventy-fours.

Captain Lucas, commanding the three English men-of-war, was for this mishap brought to a court-martial. This broke his heart; and the first parade the 12th attended on its return, was the melancholy one of depositing his remains in the cemetery near the Fort. The above, and other disquieting rumours, all meant war.

CHAPTER XIII.

THE Regiment was inspected that year (May 4th, 1798) by Major-General John Floyd, a distinguished cavalry soldier. As a Cornet he had fought at Minden.* He complimented the battalion, and expressed to Colonel Aston his belief that under such good command fresh honours were bound to come. That very day year (May 4th, 1799) saw the fall of Seringapatam, in which assault the 12th played so grand a part, but alas! without their Colonel—for he, on Christmas Eve, had been killed in a foolish duel (see "Duelling," chap. xv.) General Floyd was blessed with a pretty wife and child. The latter became, in 1820, the wife of our great Minister Sir Robert Peel, whose life ended so sadly on June 29th, 1850, caused by a fall from his horse on Constitution Hill.

A public funeral was offered, and a Peerage to his widow, but Peel's will was precise—he wished for a funeral in his parish church at Drayton, and for no member of his family to accept any reward. However, he left behind a sailor son with all the happy daring of a Nelson, who proved himself no less happy as a soldier; it was in the dark days of the Mutiny, when Sir Colin Campbell was on his march to Lucknow; Peel—his men—his guns—Sir Colin could not have done without. His death was somewhat sudden, due to that scourge, smallpox (1858).

A wave of military enthusiasm now came over all; extensive field movements, and various military operations became the order

* Many cavalry regiments bear "Dettingen" on their standards; none carry "Minden." Surely the Sackville incident did not forbid that honour. Even if so, give them "Warburg," where they all behaved so well.

of the day, caused undoubtedly by the following note from Buonaparte to Tippoo Saib:—

"Lettre du General Buonaparte à Tippoo Saib.

"Au Cairo le 6 Pluvoise an VII. 25 January, 1799.

"Vous avez deja été instruit de mon arrivée sur les bords de la Mer Rouge avec une armee unombrable et invincible, remplie du desir de vous delivrer du joug de fer de l'Angleterre.

"Je m'empresse de vous faire connaittre le desir que j'ai que vous me donniez, par la voie de Mascate et Mokka, des nouvelles sur la situation politique dans la quelle vous vous trouvez.

"Je desirerais même que vous pussiez envoyer à Suez ou au grand Cairo quelque homme adroit qui eut votre confiance, avec lequel je pusse conférer.

(Signe) "BUONAPARTE."

General Harris joined the Army* on March 1st, 1799, with a new Colonel (Shaw) for the 12th, whom they nick-named "Sour Krout"—a zealous, brave old man, but nothing pleased him. All now was in readiness for commencing the siege of Seringapatam, and on the night of April 3rd, General Baird, with the 12th, 74th, 94th (Scotch Brigade) and two Sepoy battalions assembled, with intent to clear the ground in front, where several topes gave great annoyance by their concealed musketry fire.

The night was very dark; so dark in fact that one of the battalions of Sepoys, having lost their way, had moved to the front, and become exposed to the attack of a very superior number of the

* Seringapatam, 1799, under Lord Harris:—

30,000 fighting men.
300,000 followers.
400 elephants.
1,000 camels.
1,500 bullocks.

For the Tirah Expedition, 1896, under the able General and Commander-in-Chief Sir William Lockhart, a force nearly similar in numbers, viz., 30,000 to 33,000 men was employed. The demand for transport was as under:—

16,200 camels.
45,500 mules.
12,600 bullocks.

enemy; the 12th hearing the firing, and troops coming on towards them, prepared to charge, when it was discovered that the advancing foe was the Sepoy Regiment retiring. Another incident happened owing to the darkness of the night; this was when the same party, viz., the 12th and the two Sepoy battalions were returning, as they thought, to their lines. An aide-de-camp assured General Baird, who was in command, that the troops instead of marching *from*, were marching *on* the enemy.

The guide on being referred to was obstinate, and persisted he was right; the aide-de-camp (Major Lambton) declared that by the stars he could tell that, instead of moving to the southward, the troops were marching directly *north*. Baird produced a pocket compass, and putting a firefly on the glass, ascertained that his march was erroneous, and his guide entirely astray. Fortunately he had time to remedy the mistake; jocularly observing that "in future he should put more faith in the stars than he had done formerly."

The object of this night attack had not been quite successful; so on the following evening two columns, one under Shaw, of the 12th, and the other under Wellesley, of the 33rd, set off again to clear these topes. Ill-luck attended Wellesley. Some fright brought on unsteadiness, and then there came a panic. The Colonel tried to find his men—the men their Colonel. It ended by Wellesley, in a fit of wild distraction, seeking out General Harris to tell him what had happened.

This failure was somewhat severely canvassed at headquarters by General Harris, and next morning, when a renewed attack on the wood was ordered, General Baird was asked to lead. This generous officer replied: "Might Colonel Wellesley once more try his fortune? He has failed, not through the want of skill and bravery, but from force of circumstances; and if I supersede him, it may be prejudicial to him hereafter."

General Harris, after some reflection, acquiesced, and sending for Colonel Wellesley, offered him the command again. The Colonel, somewhat mortified, readily consented, and soon this hornet's nest within the tope was quite obliterated.

Napoleon, we remember, was in a similar plight this very day at Acre, when, owing to the sniping of Sir Sydney Smith, he had to pack up his traps and run.

It was on May 2nd when the old masonry, unable to support this well-served and well-sustained cannonade, showed signs of yielding. Masses of the wall came down into the ditch, and a breach in the *fausse braie* was reported practicable ; and on May 3rd the face of the bastion was in such a state of ruin, that preparations were made for an immediate assault.

"Sir,—The breach being reported practicable, the Commander-in-Chief desires that the assault may be made this day at 1 p.m.

"I have the honour,

"Headquarters Camp, (Sd.) "Barry Close,

"May 4th, 1799. "Adj.-General."

Orders to that effect were given next morning to Major-General Baird, who had volunteered to command the storming party. A glorious revenge was now to be achieved by Baird, for he who was to lead on that resistless soldiery, by whose bayonets the life and throne of Tippoo should be extinguished, had pined in hopeless captivity, tenant of a dungeon in that capital which he was to enter in a few hours a conqueror ; for in the melancholy slaughter of Colonel Bailey and his troops by Hyder Ali, on September 10th, 1780, Baird, then a Captain, was desperately wounded, made prisoner, hurried to Seringapatam, and there subjected to treatment that, even at a period remote from the event, cannot be read without producing a thrill of horror and disgust.

Of the many who shared his captivity, few remained to narrate their sufferings. Disease, starvation, poison, and the bowstring ended their miserable lives ; fate ruled it that Baird should survive ; and after disease had failed to rob him of life, or temptation to deprive him of his honour (for at this time Hyder sent some of his principal officers to induce the English to enter his service, offering them three times as much pay as they received in our army, and as many horses, palanquins, and wives as they chose), he was spared to lead this band of soldiers to exterminate a tyrant and prostrate a dynasty.

The arrangements for the assault were all completed on the previous evening; ladders, fascines, etc., were conveyed into the trenches unnoticed by the enemy, and before daybreak the storming parties marched quietly in, and lay down until the order to assault was given.

One o'clock came. The city at that hour was perfectly quiet, while the trenches to all appearance contained nothing but their ordinary guards. Baird now appeared, and at the word "Forward!" the forlorn hope rushed on, closely followed by the supporting columns; and in six minutes the British colours were flying above the breach.

The storming parties now filed off right and left, and pressed on. The right column halted on the east cavalier, to give the men breathing time after violent exertion under a burning sun, before assailing the palace, where it was believed Tippoo had retired.

The British, after resting, gained ground, the garrison in every direction flying; while a stray shot, and occasionally a wild huzza! told that the victors were everywhere advancing. A random shot struck the Sultan. He pressed his horse forward, but his passage was impeded by a mob of runaways. His horse next was killed, but his followers managed to disengage him, dragged him exhausted from beneath the fallen steed, and placed him in his palanquin.

But escape was impossible; the British were on him. The bayonet was unsparingly at work, for quarter at this moment was neither given nor expected. Dazzled by the glittering of his jewelled turban, a soldier belonging to the Light Company of the 12th dashed forward and caught the Sultan's sword belt. Tippoo cut boldly at his assailant, and inflicted a trifling wound; the soldier, irritated by pain, drew back, laid his musket to his shoulder, and shot the Sultan dead.

The taking of Seringapatam gave occasion for a remarkable exercise of juvenile talent in a youth of nineteen, who was studying art in the Royal Academy. He was then simply Robert Ker Porter, but afterwards, as Sir Robert, became respectfully known for his "Travels in Persia," while his two sisters, Jane* and Anna Maria, attained a reputation as prolific writers of prose and fiction.

There had been such a thing before as a panorama, or picture giving details of a scene too extensive to be comprehended from one point of view; but it was not a work entitled to much admiration. With marvellous enthusiasm this boy artist began to cover a canvas of 200 feet long with the scenes attending the capture of the great Indian fort, and strange to say, he had finished it in six weeks.

Sir Benjamin West, President of the Royal Academy, got an early view of the picture, and pronounced it a miracle of precocious talent. When it was arranged for exhibition, vast multitudes, both of the learned and unlearned, flocked to see it. "I can never forget," says Dr. Dibdin, "its first impression upon my own mind. It was as a thing dropped from the clouds—all fire, energy, intelligence, and animation. You looked a second time—the figures moved and were commingled in hot and bloody fight. You saw the flash of the cannon, the glitter of the bayonet, and the gleam of the falchion. You longed to be leaping from crag to crag with Sir David Baird, who is halloaing his men on to victory! Then again you seemed to be listening to the groans of the wounded and the dying; and more than one female was carried out swooning. The Oriental dress, the jewelled turban, the curved and ponderous scimitar—these were among the prime favourites of Sir Robert's pencil, and he treated them with literal truth. The public poured in thousands for even a transient gaze."

For here we see our heroes fighting with that degree of splendour and glory perhaps unrivalled in the past, and to be unequalled in the future.†

---

* "Scottish Chiefs," "Thaddeus of Warsaw," etc.

† Sir Robert Ker Porter, K.C.H., afterwards accompanied Sir John Moore in 1808-9 as military artist. He died at St. Petersburg, May 3rd, 1842.

A key to the above picture is here given, showing our heroes in the parts they acted.

THE

GRAND HISTORICAL PICTURE

OF THE

Storming of Seringapatam,

By the BRITISH TROOPS *and their* ALLIES, MAY 4th, 1799

PAINTED BY ROBERT KER PORTER.

DESCRIPTIVE SKETCH.

THE center object is General Baird surrounded by his Staff, and calling his men to follow Serjeant Graham on the *Forlorn Hope*, who having obtained the Colours from the Enfign, planted them on the breach, and as he gave the third huzza of victory, an Indian with a piftol fhot him through the heart. At the foot of the baftion is a party of Tippoo's Soldiers, who are repulfed by the Grenadiers of the 74th.—Lieut. Prendergaft appears mortally wounded, and Lieut. Shaw lies among the flain. A little to the right is Cap. McLeod, who at the onfet was wounded thro' the lungs, being led off by a ferjeant of the Meuron regiment. On the baftion to the right, ftands Col. Sherbrook animating his men to the affault. Croffing a branch of the Cavery is the remainder of the 73d; and behind them the brave Col. Dunlop borne off wounded, between two grenadiers. In front of the mortar battery is Major Allan, with Col. Dallas and Major Batefon; near them ftands Col. Gent. of the engineers, giving directions for the removal of a wounded artillery man. In the distance is a view of General Harris and the British camp. To the left of the grand breach, lies Capt. Owen of the 77th, fupported as he is dying on one of the *Tyger guns*, taken from the enemy, he is upheld by an artillery man, who points to the Indians from whom he received his death, while the grenadiers at the point of the bayonet are preffing to avenge it. On the ramparts ftands Tippoo himfelf, directly above the gateway under which he fell. A French officer a little further off, is pointing to the fcene on the centre baftion; and on the bridge is the defperate conflict between the Company's troops and Hyder Alley's grenadiers; Lieut. Lalor of the 73d, who led a party of the Sepoys, is wounded and ftruggling with his enemy, as he is falling into the water. In the rear of the artillery men, are the Native troops preffing forward to the action, headed by Lieut. Col. Mignan, and flanked by the gallant Brigade-major Pafley

The Place of Exhibition is rendered agreeably warm by large Stoves, in which Fires are conftantly kept day and night.

N B Just Published, and to be had at the place of Exhibition, NARRATIVE SKETCHES of the CONQUEST of MYSORE, neatly Printed in a Pocket Octavo size

---

THE PUBLIC

Are refpectfully informed, that this *PAINTING* is designed from the most authentic and correct information relative to the SCENERY of the PLACE, COSTUME of the SOLDIERY, and various Circumstances of the ATTACK. It is executed upon a Scale comprehending 2550 Square Feet of Canvas, and contains feveral hundred Figures, with Portraits of the British Officers who diftinguished themfelves on that Occasion

The above engraved Outline includes ONLY THOSE OBJECTS *in the Picture to which a* PARTICULAR REFERENCE *is made in the* EXPLANATORY COLUMNS, *and is given merely to assist the Eye in tracing the* RELATIVE SITUATIONS *on the Canvas*

Open EVERY DAY, from *NINE* o'Clock till *DUSK*, in a New BUILDING, erected for that purpofe, *SOUTH-PARADE*, *Belfaft.*

ADMITTANCE One Britifh Shilling.

.·. It is intended to publish, by Subfcription. THREE LARGE PRINTS from the above PAINTING; Propofals for which may be had at the place of Exhibition, where Subfcriptions are alfo received

DOHERTY AND SIMMS, PRINTERS, 8, HIGH-STREET, Belfaft.

To FACE PAGE 142.

Baird and his soldiers slept that night conquerors of Seringapatam. Major Allen, Captain Buckeridge, Lieutenants Percival, Gahan and Shaw, and Surgeon Bagot, of the 12th, were all killed; Gordon and Neville died of fever.

On recovering the body of Tippoo, it was buried with great pomp alongside his father Hyder Ali, in the mausoleum of his ancestors—but amidst a tremendous storm of wind, rain, thunder and lightning, so much so that the soldiers became impressed with the conviction that His Satanic Majesty was present. No park of artillery could have vomited forth such incessant peals as the loud thunder that exploded close to their ears. Heaven and earth seemed absolutely to have come into collision, and no bounds set to the destruction.

After encamping a short time near Seringapatam, the regiment marched into garrison at the captured fortress, and there remained until a daring freebooter named Dhoondia, with 5000 horse, threatened the frontier of Mysore, and naturally occasioned immense alarm. No time, however, was lost in despatching a sufficient force to crush him altogether, or compel him to retire; the 12th Foot formed part of this force, which was under the command of Colonel Wellesley. By forced marches Wellesley succeeded in coming up with Dhoondia while he was encamped, as he imagined, in perfect security.

The fellow, naturally daring, took up a strong position, and boldly waited for the British assault.

Colonel Wellesley led the charge; it was admirably made, and the marauder's fate was decided. His cavalry were cut to pieces or dispersed, Dhoondia himself sabred, and his body, secured upon a gun, was brought in triumph to the camp.

Thus perished the king of "the two worlds," for such was the title by which the freebooter was styled by the banditti.

"Camp near Yarriagoranelly,

"June 3rd, 1799.

"The colours or standards taken by the following corps from the enemy during the late service to be sent on without delay to the Adjutant-General, in order to their being lodged at Seringapatam until they can be forwarded to the Presidency.

"By his Majesty's 12th Regiment, 8 colours; 74th Regiment, 3 colours.

(Sd.) "Barry Close,
"Adjutant-General."

CHAPTER XIV.

"WAR" is defined by Johnson as "the exercise of violence," and as such had told upon Lieutenant Bayly, Lieutenant Seton and two others. This meant a change of air to England for a time. They embarked upon the "Ceres"—passage money, £300. The four just made a rubber:—

> "A good old English game's the thing;
> Where still is reverenced the King!
> Next him the Queen, and third the Knave;
> Meaning true servant, not vile slave!
> The ace so paramount, may mean
> Prime Minister; who still is seen,
> Tho' without sceptre, crown or sword,
> The strongest trump upon the board."

When off Malacca, the "Ceres" anchored to re-victual and repair. A trifling incident now occurred which had a momentous result, for Lieutenant Bayly, whilst on the foreshore one morning when the breakers were high, noticed a boat approaching from off a ship, making what is called "heavy weather." It was just when touching shore that the boat upset, and a woman screamed for help. This certainly was a leap in the deep, or dark, for Bayly—he returned with his destiny in his arms, and within three years, at another watery place called Bath, "Love—all powerful Love" overcame every prudential consideration; for he led the same lady to the altar, and in eleven months he was the proud father of twins. "Beware of twins." For about the same time, an interesting case was taxing the talents of the Attorney and Solicitor-General in the Dublin Law Courts.*

In the midst of the China Seas, the fleet was overtaken by a tremendous typhoon, in which the "Talbot," of 1200 tons, foundered during the night. "I fear our time is come," were the words of the captain of the "Ceres" on entering the saloon at break of day; for amid the raging sea, though the wind had dropped, the vessel was drifting on to a fatal shoal called the "Scarborough." However, when things looked at their worst, the first mate shouted, "A breeze aloft, sir! A breeze! the head sails are filling!" and thus they were spared to reach Canton in safety.†

And thus the voyage continued; but not without the pain of having to report a very sad accident to Lieutenant Seton—for when off Falmouth he was in the act of stringing a bow, when by some mishap the cord recoiled with prodigious force, tearing out his left eye and lacerating his face. The poor fellow suffered agonies, only to be spared for a few years, when he was killed at the battle of Toulouse, 1814.

It was on entering Portsmouth Harbour, that reinforcements for Abercrombie in Egypt were leaving. The usual compliments

* A gentleman, possessed of £15,000 property, died in Dublin, leaving his wife pregnant. He made his will shortly before his death, and left the above sum in the following manner:—In the event of his lady being delivered of a son after his decease, he bequeathed him £10,000, and the remaining £5000 he willed to the mother; but if a female child, then £10,000 were to go to the mother, and the daughter was to have £5000. It so happened that the lady was delivered of twins—a boy and a girl; and the question arose whether the boy was not to have £10,000, and the daughter the £5000, as these were specific bequests, and the bequest to the wife only made contingently.

† Somewhat similar to what happened in the harbour at Apia, Samoa, on March 16th, 1889. Captain H. C. Kane writes: "The 'Calliope,' I thank God, is left afloat. I called on the Staff Engineer, Bourke, for every pound of steam he could give us, and slipped my one remaining cable, and then could only make one knot against it. Out of seven warships, four went down (130 drowned), two stranded, 'Calliope' safe." So Captain Kane and his Engineers deserved well the thanks of the Admiralty. Better still, R. L. Stevenson has put in vivid language the heroism of the men.

and huzzas passed between the two vessels—the outward and the homeward bound, the one to seek the madding crowd in strife; the other to "Home, Sweet Home."

It was to drive the French from Egypt that orders had been sent to England, Minorca and Gibraltar—at which latter places a large force was assembled—to move on to Egypt, and there to be joined by a force under Baird *via* the Red Sea.

> "Like a rock unmoved, a rock that braves
> The raging tempest and the rising waves,"

and with a discipline mingled with a courage that baffles description, arrive our soldiers and sailors from East and West to meet in Egypt.

The fleet entering Marmorice harbour from Minorca, as told by G. S. Parsons, Lieutenant, R.N., is as follows:—

"Comparative darkness came on with a suddenness that I never before had observed, and the gusts of wind were terrific. During this elemental war the British Fleet under Lord Keith, and the Army under Sir Ralph Abercrombie, closely crammed in men-of-war *(armes en flute)* and transports, to the number of 200 sail, were carrying a heavy press of canvas to claw off a lee shore. That shore was Caramania in Asia Minor, a most mountainous, well wooded, black-looking coast. We were in search of Marmorice harbour, the appointed rendezvous of the Egyptian Expedition; and the Asiatic pilots, frightened at the dangerous position of the Fleet in this tremendous weather, lost the little knowledge they had formerly possessed of this unfrequented and frowning coast, whose mountains towered high above the clouds, and on which no vestige of human life could be seen.

"Every glass, in the clearance between the squalls, was eagerly turned upon the precipitous shore, upon which the heavy waves beat with most terrific splendour.

"It was self evident to the meanest capacity that unless the harbour should be entered before night, the transports filled with British warriors would be wrecked on the lee shore, with no chance of assistance. The men-of-war, by dint of carrying sail, might claw off, but the great majority of this fine army would in a few hours become food for the deep, or the ferocious and ravenous tenants of the vast forests, that seemed interminable to our straining sight.

"The Admiral, on whom all the responsibility rested, endeavoured to assume the calmness of tone and manner that the honesty of his open nature would not brook; his agitation was visible in the contortions of his venerable countenance and the sudden starts of his nervous system.

"'Fire a gun and hoist a signal of attention to the Fleet,' said his Lordship.

"'They have all answered, my Lord,' said the officer of the signal department.

"'Now, Mr. Staines (he died Captain Sir Thomas Staines, K.C.B.), be particular; ask if anyone is qualified to lead into Marmorice.'

"As the negative flag flew at the masthead of the men-of-war, every countenance proportionately fell. At length, with heartfelt joy I proclaimed that one of our sloops had hoisted her affirmative.

"'Who is she, youngster? Boy, do not keep me in suspense!'

"'The "Petrel," my Lord!'

"I saw an ejaculation of thankfulness rise warm from the heart to the lips of Lord Keith, as he piously raised his eyes and pressed his hand to his heart.

"'Signal for the Fleet to bear up, make more sail, and follow the "Petrel,"' said Lord Keith.

"'Captain Inglis may be depended on!' And we shook out a reef, and set the main-top-gallant sail, which soon closed our leader in the 'Petrel.'

"As we approached this mountainous and novel land, the idea (and it was an astounding one) seemed to dwell on and occupy the most unreflecting mind, that should Captain Inglis be wrong, every ship, with twenty-five thousand men, would be the sacrifice of such error.

"Lord Keith ordered the signal of 'Attention' with the 'Petrel's' pendants.

"'Captain Inglis, your responsibility is awful,' said the telegraph. 'Are you perfectly certain of the entrance of Marmorice?'

"'Perfectly sure,' said the answer; 'and right ahead!'

"'Signal officers on the fore yard with their glasses!' said the Admiral; and slinging our telescopes we ascended.

"Indeed it was time; for now the roar of the waves as they broke on the coast, throwing their spray on high, conveyed a dismal idea of our impending fate.

"'A narrow entrance ahead!' called the Signal Lieutenant, Staines.

"'Do the Midshipmen make out the same?'

"'We all of us discern it, my Lord!' shouted the whole.

"'God be praised for this great mercy!' said his Lordship.

"The entrance of Marmorice now became distinctly visible to all on deck, from the contrast of the deep, still water to the creamy froth on the shore; and the signal for the port in view, and the men-of-war to haul their wind until the merchantmen had entered the channel, was flying at the 'Foudroyant's' masthead as she shot into the gut of Marmorice.

"Scarcely had we moored, when the heavy masses of clouds that had rested on and capped the high land, now opened upon us in earnest, and the forked lightnings darted among the Fleet with fatal effect.

"The gale increased to a perfect hurricane, and blew from all points of the compass; the flakes of ice, for they were too large to be called hail, came down with such prodigious force as to destroy man and beast; and whoever witnessed that storm could entertain no doubt of a special Providence in the affairs of men.

"We were all safe moored, and the heart expanded in thankfulness to the Eternal Power that had watched over our safety."

After a short delay at Marmorice, the English troops—some 11,000 strong—made good their landing at Aboukir Bay early in March, 1801, and on March 21st defeated the French in the memorable action before Alexandria, where their gallant chief, Sir Ralph Abercrombie, received his mortal wound.

The landing of our troops in Egypt forms again quite an episode in the annals of war. It shows the British soldier firm, dignified, invincible; for upon an amphitheatre of hills and hillocks, and on which cannon at frequent distances all around were placed, as well as men to annoy us, did our brave army advance, in spite of the saying of the French General, "Nothing with life could be thrown on his shores but a cat." The French suffered our boats to advance unmolested, until the whole division was within with the arc of the circle already described. Then opened such a hail-storm of shot and shell upon our boats, that our fine men, cooped up in them like sheep, their hands across their breasts, could stand it no longer. With Major Ogle at their head, they leapt from their boats, and either swam or waded ashore; and when visible to the Fleet a shout went up, "They are on shore! Huzza, my boys! They are on shore!" And as this cry passed from ship to ship, it caught like wildfire. The whole Army shouted "Huzza! Huzza!" and on shore they went to storm the inaccessible hill, with General (Sir John) Moore at their head.

The French were staggered at this wonderful audacity. They seemed to hesitate and look and turn aside, then stop and look again; not firing as they might have done, with all the steadiness and advantage of their situation; but having no feeling but a sense of their danger, they ran. Our brave soldiers gained the summit, took the field pieces, and pursued the enemy down the back of the hill to complete their dispersion.

It was on the 21st when the French mustered in every available force to drive us back to sea and ships. The hostile armies soon became intermingled. The 28th, 58th, 23rd, 42nd and Minorca Regiment (now the Manchesters) were one and all engaged in front, flank and rear. The 28th still wear a badge on front and back of helmet to record the fact of their back to back position. Anthony Lutz, a private in the Manchesters, took singlehanded the Standard of the French Invincibles.

The brave Abercrombie, well over sixty years, was no exception; he engaged himself with a young French Dragoon and took his sword, only to be shot shortly after by a musket ball close up to the hip. "Father, where are you wounded?" for the son had been told of what had happened, and galloped off to help; but, alas! the wound was mortal.

The French withdrew, dispirited and crushed. The English as victors mourned over their Chief, for after three days' suffering, Sir Ralph died on board the "Foudroyant," on which ship his body was carried to Malta and there entombed.

The command now devolved on General Lord Hutchinson. The French withdrew inside the walls of Alexandria, in which place and in Cairo the bulk of their army was massed. Baird, with his 4000 Europeans and 4000 Sepoy *Volunteers*, left Bombay in December, 1800. From the delay caused by the north-west monsoon, he decided to land at Cosseir (instead of Suez), in the Red Sea, where despatches awaited him from General Hutchinson pleading urgency, as he wanted Baird's force for the siege of Cairo and Alexandria.

Baird at once made all arrangements—5000 camels and water bags innumerable were purchased, and small parties were sent on, and encamped at the different wells along the Desert route to deepen and cleanse them, and to dig fresh ones if requisite, besides making watering places for camels and depots for fresh meat where flocks of sheep were collected.

Two gallons of water were allowed daily, and the only rations issued on the marching days were a liberal supply of rice and a pint of wine. The marches commenced at sunset, when half a pint of wine was issued. The rice was carried in their mess tins, whilst the water in which it had been boiled was mixed with the other half-pint of wine and carried in their water kegs.

In this way the desert between Cosseir (June 21st) and Kinneh was crossed with scarcely a casualty. At Girgeh, on July 29th, they learned that the French had evacuated Cairo on June 28th. The force now took river boats, and reached Cairo—a distance of 400 miles from Girgeh—on August 28th.

Sir James McGregor, then a young surgeon, showed his great ability and aptitude and soon rose to highest rank—Director-General,

Army Medical Department. It is to him much of the success was due.

General Kléber now wished to enter into a capitulation of the country with the Grand Vizer and Sir Sydney Smith, who commanded the Mussulman Fleet. Lord Keith and the English Cabinet forbade this, and insisted on the French army surrendering to us as prisoners of war and embarking on English vessels for their return to France, which accordingly was done. So ended the French occupation.

Cesar, who was coachman to Napoleon; his skill in driving six most beautiful horses in the narrow streets of Cairo and Boulac; and the wonderful carriage attached to same, which Napoleon used in his transit through Syria to St. Jean d'Acre, remained to the Egyptians as the greatest wonders of their French visitors.

Thus the century closed. Napoleon, in spite of everything, reigned supreme in France.

> "Yet live, we give thee one illustrious day,
> A blaze of glory ere thou fad'st away."

Wellesley was recognised as a man of mark and great promise.

Nelson was beloved and trusted as the guardian of our seas.

The new century augured well, as Lord Cornwallis for England, and Joseph Buonaparte for France, met at Amiens to settle some preliminaries of peace. They met again in October, 1801, to sign, which stayed the clash of arms and sent grim-visaged War off the stage for awhile.

Then enters Peace. Her ways are ways of pleasantness—which Bayly undoubtedly found out to be so, for it gave him time to narrate to his friends all that he had seen and done in the past five years—besides enabling him to enjoy life to the full in his home surroundings. His first visit was one of respect and etiquette to General William Picton, Colonel of the 12th, who confronted him at once. "Well, Lieutenant, I am glad to see you; and to learn that the 12th Regiment again distinguished itself at Seringapatam. Now tell me about poor Colonel Aston and of his quarrel with my nephew John."

The two brothers Thomas and John Picton, both were in their uncle's regiment (the 12th). Thomas, when a Captain, elected to go on half-pay, and for upwards of twelve years led a country, but not an idle life. In 1793, when war broke out and finding himself unemployed, he went to the West Indies, and shortly after Sir Ralph Abercrombie became the Governor. Now two kindred spirits met. It was not until the Peninsular war, in those days of bloodshed around Badajoz, Ciudad Rodrigo, and Vittoria, that his fighting powers were seen—for Picton with his division meant certain victory. It is said that Picton was never exactly amical to Wellesley.

Certainly in the Peninsula there were many differences between the two. On one occasion, when Badajoz was hard to win, and 2000 men had already fallen, and the roar of conflict was at its height, there arrived an officer to report the capture of the castle. "Who brings that intelligence?" said Wellington, looking up. "Lieutenant Tyler," was the reply. "Ah, Tyler! Well, but are you sure, sir?" "Yes; I entered the castle with the troops, and have but just left it." "Whose troops?" "General Picton's, sir." No further query, for Wellington knew that if Picton was there, Badajoz was won.

"Well done, Fifth Division!" cried the Duke. "Well done, *Fifth* Division, indeed!" said Sir Thomas; "it is my division, the *Third*." "Oh, I thought it was the Fifth," cried the Duke. "You thought, indeed!" cried Sir Thomas, and rode proudly away.

At Burgos further altercations passed, and Sir Thomas felt as if they wished to force him out; but the Ministers and the Duke knew his worth too well. On the escape of Napoleon from Elba, Picton was the first General Officer employed, and about the first to yield up his life in the great fight of Waterloo.

"I met poor Picton one day in Pall Mall. 'What's the news, Sir Thomas?' 'Why, I have just now heard,' said he, 'that Buonaparte has escaped from Elba and is now in France.' I expressed my surprise and incredulity. 'Well,' said he, 'you have it as I have had it, and that is all I know about it.'

"The next time I saw him was in the front parlour of his lodgings in Edward Street, Portman Square, lying dead in his coffin, his sword scabbard bruised with shot, his boots uncleaned and stained with the yellow mud of Waterloo, and his accoutrements lying about in confusion. He was buried in St. George's Chapel, Uxbridge Road—his brother, the Rev. E. Picton, General the Hon. Stewart, and his faithful aide-de-camp Major Tyler, being his few friends assembled."*

* Memoirs, Captain Elers, 12th Regiment.

CHAPTER XV.

DUELLING.

"He must, at length, poor man! die dully at home, when here he might so fashionably and genteelly have been duelled or fluxed into another world."

THIS evil practice died hard.* Shakespeare, with his humour, was unable to kill it. Napoleon was a sworn foe to it: "Bon duelliste, mauvais soldat;" and when the King of Sweden challenged him, he replied "I will order my fencing master to attend as plenipotentiary," and to Sydney Smith, who challenged him off Acre, "Not even if you were Marlborough."

The best check this wanton and silly ordeal ever received was in the island of Ceylon. There the person who gave the offence received the fire of his adversary, and then apologised to him; but turning on the second of his adversary, addressed him thus: "But you, sir! I have an account to settle with. You insulted me in the manner in which you gave your friend's message. Be so good as to take his place and give me satisfaction for the insult." He did so, and was shot dead the first fire. "Bon duelliste, mauvais soldat;" for there were many—but take one for example.

In the years 1796-97, when the 12th Regiment and many other units were resting at the Cape† *en route* for India, so frequent was this direful practice, that an Order forbidding duelling in the Governor's garden was issued, "as being dangerous to the public." This order had no effect in preventing or even checking same, and other meeting-places were easily found.

---

* Lieutenant Munro was tried for murder in 1847, for shooting Col. Fawcett—about the last military duel.

† At the Peace of Amiens, 1802, the Cape was ceded to the Dutch, and retaken by the English under Sir David Baird in 1806.

Captain Keating thought fit to quarrel with a young officer of the 12th Regiment. The former was well known to be a professional duellist, able to strike a shilling, or snuff a candle, at twelve paces. The second to Captain Keating, knowing this, diminished the charge, so that the intended wound might not prove fatal. The bullet, as expected, hit the young man in the side, and not proving fatal, Keating, with a wicked oath, exclaimed: "If you had only put powder enough in the pistol, I should have killed him." For this remark he was afterwards cut.

But to continue. In the *Times* of November 7th, 1805, with its leader on the most decisive victory that has ever been achieved (Trafalgar), and its grief, for "if ever there was a man who deserved to be praised, wept and honoured by his country, it is Lord Nelson," on the other side we read in the Law Report:—

"The King *v.* Thomas Keating and J. Wylde.

"Mr. Garrow applied for a criminal information against these gentlemen—the former is a Captain in the 45th Regiment, the latter an assistant surgeon of the same regiment—for endeavouring to provoke Mr. R. Foxcroft, an attorney at Nottingham, to fight a duel.

"Rule Granted."

In later pages, at the capture of Mauritius, the same Mr. Keating comes off quite as the "mauvais soldat." In spite of all, however, we read in a "Voice from St. Helena," that on July 27th, 1816, Colonel Keating, late Governor of the Isle of Bourbon, had an interview with Napoleon which lasted for nearly an hour.

The duel resulting in the death of Colonel Aston was caused greatly by the low behaviour of an officer, Lieutenant Hartley, just exchanged from the 36th into the 12th Regiment; for, presuming on his slight acquaintance with Colonel Aston (then on leave in Madras), he wrote his own version in a private note concerning some transaction between himself and the Paymaster, Major Allen.

The Colonel, in reply, simply stated: "From what you say, perhaps, the Paymaster has not treated you liberally." This private note Lieutenant Hartley showed to every one, creating mischief everywhere—so much so, that Major Picton, whilst in temporary command, convened a Court of Enquiry, which found that the Paymaster's accounts were perfectly clear with Lieutenant Hartley, and consequently did not deserve the construction placed upon them by the Colonel.

Surely Lieutenant Hartley was the man to become the target, as was done with the second in the previous duel at Ceylon.

The Colonel then reprimanded Major Picton for convening this Court in his absence; Major Picton, feeling aggrieved, sent a challenge to his Colonel, and the usual meeting followed. Major Picton's pistol missed fire, and he threw it down on the ground in a rage. The Colonel told him to "Try again!" This, of course, was not allowed; and after shaking hands, the Colonel said: "Well, Picton! You and your friend must come and dine with me to-night, and all must be forgot."

On the following day, Major Allen, the Paymaster, asked satisfaction for the reflection cast on his character by the word "illiberal." "Certainly!" replied the Colonel; "but allow me to say you have been very tardy in demanding it." Then, sending for the Adjutant, he desired him to go to all the officers and to say, "If any one of them felt aggrieved he was to say so, and he (the Colonel) would fight them one after the other, and finish the business altogether."

The meeting took place. The distance was measured, and Major Allen fired. The Colonel remained erect, and the seconds concluded that the ball had passed him; but dropping his pistol arm, he exclaimed, "I am wounded, but it shall never be said that the last act of my life was that of revenge." This was at Christmas time, and thus lay Colonel Aston—

"Fallen from his high estate,
And weltering in his blood."

When the heavy roll of the muffled drums announced the funeral of their much lamented chief, Major Allen, the portly, full-faced, handsome man of six feet, appeared on parade, cadaverous, hollow-eyed, stooping, and so weak, that he could hardly follow in procession; and at the painful sounding noise of the coffin being struck by the descending earth, he fell fainting. His dubash then stated that his master had not touched food since the fatal duel, and had only asked for water, muttering, "Oh God! Oh God! I have killed my best and dearest friend."

The two Majors were tried by Court Martial and admonished. Major Allen was afterwards tried by Civil Court and acquitted, but without the word "honourably." He was never seen to smile again, and was the first to fall at the siege of Seringapatam.

Perhaps the sad result to poor Colonel Aston's wound might be due, or at least been hastened, by the attachment of some foreign body to the bullet, such as Mr. Macklin, in the following duel, wished to guard against:—

"Macklin, the great Irish actor, anxious to ward off his son from the hazards of his own profession, obtained for him a cadetship in the East India Company's service.

"Macklin fils, like his father, was somewhat eccentric, for in his first quarrel which led to a duel, he appeared at the rendezvous in a loose great coat, which covered him from head to foot. When all was ready, he cast off this garment—and there he stood perfectly naked; giving for his reasons that most of the wounds that proved fatal in India, were due to the fact that the bullet carried some of the woollen or linen clothing along with it.

"'Now I determine to fight naked, and you may do the same,' said he to his opponent. The seconds could not allow of this indecent exhibition, so the duel was stopped, and the quarrel mended."

How many of these frivolous affairs could have been settled as Touchstone has it: "If you said so, then I said so," and they shook hands and swore brothers. Your "if" is the only peacemaker; much virtue in "if."

Even the base and wicked murderer awaiting death on the scaffold, and offering up perhaps his first and only prayer, sought virtue in the "if." "Oh God! if there is a God—save my soul if I have a soul!"

In France, perhaps, they manage these affairs of honour a shade better. On the eve of Hohenlinden, young Ségur, who we first met on the 18th Brumaire, and was now serving under Macdonald* and Moreau, met his Colonel, who evidently had been dining too well. Ségur was thrust aside from off the footway by an unpleasant push, and his Colonel passed on unblushingly.

Ségur passed the whole night in transports of rage, and at break of day, seeing his Colonel walking by himself on the plain, ran to him, tendered his resignation, giving him to understand that immediately afterwards — having become his equal — he should demand satisfaction for the insult of last night. The Colonel evidently had no recollection of the incident, for, all surprised, he measured his young antagonist from head to foot with a glance of wonder and astonishment, and told him he could not in presence of the enemy resign his commission; but he himself felt sorry if he had inadvertently been guilty of any insult. Thus the quarrel ended, and both men went forth to Hohenlinden.

"On Linden, when the sun was low,
All bloodless lay the untrodden snow;
And dark as winter was the flow
Of Iser rolling rapidly.

"But Linden show'd another sight
When the drum beat at dead of night,
Commanding fires of death to light
The darkness of the scenery."

---

* Macdonald was of a Scotch family which had followed the exiled James II. to France. He was born in 1765 and died in 1831. He fought at Jemappes in 1792, and Hohenlinden, 1799. Napoleon created him Marshal at Wagram. His services at Lautzen, Bautzen and Leipsic were brilliant, and Napoleon, on his abdication, gave him the sword that Murad Bey had presented to him in Egypt. He became a member of the Chamber of Peers in 1816.

Napoleon sums up Hohenlinden thus: "Sans doute la bataille de Hohenlinden fut très-glorieuse pour le General Moreau, pour les généraux, pour les officiers, pour les troupes Françaises. C'est une des plus decisives de la guerre; mais elle ne doit être attribuée à aucune manœuvre, a aucune combinaison, à aucun génie militaire."

George II. gave strong sanction to duelling. He himself challenged his brother-in-law, Frederic the Great, to a hostile meeting. When two officers under Lord Albemarle's command—Ensign Campbell and Lieutenant Ferguson—quarrelled, and Campbell knocked Ferguson down, the comments on the Court Martial that followed read thus:—

"In reference to the ensuing Court Martial, the Secretary of War (Henry Fox) writes to Lord Albemarle, November 27th, 1746: 'Mr. Ferguson is justly acquitted of the charge against him; but his complaining to a Court Martial, instead of resenting in another manner the usage he had received from Campbell, it must be supposed, will necessarily prevent the officers of his regiment from rolling with him. His Majesty particularly asked if they had not their swords on when this happened, and bids me tell your Lordship, that as an officer—not as a king—it is his opinion that if Campbell is pardoned, a hint should be given to Ferguson that he must fight him or be broke.'"

Duels on land, *i.e.*, the Cape, had already been proscribed by the Governor as dangerous to the general public, so perhaps when M. de Grandpré and M. le Pique had agreed to fight (Paris, 1808), they decided it must be from a balloon, and in regions affording more elbow room.

Each combatant, with his second, was armed with a blunderbuss, for the conditions were to fire at the balloon, and not the car. The first signal was to cut the rope, and let the balloons ascend; a second signal then was given to commence the fight. At the latter, M. le Pique fired and missed. M. de Grandpré fired and hit, when M. le Pique and his second both descended, and were dashed to pieces. M. de Grandpré alighted some leagues from Paris.

Now that we have flying chariots through these fields of air, God forbid that their sole function shall be for horrid war: to watch the mysterious grouping in the sky of nations with their air ships, sending wireless messages to their submarines below—we who live between must have a sorry time.

CHAPTER XVI.

NAPOLEON—now absolute in France and all-powerful abroad—thought fit to demand from England the evacuation of Egypt and the surrender of Malta.

"Malte on rien," said Napoleon to Lord Whitworth. "Jaimerais meux vous voir en possession de Montmartre que de Malte." Whitworth on this demanded his passport, which meant "Amen to Amiens," and war was declared on May 13th, 1803.

Pitt at the time was not in office, but the cry (like in his father's time) soon became: "Si ce ministère dure, la Grande Bretagne ne durera pas." His re-appearance in the House of Commons (says Lord Rosebery) was unique. Two hundred Members of the House had never heard him, many of whom had never seen him. His influence at once exceeded all belief, and he was welcomed back as Prime Minister on May 18th, 1804—the same day as Napoleon was proclaimed Emperor of the French and was *en evidence* at Boulogne, preparing for the invasion of these shores.

Napoleon commenced this second burst of war (1803) by ordering the arrest of all English people travelling in France. Sir John Abercrombie, who had nursed his father Sir Ralph, when mortally wounded at Alexandria, was the first so caught.

The ridiculous pretext offered by Napoleon was that, as all Englishmen were bound to serve in the Militia, he was authorised to consider them as military men; but this was a pitiful excuse, for the order of arrest even included women and clergymen, many being over sixty years of age.

Verdun was the town selected for their prisonhood, and the captives soon found means to enliven their existence. Theatricals, races, balls and masquerades became the order of the day, and the "Detenu's Song" by Lawrence remains a poem far above the average. To others it went harder. In the Obituary, February 19th, 1805 :—

"At Verdun, in France, Lieutenant William Fitzgerald, second son of the Vice-Provost of Trinity College, Dublin, late of the frigate 'La Minerve,' which on the 2nd of July, 1803, went aground in a fog near Cherbourg, and was taken by the enemy, but not till after a gallant resistance while resistance was of any avail. He and his brother, a young midshipman only twelve years old, who shared in the danger of this disastrous event, were marched several hundred miles into the interior of the country, to different places of confinement, and ultimately to Verdun, where the despair of an exchange of prisoners, and consequently of any farther opportunity of distinguishing himself in the service of his country during the remainder of the war, is supposed to have preyed upon his spirits and to have produced a decline, which at length released him from his earthly captivity."

In later days, Napoleon was reminded of his unpleasant edict by a Lieutenant on board the "Bellerophon," when *en route* to his exile home at St. Helena. He remarked to this Lieutenant that he was somewhat old for his rank. "Ah!" replied the Lieutenant, "I was imprisoned by you at Verdun for some years ; that is what checked my promotion." Napoleon shrugged his shoulders and looked away, displeased.

Napoleon and Josephine once, when posting through the country, stayed for a night or two in Verdun (in 1803).

Josephine made the coffee, and the landlord of the hotel gave away another family secret, viz., that Monsieur and Madame slept not only in the same apartment, but in the same bed. These arrangements proceeded from fear of assassination, and perhaps not from affection, for Napoleon at this time was distinctly unpopular.

Robert Emmet was also a traveller in France when the Peace of Amiens ceased to be; but he was kept by the First Consul not as a foe but a friend, and only left for Ireland when primed with the plans and intentions of Buonaparte against England.

Full of ardour and enthusiasm, Emmet had no misgivings—he spent his days and nights in devising plans of attack. His projects were too visionary, his pecuniary means scanty, and many of his colleagues false; but still, on July 23rd, 1803, he sallied forth dressed in green and gold at the head of 120 men, all drunk. Pillage, outrage and murder followed. Lord Kilwarden was killed. Emmet at these proceedings abandoned everything in disgust, and might have fled the country, but the bidding adieu to his betrothed (Miss Curran, daughter of the great wit John Curran) gave time for his arrest.

His speech why sentence of death should not be passed is a masterpiece of impassioned oratory—an inspiration to all actors. He was hung in September, 1803.

This scene so affected the poor girl (although she married later) that it caused her to droop and die. The same gloom settled over the mind of Curran; he resigned all his appointments, and sought travel abroad. The country for a time rang with his speeches and witticisms; they all came so readily. Only to quote one: when Lundy Foot, the great tobacconist, asked Curran for a Latin motto for his coach, out sprang the answer "Quid rides."

The theatres of Paris could no longer stave off their abuse of the Great Man, and chose the week of the famous trials when Moreau, Georges, Pichegru, etc., were all indicted for treason. The badinage and wit partook somewhat of the following:—

Oranges were flung from the gallery to the stage amidst cries of "Peel them! Peel them!" At first no attention was paid to the noise, but as it continued, the whole audience joined in until the oranges were peeled, when inside each was a portrait of Louis. "Gardez le Louis et jettez l'ecorce" was the fresh demand.

The prisoners, at the close of each day, were conducted back to their cells between files of soldiers. Many a time were the whispers of the men loud enough to be heard, especially to Moreau: "Mon General, voulez vous de nous?" "Non," was the answer, "je n'aime pas le sang." Had he but given the word, probably Buonaparte and he might have changed their residences—Buonaparte to the Temple; Moreau to St. Cloud.

Moreau, in his defence, made a noble speech; so much so that the audience rose and clapped their hands, which is as unusual in the Courts of France as it is in those of England. This, when told to Napoleon, so enraged the tyrant that he raved for a death sentence. He was then quietly told by General Moncey, the Commandant of Gendarmerie, that disaffection already prevailed amongst the corps, and if anything happened to Moreau they would set fire to the four corners of Paris—so Moreau was only exiled. He was killed at Dresden fighting against Napoleon, 1813.* Louis XVIII. in the following year placed the Field-Marshal's baton on his tomb.

Pichegru, another fine soldier who by his wise dispositions had won many battles for his country (some of which England remembers to her cost, viz., Nimequen, Dordrecht, etc., under the Duke of York), was supposed to have strangled himself in prison. As a proof of his strangulation, the body was to be taken from the prison to the Criminal Court, for the surgeon to report the cause of death as self-inflicted; but at the hour appointed there was no body, for Pichegru never intended to strangle himself. The next day a fresh order was given for the judges to attend, when the body of Pichegru *did* appear, for he had been strangled in the night "by order." And had not Captain Wright, of our Navy, disappeared under somewhat similar circumstances? Wright had received orders, or say permission, to convoy over some French Royalists, and land them on the shores of France. If this licence was indiscreet on the part of England, it offers no excuse for the murder of this officer by order of Napoleon.

---

* This fact was first made known to the French by a little dog whining outside the hut where dead Moreau lay. On the dog's collar were the words: "J'appartiens au General Moreau."

The murder of the Duc d'Enghien appalled and frightened everyone. He was entrapped at Baden across the Rhine, on March 15th, 1804, by Canlincourt and some French soldiers, who had orders to try, and to shoot him, at Strasbourg. The Prefect of that town, however, plainly said that the people of Strasbourg would never suffer such an execution—so the unfortunate Duc was driven in haste 450 miles to Paris, where a court-martial at once awaited him, and he was shot the same evening (March 21st).

The next morning Paris was in stupor when she heard it. Fouché even said, "C'etait un coup du fusil inutilement laché. Cambacérès." The Second Consul exclaimed, "Vous êtes devenus bien avare du sang des Bourbons;" and Josephine, who chided her spouse, received in reply, "Laisse moi tu n'est qu'une femme, et tu n'entends rien a la politique."

"Should vice expect to 'scape rebuke
Because its owner is a duke."

Even these crimes did not suffice, for Napoleon sent an insolent proposal to Louis XVIII., then residing at Warsaw. He styled himself the First Consul, but before assuming Imperial dignity he wished for Louis XVIII. to relinquish all rights to the throne of his ancestors, or in other words to abdicate in his favour.

The bearer of this message received a somewhat curt reply: "His Majesty did not wish to confound Mr. Buonaparte with those who had preceded him, or to think of him as one likely to succeed him." He was likened to the mule, without pride of ancestry, without hope of posterity.

Finally, perhaps, no phial of hate could contain more than the following letter:—

"Brompton, October 31st, 1803.

"Sir,—

"I yesterday partook in the Sacrament of the Lord's Supper at Brompton Chapel. The service was well conducted, and the congregation appeared devout; but I noticed a breach of *female decorum*, which I trust you will permit me pointedly and publicly to reprehend. . . .

"Yesterday, sir, in Brompton Chapel, a well-dressed aged female presented herself before the altar of the God of Christians, bearing on her breast, in the *most ostentatiously indecent manner possible*, a flaunting medallion of the most notorious Anti-Christ that the Almighty, in His wrath, ever permitted to pollute the globe. I saw the officiating Minister (the Rev. Weeden Butler, jun.) was extremely shocked at so glaring an instance of eccentric impertinence: but how was he to act? The Sacrament was administered, evidently with a hesitating and irresolute hand; and the young clergyman seemed relieved from the horror when the foul profanation was removed.

"A FRIEND TO DECORUM."

CHAPTER XVII.

WHILE these events (so much resembling a second Reign of Terror) were happening in France, the Bengal Army, under Lake and Wellesley, were engaged in the great Mahratta War.

Nadir Shah, a Persian monarch, looted Delhi in 1740 of all its treasures, amounting in jewels and money to many millions sterling, and left the city defenceless just with a puppet king. It was then the Mahratta power stepped in, and Scindia took to rule. The interlopers brought the French, who soon intrigued against the English, as was only too apparent in the Mysore War, 1799.

Wellesley made his theatre of war in Berar, and Lake in Gwalior—time, September, 1803. The former, with a force much harassed by forced marching and the heat, caught sight of his enemy in full strength near the village of Assaye on September 23rd, where the 74th and 78th and 19th Light Dragoons had at one time to sustain a heavy onslaught, and to rescue one another. Colonel Maxwell, of the 19th, was slain, and hardly an officer in either corps remained unwounded. Their behaviour was splendid, and soon a decisive victory was assured. Other fortresses, Asseerghur* and Argaum (November 29th, 1803), soon fell to escalade.

---

* This old Mahratta stronghold, perched seven hundred feet above the plain, and large enough to contain a regiment, is quite worthy of a visit. Being detained in a Rest Camp at Khundwa in 1872, I sent proposals for a friendly game of cricket, which offer was accepted. My men, to me, were total strangers, certainly in a cricket sense, but trusting to our village greens, I thought they might know enough. The scores in this test match have never been recorded, and I hope never will. So if no centuries of runs, there were centuries of time—

Advancing from Cawnpore, Lake took Alighur by storm in early September; then marched on to Delhi*, where in a stiff battle he drove off the Mahrattas; and entered the city on September 8th, 1803. He then released the puppet king from his cruel thraldom, and left Colonel Ochterlony in charge as Resident to protect the re-instated monarch. Lake next captured Agra, and shortly afterwards defeated at Laswaree the whole of Scindia's army, *i.e.*, the troops he had marched up with from the Deccan. After this great victory, Lake's future lay, in successes and reverses, with a Rajah named Holkar.

Lake, in his despatch dated May 10th, 1804, mentions the flight of Holkar after a warm engagement, also the success of Colonel Don in capturing by assault a strong fort, Rampoora, on May 16th. Here Sergeant O'Loughlan, Major Doveton, Captains Wood and Ruhan, and Lieutenant Hay received special mention. In August, Colonel Monson reported the capture of Hinglaiz Ghur, after which Monson set out to co-operate with Colonel Murray in bringing Holkar to bay elsewhere; but unfortunately a great disaster veiled the sun awhile.

The detachment under the command of Colonel Monson, which consisted of 3000 men, was perpetually harassed in its march by a superior enemy. His situation was aggravated by a want of supplies; by continued and heavy rains, which impeded his march and rendered the conveyance of the guns almost impracticable; and lastly, by treachery, some of his native officers having entered into a correspondence with Holkar.

---

for there lay a splendid gun, prized by Sir Arthur when he scaled the place. It is supposed to be *en route* to Woolwich—when tackle can be found, first to get it down the hill, and then to roll it home. Next came the caves of Ajunta, where some most interesting frescoes throw the art of painting back into an unknown period; and close by was the whitewashed mosque, where Sir Arthur wrote his despatches proclaiming his victory at Assaye. Our visit lasted three days; we returned quite delighted.

* Lake made frequent mention of the undaunted bravery and steadiness of His Majesty's 76th, then commanded by Captain Boyes, and on this occasion it was never more conspicuously displayed. His grandson, Captain W. J. Boyes, was Adjutant of the 12th from 1865 to 1870.

An extract from a private letter describes the situation :—

"October 11th, 1804.

"I am one of two officers who were saved from a whole battalion. The 2nd Battalion of the 2nd Regiment is even worse, one only of their officers having escaped. One battalion of the Irregular Horse went over to the enemy at the beginning of the battle. A company of Sepoys were following its example, when a young lieutenant (I don't know his name) went up to the Zeminda (a native officer) and shot him through the heart; which so much surprised and astonished the Sepoys that they immediately altered their intention, and swore to follow the valiant lieutenant, which they afterwards did."

After this, shall we say, perfidious success, Holkar collected all the men he could, and laid siege to Delhi, but Ochterlony made a grand defence, determined to hold his own until relieved by Lake (October 16th, 1804).

The East India papers give the following: "The fortifications were in the most ruinous state, and the native troops deserted in great numbers; but British valour prevailed after a siege of nine days, in the course of which the enemy made every effort to storm the place. The city of Delhi is ten miles in circumference, and never before sustained a siege."

According to other official papers, the French were all this time corresponding frequently with Holkar. A French chasse-maree, after a desperate resistance, was captured off the Malabar coast, crammed with officers and munitions of war. Then, from Mauritius, the French officers came over in shoals; they landed on the coast of Cutch and traversed the borders of the desert as far as Jelour, then, taking an eastern direction, they passed into the Mahratta country just above Dunderpoor.

On November 13th, General Fraser routed Holkar at Deeg. The 76th and 1st Europeans (101st) again were irresistible, but when the evening came, General Fraser was dead, and 3 Captains, 17 Lieutenants, 1 Ensign, 12 Sergeants, 22 Corporals, 218 Privates.

After the relief of Delhi, Lake lost no time, and was soon in pursuit of Holkar—a prelude to a grand cavalry fight. For on November 18th, Lake reported his great cavalry fight to the Governor-General:—" Since I left Delhi on the 31st ult., the troops have daily marched a distance of twenty-four miles, and during the day and night previous to the action they marched fifty-eight miles; and the distance to which they pursued the enemy must have exceeded in all seventy miles." The 8th Dragoons, 27th, 29th, the Native Cavalry, and the Royal Horse Artillery, for their mobility, were especially praised.

In the General Order of Lord Lake, in which he thanked the officers and men who conducted themselves so gallantly in the action of the 17th of that month, with the cavalry of Holkar, he particularly noticed the services of Colonel Macan, Lieutenant-Colonel Vandeleur, and Major Need, of the 1st Foot and 2nd Brigades Cavalry; Lieutenant-Colonel Toone, who commanded the Advanced Guard; Captain Abercrombie, officers and men of the 8th Dragoons; Captain Philpot, officers and men of the 27th Dragoons; Major Wade, officers and men of the 29th Dragoons; Captain Welsh, officers and men of the 1st Native Cavalry; Captain Elliott, officers and men of the 4th Native Cavalry; Captain Swinton, officers and men of the 6th Native Cavalry; and Captain Browne, officers and men of the Horse Artillery. He added "He cannot avoid to embrace this opportunity of noticing the uncommon perseverance and praiseworthy conduct of the Reserve, which effected a march of upwards of three hundred miles in seventeen days, and which bore the unusual fatigues, to which it has been so long exposed, with the utmost cheerfulness. He begs Lieutenant-Colonel Don, commanding the Reserve, to accept his best thanks, and to signify to the officers and men of the detachment of artillery, to the officers and men of the flank companies of His Majesty's 22nd Regiment, and 2nd Battalion of the 21st Foot, the great satisfaction he has derived from the zeal, alacrity, and patience which they have uniformly manifested."

The reader will notice that, in the above Order, a Major commanded a brigade of cavalry, whilst the Regiments had only Captains for their leaders, and all thoroughly went well. This remark is in no way meant against the Colonel; for without the good Colonel, who could instruct? But it is meant against the retention of the Pompadore Mount Razor type, whose vagaries do such damage to the growth and development of that splendid race of Captains so eulogized by Lake, and of our gallant Subalterns, who have pulled the English Army out of many a tight place. Was it not Lord Cardigan who, as a Colonel, so disgraced England by the following:—

"April 11th, 1841—The Earl of Cardigan causes 100 lashes to be administered to Private Rogers, of the 11th Hussars, in the riding school at Hounslow, immediately after divine service, and before the men could return to barracks."

Then, after the Crimea, he has to refute charges of cowardice:—

"CARDIGAN *v.* GENERAL CALTHORPE.

"The main charge made against his Lordship was one of cowardice, the General alleging that his Lordship, although well known to be a consummate horseman, had allowed his horse during the charge to take him to the rear. Lord Cardigan applied repeatedly to General Calthorpe to retract his statement, but without any other result than a note to the following effect in the second edition: 'The author had relied on statements furnished by officers actually engaged in the charge; but as the excellence of Lord Cardigan's horsemanship is unquestionable, the idea that his horse ran away with him is no doubt erroneous.'"—June, 1863.

This trick of horse and rider is very well at times, and sometimes most amusing; but between War and Peace there is a difference. At Aldershot, whenever a field day went too long, I often noted there was sure to be a runaway, where equally good riders were concerned. Roddy Owen sometimes was one—his horsemanship was superb. He, poor fellow, died of cholera at Uganda, 1897. But by some strange coincidence, these runaways occurred within half-an-hour of the express train leaving Aldershot for London, and on making tender enquiries as to the result—the horse was in the stable, and the rider a pedestrian in town.

One scene more from the theatre of war: for from Bhurtpore Lake had been told by the Rajah that he would pay all the expenses of the war if Lake would but desist. "If not," said the Rajah, "I will defend it to the last extremity, when I will blow up the fort, with my wife and children, sooner than you shall have it." Lake had to withdraw before his object was obtained. The rewards came duly.

"Putney Hill,

"August 30th, 1804.

"My Dear Wellesley,—

"The letters which you will receive by this conveyance, will inform you that the King has conferred a peerage on General Lake*, and an extra red ribband on your brother, General Wellesley.

"W. Pitt."

Sir John Craddock arrived in India about this time as Commander-in-Chief in Madras. He brought out the decoration for Major-General Wellesley, asked of his servant to bring his master's coat to him, and placed the Star of the Order upon it, before he got up in the morning, and then enjoyed Sir Arthur's surprise.

Sir Arthur, after nine years of distinguished Indian service, longed to gaze once again on the white cliffs of Old England. Before this wish was realised, he had to accept the invitation of the Field Officers and Captains of His Majesty's Regiment to a banquet at Madras, with Lord William Bentinck† in the chair. The evening was very festive.

---

* Lord Lake took for his supporters a 76th and a Sepoy. He was a Guardsman, and had served in the same campaigns as the 12th, viz., Minden and the Seven Years' War. He died in England in 1808, from a chill caught at the Court Martial assembled for the trial of General Whitelock at Chelsea Barracks. Whitelock was charged with surrendering Buenos Ayres. In 1805, and again in 1807, the British besieged Buenos Ayres. The mother country—Spain—told the natives to defend themselves as best they could, so they took to arms and twice expelled the invaders. It was on the latter occasion (1807), when Whitelock surrendered, that the charge of incapacity and cowardice was brought against him. He was acquitted of cowardice.

† Governor of Madras; afterwards a General with Sir Arthur in the Peninsula; next Governor-General of India.

Sir Arthur, on arriving home, soon resumed political life. In 1790 he had sat for Trim, and was now elected as Member of Parliament for Newport, Isle of Wight, shortly to be followed by office as Chief Secretary for Ireland. At this time, Napoleon, Ney, Lannes, and Davoust, with 160,000 men and 2300 petites batiments, were at Boulogne, waiting for a fleet to escort them over, for says Napoleon: "Que nous soyons maîtres du detroit pendant six heures, et nous serous maîtres du monde."

It was on May 18th, 1805—perhaps the hour when Pitt entered the House of Commons as Premier of England—that the guns across the water were saluting Napoleon as Emperor, and his family Imperial Highnesses. An Imperial Eagle was hoisted in place of the Fleur de Lys of the Bourbons and the Coq Gaulois of the Republic; and eighteen Generals were given their Marshal's baton.

The 14th of July was a grand fête for the distribution of the Legion d'Honneur, and in December came a grand parade for the regiments to receive their Eagles. "Les aigles imperiales voleront de victoire en victoire," said Napoleon; but do what he could, he could not find the Eagle to fly across the Channel. The Eagle meant his fleet, which could not be made to concentrate and manœuvre with the same accuracy as troops on land. "Wind and weather permitting" never entered into Napoleon's calculations.

The much wished for fleet had already met with one mishap, for Admiral Latouche Treville (Commander-in-Chief) died quite suddenly off Toulon on August 1st, 1804. In his possession at the time were sealed orders for the squadrons of Toulon under Villeneuve, of Rochefort under Missiessy, and of Brest under Gantheanne, to rendezvous off Antilles (West Indian Isles) and await the arrival of the Spanish fleet from Ferrol and Cadiz, and then for the United Armada to make for English waters.

The command then passed into the hands of Villeneuve.*

* Villeneuve (born in 1763), was the second in command at the Nile. After Trafalgar he escaped in the "Guillame Tell," was taken prisoner at Gibraltar, and sent to England on board the "Victory." He was liberated in 1806, but was not allowed to return to Paris by order of Napoleon. He was found dead in his bed at Rennes, 1806.

Admiral Missiessy was the first to reach Antilles; there failing to meet his sister ships, he returned to Rochefort.

Admiral Gantheanne could not get out of Brest, for our Admiral Cornwallis had securely blocked him in.

Villeneuve, with seven of the Spanish fleet, reached the Antilles. Orders were soon sent him to return and raise the blockade at Brest, and to enter at all cost the English Channel and keep it free for the flat-bottomed boats to cross.

Admiral Villeneuve failed. He fought an indecisive action with the English under Sir Robert Calder off Cape Finisterre, when, considering himself sufficiently beaten, he entered Ferrol and Cadiz, where the English fleet safely kept him.

When the news of Villeneuve's failure reached Napoleon at Boulogne on August 13th, 1805, Daru* was summoned, and the Emperor—wild and black as thunder—paced up and down the room for nearly an hour. Then, venting his anger in a torrent of oaths, reproaches, and painful words, he pointed to a desk loaded with papers, and said, "Seat yourself there; write!" and immediately, without transition, without apparent meditation, and with his sharp, short and imperious accent, he dictated without hesitating, the plan of the campaign from Ulm to Vienna.

Immediately the Grand Army of seven Corps—each under a Marshal—wheeled about, broke into a hundred columns, and hurried towards the Rhine.

The camp at Boulogne lasted from 1803 to 1805, during which time Napoleon's sword had been dangling in the scabbard. His pen, however, had been employed in Paris for the better government of France. The Code Napoleon of 1803 is still their rule and

* Daru (born 1767, died 1829) was a born administrator and author. He entered the Commissariat at the age of seventeen. He could, amidst the duties of the difficult post of Commissary Chief to an Army, write "Odes and Epistles of Horace" and "Poems to the Alps." "Daru is fit for anything," said Napoleon; "he has judgment, intellect, a great capacity, a body and a soul of iron."

guide. Unsurpassed as a soldier, Napoleon was equally unsurpassed in all other affairs of State.

But to return. Now, bursting forth like a thunder clap, Napoleon wished to clear the air everywhere. Russia would not recognise him in his new title of Emperor, and sent her forces to Corfu, to be in touch with the English in Sicily and Malta, and to support the reigning House of Naples—of whom Napoleon well remembered that an Englishman was their Prime Minister, for had not Acton (grandfather of Lord Acton of Cambridge) permitted Nelson to water and provision his fleet at Syracuse previous to the battle of the Nile?

Acton's position soon became untenable, and he was forced into resignation. He was replaced by one Hugh Elliot, a younger brother of Lord Minto (whom we shall soon meet as Governor-General of India), and France sent to Naples a M. Alquier. These two were of the same age—both born in 1752. The former had been trained for English diplomacy, and enjoyed a splendid record; whilst the latter was a Jacobin and regicide well versed in the ways of French intrigues. The real tension came on Christmas Day, 1805, when Napoleon issued a bulletin: "La Dynastie de Naples a cessé de regner," and vowed vengeance.

CHAPTER XVIII.

"Their lives did not end when they yielded their breath:
Their glory illumines the gloom of the grave."

THE same years—1803-4-5—had been to England a time of anxious watching. Pitt and Nelson stood out as beacons in full effulgence, a brilliancy which, alas! was soon to go, for death demanded both in this critical year, 1805-6.

Patriotism and port wine had kept Pitt alive for forty-seven years—twenty-two of which he had ruled England; and now (1803), when Warden of the Cinque Ports, he raised and drilled a corps of 3000 Volunteers. Somehow, in the Charter of Enrolment of this corps, appeared accidentally in our massive English style, that "This Corps is on no account to quit the shores of England except in case of invasion."

Volunteering and enlisting was in force everywhere, and Captain Bayly, 12th Regiment—now a Recruiting Staff Officer at Bath and Wells—mentions in his diary that he had to find 200 substitutes for those unwilling to serve. To cry off meant a fine of £40 to £60, the greater portion of which had to be paid over to the Military Chest at Chatham. At one time Bayly found himself possessed of several thousand pounds. How was he to secure same?

The idea of sewing the bank bills into the lining of his regimental cap came uppermost; this accordingly he did. The cap had now to be worn or watched morning, noon and night; and it was this vigilance that attracted the attention of an old rogue of a Sergeant. The moment came when the cap was left unguarded, and the

Sergeant seized the bills and absconded. Bayly's horror on discovering this was great.

His landlady said she had seen a fat Sergeant enter his room; and further enquiries told him that the same man had left by coach for Bath; so mounting a beautiful young fleet bay mare, a present from his father, he galloped thirteen miles in thirty-five minutes. The mare, on entering Bath, gave a sudden plunge forward and fell dead at his feet. The object of his search was not there; he had quitted the coach a mile from Bath on to the Bristol road.

It was whilst arranging to continue the pursuit, that the Bristol coach rattled through the town for London, and there, in the back seat, was the Sergeant. "Stop coach! Stop coach! A thief! A thief!" soon brought the miscreant down, when all the missing notes were found secreted in his neckcloth.

The Sergeant was brought to a Court Martial, broken and flogged. The dead mare—valued at least 100 guineas—was hard to replace.

Two hundred substitutes from one town alone, when Napoleon was facing us at Boulogne, hardly meant roaring patriotism; but in less than a month, and from another part of England, the hero of the Nile, a quiet looking gentleman, started from his country seat at Merton to drive to Southsea on the coast—and he is about to tell us, in language as lasting as the memory of England, that there is a time when

"England expects that every man will do his duty."

In his Lordship's memorandum book is written:—

"Friday night, September 13th, 1805.

"Friday night, at half-past ten, drove from dear, dear Merton, where I left all which I hold dear in this world, to go to serve my King and Country. May the great God whom I adore enable me to fulfil the expectations of my Country! and if it is His good pleasure that I should return, my thanks will never cease being offered up to the Throne of His Mercy. But if it is His good Providence to cut short my days upon earth, I bow with the greatest submission: relying that He will protect those so dear to me that I may leave behind. His Will be done! Amen, Amen, Amen."

One is tempted to conjecture how far the superstition of a Friday (especially the 13th) entered into Nelson's mind. For the sake of his sailor's superstition he would not have sailed on a Friday. Would the God of Battles have spared him, had his drive from Merton been delayed for ninety minutes?

The following interesting extracts are faithfully copied from his Lordship's memorandum book, which was written entirely with his own hand:—

"Saturday, September 14th, 1805—At six o'clock arrived at Portsmouth, and having arranged all my business, embarked at the bathing machines with Mr. Rose and Mr. Canning, who dined with me. At two, got on board the 'Victory' at St. Helen's, with 'Euryalus' in escort.

"Wednesday, September 25th, 1805—Light airs southerly. Saw the rock of Lisbon S.S.E. ten leagues. At sunset the Captain of the 'Constance' came on board, and sent my letters for England to Lisbon, and wrote to Captain Sutton* and the Consul. The enemy's fleet had not left Cadiz the 18th of this month, therefore I yet hope they will wait my arrival.

"Saturday, September 28th, 1805—Fresh breezes at N.N.W. At daylight bore up and made sail. At nine saw the 'Etna' cruising. At noon saw eighteen sail. Nearly calm. In the evening joined the Fleet under Vice-Admiral Collingwood. Saw the enemy's fleet in Cadiz, amounting to thirty-five or thirty-six sail of the line.

"Sunday, September 29th—Fine weather. Gave out the necessary orders for the Fleet. Sent 'Euryalus' to watch the enemy with the 'Hydra' off Cadiz.

"Wednesday, October 9th—Fresh breezes easterly. Received an account from Blackwood that the French ships had all bent their

* Of His Majesty's ship "Amphion," then in the Tagus.

top-gallant sails. Sent the 'Pickle' to him with orders to keep a good look out. Sent Admiral Collingwood the Nelson truth. At night wind westerly.

"Monday, October 14th—Fine weather: westerly wind. Sent 'Amphion' to Gibraltar and Algiers. Enemy at the harbour's mouth. Placed 'Defence' and 'Agamemnon' seven to ten leagues west of Cadiz, and 'Mars' and 'Colussus' five leagues east of the Fleet, whose station is from fifteen to twenty west of Cadiz, and by this chain I hope to have a constant communication with the frigates off Cadiz.

"Wednesday, October 16th—Moderate breezes westerly. All the forenoon employed forming the Fleet into the order of sailing. At noon fresh breezes W.S.W. and squally. In the evening fresh gales. The enemy as before by signal from 'Weazle.'

"Thursday, October 17th—Moderate breezes north-westerly. Sent the 'Donegal' to Gibraltar to get a ground tier of casks. Received accounts by the 'Diligent' storeship that Sir Richard Strachan was supposed in sight of the French Rochefort squadron, which I hope is true.

"Friday, October 18th—Fine weather: wind easterly. The combined fleets cannot have finer weather to put to sea.

"Saturday, October 19th—Fine weather: wind easterly. At half-past nine the 'Mars,' being one of the look-out ships, made the signal that the enemy were coming out of port. Made the signal for a general chace S.E. Wind at south, Cadiz bearing E.S.E. by compass, distance sixteen leagues. At three the 'Colussus' made the signal that the enemy's Fleet were at sea. In the evening made the signal to observe my motions during the night; for the 'Britannia,' 'Prince,' and 'Dreadnought' to take stations as most convenient; and for 'Mars,' 'Orion,' 'Belleisle,' 'Leviathan,' 'Bellerophon,' and 'Polyphemus' to go ahead during the night, and to carry a light, standing for the Straits' mouth.

"Sunday, October 20th—Fresh breezes S.S.W. and rainy. Communicated with 'Phœbe,' 'Defence,' and 'Colussus,' who saw near forty sail of ships of war outside of Cadiz yesterday evening; but the wind being southerly they could not get to the mouth of the Straits. We were between Trafalgar and Cape Spartel. The frigates made the signal that they saw nine sail outside the harbour. Sent the frigates instructions for their guidance; and placed the 'Defence,' 'Colussus' and 'Mars' between me and the frigates. At noon fresh gales and heavy rain; Cadiz N.E. nine leagues. In the afternoon Captain Blackwood telegraphed that the enemy seemed determined to go to the westwards;—and *that* they shall *not* do if in the power of Nelson and Bronte to prevent them. At five telegraphed Captain Blackwood that I relied upon his keeping sight of the enemy. At five o'clock 'Naiad' made the signal for thirty-one sail of the enemy N.N.E. The frigates and look-out ship kept sight of the enemy most admirably all night, and told me by signal which tack they were upon. At eight we wore and stood to the S.W., and at four wore and stood to the N.E.

"Monday, October 21st—At daylight saw enemy's combined Fleets from E. to E.S.E. Bore away. Made the signal for order of sailing and to prepare for battle. The enemy with their heads to the southward. At seven the enemy wearing in succession."

On the signal "Bear down upon the enemy in two lines," the British Fleet set all possible sail. The lee line, consisting of thirteen ships, was led by Admiral Collingwood in the "Royal Sovereign," and the weather line, composed of fourteen ships, by the Commander-in-Chief in the "Victory."

Shortly before the enemy opened fire, the signal "England expects every man will do his duty" was spread and received throughout the Fleet with great enthusiasm. At fifty minutes past eleven the "Victory" was in the thick of it.

Lord Nelson and Captain Hardy walked the quarter-deck for some time, the former remarking "This is too warm work, Hardy, to last long," for the "La Fougereux," "Redoubtable" and "Temeraire" were alongside—forming as compact a tier as if moored together.

When the guns were run out, their muzzles came into contact with the "Redoubtable's" side. Here, then, was seen the astonishing spectacle of the fireman of each gun standing ready with a bucket full of water, which as soon as his gun was discharged he dashed into the enemy through the holes made in her side by the shot, to prevent fire.

It was from the "Redoubtable" Nelson received his mortal wound. In his death agony he heard that fourteen ships had surrendered. "That is well, but I bargained for twenty." He then exclaimed, "Anchor, Hardy! Anchor!" and raising his head as if to whisper, slowly said, "Kiss me, Hardy. Do not bury me at sea."* Thus England's greatest hero died at 4.30 p.m. on October 21st, 1805.

There was no lead on board to make a coffin, so a cask called a "leaguer" was chosen and filled with brandy. On December 11th the "Victory" arrived at Spithead, when the body of Nelson was taken from the cask in perfect preservation; he was then dressed in full uniform and conveyed to Greenwich Hospital, there to await the funeral at St. Paul's.

> "The fiery spirit working out its way,
> Fretted the puny body to decay."

Alas! and may the same be said of Pitt. He lived through the enthusiasm and grief for Nelson; but the news of Ulm and Austerlitz—the former fought on the self same day as Nelson died—fretted *his* puny body to decay, and with the words of "Oh! my country" on his lips, he died on January 23rd, 1806.

---

* All sailors have a strong objection against the burial at sea.

"A sailor was supposed to be dead; they were preparing him for burial, when by accident the needle pierced him in the nose; he instantly revived; and to avoid the recurrence of such a mistake they have ever since passed the needle through the nostrils, as the ultimate criterion of life or death."

The church register of St. Helens records that in 1808—commencement of Peninsular War—there were eight bodies washed ashore, sewn in hammocks. A recent Act of Parliament compelled burial for same in the Parish Church. In 1809, five men were so buried; in 1810, four; in 1811, three; in 1812, six.

CHAPTER XIX.

WHAT had been the fortune of the Grand Army since August 29th, 1805, when Napoleon manœuvred them from Boulogne into Eastern Europe? Like an avalanche they had brought ruin upon ruin, rout on rout up to the walls of Vienna.

Comte de Ségur, whose acquaintance we have previously made, had been selected by Napoleon for his personal staff. "Your duty will be to command my body guard: you see the confidence I place in you; you will justify it. Your merits and your talents promise a rapid advance." The personal staff also had a passport into the domestic circle; thus we read of the farewell to Ségur by Josephine and of other incidents in the campaign.

"Go; my prayers go with you; and be as happy as the army of France," were the words spoken by Josephine to Ségur, when she left Napoleon on October 1st. "Never doubt it, for the Emperor has just announced to me that the enemy's army will infallibly be made prisoners within eight days."

Mack was completely turned on the 8th, and Ségur had to arrange for the capitulation on October 21st. The battle of that day commenced with a skirmish between the Emperor's personal escort and a few Uhlans. Without waiting for the result of this movement, the Emperor turned into a neighbouring farmhouse for a little sleep and shelter from the weather. Here Ségur found him slumbering on one chair, and a tired and wounded drummer on another. Soon a number of Generals arrived, waiting for orders, and amongst them Lannes, who hurrying in, exclaimed, "Sire! what are you doing here? You are sleeping; and Ney quite alone is struggling against the whole Austrian army!" "And why did he engage?" replied the Emperor. "I told him to wait, but he is always the same; he must fall on the enemy the moment he catches sight of them." "Good God!" replied Lannes; "but one of his brigades is repulsed. I have my Grenadiers at hand. We must go to him. There is not a moment to lose!" And he carried off Napoleon, who soon was here, there and everywhere.

Similar again is the *rendezvous* in the peasant's hut on the eve of Austerlitz, when the conversation turned on dramatic poetry. Junot, who prided himself on some literary attainment, cited some new tragedy. Napoleon, as if he had forgotten the Russian army, the war, and the battle of the morrow, entered fully into the discussion, and alluded to the tragedy of "Les Templiers" being a failure. He had told the author so, probably after the heckling he had received at the theatre on the night of Moreau's trial.

The argument becoming warm, Napoleon rose and said "En attendant allons nous battre!" When Murat, Lannes, Bernadotte, Soult and Davoust met Napoleon the next morning, he simply said, "Within half-an-hour the whole line must be en feu; Allez." Bernadotte he distrusted.

The battle was over by 4 p.m. (December 2nd, 1805). The two Emperors met the next morning, when a suspension of arms was granted, for the French army was not quite equal to a continuation of hostilities. Napoleon's last words to Francois II., as the meeting broke up, were "So your Majesty promises me not to recommence the war?" Francois replied that he swore it, and would keep faith.

Napoleon on mounting his horse, exclaimed "We shall soon see Paris again; the peace is as good as signed." Undoubtedly it was his wish for hostilities to cease altogether for a while. Therefore for Prussia to demand war the following year—for the simple reason that Bernadotte had violated their territory—when in a state of thorough unpreparedness, was a foolish stroke, and one to be regretted.

Napoleon saw at once that this war would be neither costly or long. He left Paris on September 24th, 1806, and was able to announce the annihilation of the Prussian army on October 15th. Anerstadt and Jena were the two battlefields. The Emperor was quartered in a village inn, and sleeping off the fatigue of his small affray at Jena, when Ségur awoke him to announce the victory of Anerstadt—for there it was that the *elite* of the Prussian troops, with their most renowned Generals, their Princes and their King, had been annihilated. The glory rested with Davoust.

Between Jena and Friedland "much water," as the saying goes, "had flowed beneath the bridge." Ségur was taken in a skirmish and kept a prisoner until the peace of Tilsit.

The Duke of Brunswick* likewise the two Generals who succeeded him were killed, and upwards of 40,000 men were placed *hors de combat.* Hohenlohe led the wrecks of the army which had been collected at Magdeburg towards the Oder; there he capitulated with 17,000 men.

Blucher fled to Lubeck, where he was made prisoner. Magdeburg capitulated to Ney on November 8th, and Hameln on the 20th; and so things went from bad to worse. The Prussians seemed lost to all courage and confidence, and bereft of understanding.

Davoust was given the post of honour, and his corps was the first to enter Berlin, followed by Angereau. Napoleon stayed at Potsdam for a while to visit the tomb of the Great Frederick. He ordered the sword, the scarf, the ribbon of his order the Black Eagle, and all the colours appertaining to this great man, to be sent to the Hotel des Invalides in Paris. He even removed the monument commemorating Rosbach to Paris—and from the bronze cannons he built La Colonne Vendome, Arc de Triomphe and La Barriere de L'Etoile.

From Berlin Napoleon marched into the heart of Poland, to bring to terms the last opponent left him on the Continent. The winter was severe, and after the somewhat mixed up battles of Pultusk and Golynin, Napoleon went into winter quarters at Warsaw, which soon became the centre of a brilliant society. The command of the

---

* His son was the first to be killed at Waterloo.

Russian Army passed from Kamenskoi—a veteran of eighty, and then insane—into the hands of Benningsen, from whom Napoleon received a nasty check at Eylau (February, 1807). Here was an opportunity for wise councils at home, but the English Government refused the urgent entreaties of Russia and Prussia for men and money. This refusal lengthened the war, and enabled Napoleon to gain a decisive victory at Friedland (June, 1807)—a prelude to the Peace of Tilsit, July 8th, 1807.

Napoleon was now master of the greater part of the Continent. He was already King of Italy, and as Protector of the Confederation of the Rhine he ruled Germany. It was from Berlin he published his celebrated Decree, declaring that the whole of the British Isles were in a state of blockade, in fact boycotted; for it now became his dream

"de dominer la mer par la terre.'

CHAPTER XX.

THE English Army at this time only gave proof of its existence, and not its power. The power came later in the Peninsular War, when its purpose was fixed.

On Christmas Day, 1805, Napoleon sent an ultimatum to the reigning house of Naples, for violating a treaty dated October 8th, 1805, in which France had consented to withdraw her troops from Naples, provided the King of Naples remained strictly neutral throughout the war. Scarcely six weeks were over, when, on November 20th, English and Russian ships appeared in the Bay of Naples, and landed a body of men in that city.

On this threat of vengeance from Napoleon, the Russian troops re-embarked and left for Corfu, and the English retired to Sicily.

Sir James Craig (the Governor of the Cape in 1796) was in command, and his second was Sir John Stuart, the same officer who commanded the Cannanore column at Seringapatam.

Naples thus left defenceless, the King and Queen fled to Palermo, and the heir apparent to Calabria, where he carried on a brisk resistance.

Sir Sydney Smith, in command of a squadron of six ships of the line, besides frigates, transports and gunboats, arrived off Naples on the night when the town was *en fête* for the coronation of Joseph (Napoleon's brother) as King of Naples (March 30th, 1806). He allowed the festivities to pass off in peace, but next day he offered his aid to the Calabrians in more ways than one.

Sir John Stuart and General Acland, with the 20th, 58th and 81st Regiments, etc., soon after landed from Sicily, and defeated the French at Maida—4000 casualties, including 1000 prisoners. The results of this victory were grand, for the Calabrian peasantry took great heart, and rose against the French, who soon thought fit to evacuate, leaving behind all the stores, cannon, ammunition, etc., which they had intended for the invasion of Sicily.

General Fox and Sir John Moore soon after arrived, to decide if a descent upon Naples was advisable or not. They advised no further expedition to the mainland, but to rest content with Sicily.

Concurrent with these events in Sicily, an expedition left the shores of England for the Cape of Good Hope. The force consisted of five thousand men under Sir David Baird (the hero of Seringapatam and of the desert march), with a proportionate Naval force under Sir H. Popham. The brigadiers were Ferguson and Beresford (a future peer for the Peninsula).

The surf ran extremely high, so Beresford, with the 38th Regiment and 20th Light Dragoons, went northward to Saldanha Bay. The rest of the army waited for the surf to subside before landing at Lespard Bay; and even then they lost thirty-five men by the upsetting of a boat, and about the same number from sharpshooters on the heights.

Sir David Baird commenced his march on January 8th, 1806, and soon descried the enemy, drawn up in the plain around the Blue Mountain. They were commanded by General Janssens, and their force amounted to above 5000, chiefly of cavalry.

Brigadier Ferguson advanced with his brigade deployed—but the Dutch held their ground, and fired right well, and only vacated when "Fix Bayonets" and the "Charge" was sounded. The force then dispersed in the direction of Holland's Kloof, where General Beresford met them, and brought them to surrender.

By the terms of capitulation, General Janssens and his army were sent back to Holland—not as prisoners of war, but in return for the complete surrender of the Colony and its dependencies.

Sir John Popham, the Naval Commander who first reported to Pitt and Melville on the weakness and defenceless state of the Cape —which made it a tempting place for others to set their flag and speculate in trade—now reported, from information received, of the weakness and disaffection of the Spanish Colonies on the Rio de la Plata. Without waiting for further instructions, he arranged with Sir D. Baird to part with General Beresford, and for a small party of troops to co-operate with him in this new enterprise. He arrived off La Plata in early June, and by the 24th, Buenos Ayres and Monte Video were in his possession. He sent to England 1,200,000 dollars, and other prizes. Great was the joy at this easy capture; but "Quickly come, quickly go." We must now relate how Buenos Ayres was re-captured.

The home authorities, dazzled by the easy conquest and the bag of dollars, winked at Popham's unauthorised departure from the Cape; but soon found occasion to make him a scapegoat.

When the Spaniards, recovered from their panic, had discovered the small numbers of their opponents, a French Colonel Liniers arrived from Monte Video with 1000 men. After a creditable resistance, Beresford and his force (71st Regiment) surrendered and were marched prisoners up the country. Soon re-inforcements from the Cape and from England, under Sir Samuel Auchmuty, with Sir Charles Stirling in the "Ardent" (who was to replace Popham) arrived. Popham was sent home for trial, the result of which was a severe reprimand.

On arrival, Sir Samuel found the troops were without artillery, stores or provisions of any kind, and in a letter to the War and Colonial Minister reported: "The enemy are armed with swords and muskets; they ride up, dismount, fire over the backs of their horses, mount, and gallop off. All the inhabitants of this country are accustomed to this mode of warfare, and every inhabitant is an enemy."

The only point offering any chance of success was Monte Video. Auchmuty had breached it over night, and orders were given for assault early in the morning; but the enemy had blocked the breach at night with hides and other obstacles, so it became a case of "Cannon to the right of them, Cannon to the left of them." The 40th, 87th and 95th were, however, not to be stopped, and the Citadel went, but not before 600 men had fallen.

The British Ministry, unaware of the recapture of Buenos Ayres, had entrusted to General Crawford 4200 men, and he, with a competent force under Admiral Murray, was to take the East Coast of South America—that is, the province of Chili—and when in possession of Valparaiso, was to establish communication with Beresford by a chain of posts *via* Buenos.

This expedition was stayed at the Cape, where Crawford received orders to join Auchmuty at Buenos Ayres. At this junction between the two Generals, the whole British force amounted to 9500 men, and this required a General of high rank, as well as of talent and judgment, for the supreme command. The choice fell upon General Whitelocke, who reached La Plata in March, 1807.

The escape of General Beresford and Colonel Packe enabled them to narrate experiences which left no doubt but that the Spaniards were determined to fight *a l'outrance.*

On the 28th June, Whitelocke made dispositions to invest the town and suburbs, and the attack commenced. He soon found out —what Auchmuty had said—that every inhabitant was an enemy; and after losing 2500 men, he accepted the terms of General Liniers to withdraw at once from Buenos Ayres.

He was tried at Chelsea (January 28th, 1808, until March 18th), when he was found guilty of incapacity, and declared unfit to serve His Majesty in any military capacity whatever. He was fully exonerated from the charge of cowardice; and the precaution of keeping his men unloaded until they reached their proper *rendezvous* was approved by the Court.

England next had occasion to distrust the Czar Alexander, for since the Peace of Tilsit he had bid for French aid in his projected raid on Constantinople; so he joined the Berlin Decree to boycott the commerce of England, forced Sweden to do the same, and tried to coerce Denmark to do likewise.

The latter hope was shattered, for Canning demanded at once from Denmark the surrender *(pro. tem.)* of her Fleet. This was refused, so Copenhagen was besieged, and the whole fleet, with stores and ammunition, was taken into British ports in September, 1807. The officers and men were left behind, and they were at once subsidized into the manning of Napoleon's own vessels.

Napoleon at this time kept a small standing camp at Dunkerque, where smugglers and spies from England had a welcome.

The mere mention of Turkey or Constantinople (May, 1807), with Europe at war, was sure to mean fuel for the flames. The Russians declared they wished to save the Ottoman Empire from the French. The French conjured the Ottomans to beware of the Russians. The Porte sent troops to Wallachia and Servia, to watch both. Sebastiani—formerly a monk, now a general—was supposed to watch the proceedings on behalf of France; whilst an English fleet (but no troops) under Sir John Duckworth entered the Dardanelles. At the same time a well found military expedition from Messina, under Generals Fox and Mackenzie*, entered Alexandria and occupied the heights of Aboukir, as of old in 1801. Generals Wauchope and Meade went to Rosetta with the 31st Regiment, but no good results, politically or otherwise, followed, and the expeditions were withdrawn.

After Trafalgar, which had given to England the Trident of Neptune or the sceptre of the world, Napoleon determined a new era should commence in the Navy of France. Accordingly ships were laid at once on slips at Antwerp, Brest, L'Orient, Toulon, and other ports, and building went on apace. He forced his brother Louis, the new King of Holland, to cede Flushing to France, when he levelled 1500 houses to make room for dockyard extension, to the cost of 66,000 francs. Anything and everything was done to enforce the boycotting of England, and he made it now his pleasure that

* Killed at Talavera.

Portugal and Spain should become pliant to his will. With what result we shall see, for England intervened.

When the *Madrid Gazette* of May 20th, 1808, announced the abdication of Ferdinand VII. in favour of the Emperor of the French, there arose a furious indignation, for all knew the ugly story of how Napoleon had obtained this abdication, which when told is briefly this :—

Ferdinand had asked Napoleon to provide him with a second wife. These negotations, which were of a private nature, had been interpreted by his parent King Charles IX., his mother, and that rascal Godoy the Prince of Peace, into a sort of plot connected with his father's life. Napoleon, with his usual subtlety, employed this scandal to efface the whole family, and to hand the kingdom over first to Murat, then to his brother Joseph.

The House of Braganza had already ceased to reign by methods somewhat similar, although Sir Sydney Smith with a British Squadron was there to help; but the Regent and the insane Queen of Portugal preferred Brazil, particularly now that their country had been divided off between France and Spain (Fontainebleau, October 27th, 1807). Napoleon fixed upon Junot (Duc d'Abrantes) to be his Viceroy.

At Cadiz the people rose at once against their pro-French Governor Solano, and would not listen to his talk; and similar scenes occurred in Carthagena, Malaga and Saragossa.

General Solano, in trying to restore order, was killed. The crisis had to be met. Lord Collingwood and his fleet, Sir Hew Dalrymple with his garrison at Gibraltar, and their own General Castanos with his patriots in camp, were all near at hand, and they would surely help to bring back their King! The people petitioned, and Lord Collingwood volunteered at once to disperse the French Fleet from Cadiz—an offer which, when told to Admiral Rosily, determined him to make the best terms he could with Spain. He surrendered on June 14th, 1808.

Similar petitions from Portugal reached Admiral Sir Charles Cotton, to which he replied in true British diction :—

"Agreeably to your desires, I send you ships, troops, arms and ammunition, and have given orders for hoisting the flag of His Royal Highness the Prince Regent of Portugal, around which the whole Portuguese nation ought instantly to rally, and take up arms in a cause so just and so glorious. To secure success, unanimity is necessary. Unite yourselves with your brave friends and neighbours the Spaniards."

(Sd.) "COTTON,

"On board the 'Hibernia,' off the Tagus, July 4th, 1808."

CHAPTER XXI.

"War, war is still the cry; war even to the knife.'

"WAR to the knife!" This was the answer of Palafox, who defended Saragossa against the French in 1808-9, when called on to give up the city. They tore down the awnings from their windows, made them into sacks, and filling them with sand, barricaded every gate, digging round each a trench. The women all helped, and the monks made the cartridges and even gunpowder—for the sulphur of the place was requisitioned; the washings of the streets gave the saltpetre; and charcoal was made from the stalks of hemp, which grew to an unusual size. It was after a desperate assault on one of these batteries, when all the gunners lay wounded or exhausted, that Augustina, snatching a match from the hand of a dead artilleryman, fired off a 20 pounder—and remained with her gun throughout the day. Her name has gone into history as "The Maid of Zaragosa," and until her death she wore, embroidered on the sleeve of her gown, a small shield of honour with "Zaragosa" inscribed upon it. She received from her Government a small pension and the daily pay of an artilleryman.

Dupont was an officer of high military reputation. He marched upon Cadiz in May, and reached Cordova in June—which town he sacked in a merciless way. The patriots from the hills then hemmed him in, and his advance was difficult. However, he managed to effect a retreat; but Castanos was on his heels, to whom he surrendered with 20,000 men.

This capitulation was received with great enthusiasm by the Spaniards, King Joseph writing to his brother: "N'avait qu' un competiteur à vaincre; j'ai un nation tout entiere. Je renonce à régner sur un peuple qui ne veut pas de moi." He asked to return to Naples.

Generals Dupont and Vedel were degraded, and some say shot. "Ils ont sali notre unforme," said Napoleon; "il sera lavé dans leur sang."

Thus came England's opportunity. She was determined to rid the Peninsula of Napoleon, and to render back to Cæsar the things that were Cæsar's. Supplies of men, money and provisions were sent to the Spanish patriots in reckless profusion, and two armies under Sir Arthur Wellesley and Sir John Moore were soon equipped for service.

The patriotism of the Spaniards in shutting the door to Joseph and opening it to Wellesley quite exasperated the great Napoleon. "Je leur avais envoyé des agneaux," said Napoleon, in speaking of the young soldiers of Dupont; "et ils les ont devorés, je leur enverrai les loups qui les devoreront à leur tour." For this, the first check to Napoleon on the Continent, had spread through Europe with rapidity, and had given joy to his enemies. Austria, who had for a time stopped all her armaments, re-commenced to arm, and showed herself inclined for war.

The force under Sir Arthur sailed from Cork on July 12th, and arrived at Mondego Bay on July 30th. General Spencer, from Gibraltar, arrived the day after, and brought the united strength to 13,000.

On landing, they at once learned of the reverse to Dupont and the scattering of his force by Castanos on July 19th, and of the general restlessness amongst the French. Laborde was in position in front of the village of Roliça, and tried to keep Wellesley in check until the arrival of Junot from Lisbon. But to keep our men in check was another thing—eager for the fray, the 5th, 9th, 29th, 60th and 95th gallantly carried all before them, capturing three guns.

The 29th had their Colonel, the Hon. G. A. F. Lake, killed; Majors Way and Thomas Egerton, Captain Hodge Pattison,

Lieutenants R. Birningham, St. John, Lucas and Stannus wounded; Captains Tod H. Birningham, Ambrose Newbold, and Thomas Langton were reported missing.

Thanks to the Army were expressed in General Orders, after the battle of August 17th, 1808:—

"Headquarters, Lourinha, August 18th.

"The Lieutenant-General was perfectly satisfied with the conduct of the troops in the action of yesterday, particularly with the gallantry displayed by the 5th, 9th, 29th, 60th and 95th Regiments, to whose lot it principally fell to engage the enemy.

"From the specimen afforded yesterday of their behaviour in action, the Lieutenant-General feels confident that the troops will distinguish themselves whenever the enemy may give them another occasion; and it is only necessary for him to recommend to them a steady attention to their preservation of order and regularity, and strict obedience to the commands which the officers may give.

"G. B. Tucker, D.A.G."

Sir John Moore, with the Reserve Division, left Spithead on July 27th, 1808, and reached Vigo on August 17th, gales and contrary winds being answerable for this tedious sail of three weeks. There he received letters telling him that Wellesley had landed on the 1st at Mondego Bay, had stormed and captured the Pass of Roliça on the 17th, marched to Lourinha on the 18th, to Vimiera on the 19th—where he halted—and on the 20th was joined by Anstruther's brigade of 2400 men, who had landed the evening before.

On the 21st, the French—led by Junot, Duc d'Abrantes, in person—were, after four hours' fighting, repulsed with heavy loss, including thirteen guns and twenty-five waggons full of stores. General Thiebault was killed, and Bernier taken prisoner—the latter being exchanged for Sir John Abercrombie, then a prisoner in France.

Sir Harry Burrard joined the army during the fight, but sensibly did not assume command. After the repulse and dispersion of the enemy, Wellesley rode up to Burrard and proposed that the right should advance to gain possession of Torres Vedras,* and that he with the left should pursue the enemy, who were retreating precipitately. Burrard replied that a great deal had been done and that he thought it inadvisable to move off the ground. General Ferguson and others all likewise urged the advantages of this advance, but Burrard was inflexible.

Early next day Sir Hew Dalrymple arrived and took over the command. The dead were buried, and the wounded sent to the ships. Early in the same afternoon an alarm was spread that the enemy were advancing to renew the fight, and all went into position again as on the day before; but the "alarm" turned out to be General Kellerman, with a body of cavalry, under a flag of truce.

A conference there and then took place—Dalrymple, Burrard, Wellesley and Kellerman attending—which decreed a suspension of hostilities, on condition that the French quitted Portugal and returned home.

This came as a great surprise to the assembled troops, for they were all thanked so warmly for their fight on the 21st—only to learn on the 22nd that all was over for the present.

Thanks were again expressed to the Army after the battle of August 21st, 1808 :—

"Headquarters, Vimiera, August 21st.

"Lieutenant-General Sir Arthur Wellesley congratulates the Army on the signal victory they have this day obtained over the enemy, and returns them his warmest thanks for their resolute and heroic conduct.

"He experienced the sincerest pleasure in witnessing various instances of the gallantry of the corps, and has in particular to notice the distinguished behaviour of the Royal Artillery, 20th Light Dragoons, 36th, 40th, 2nd 43rd, 50th, 2nd 52nd, 60th, 71st, 82nd, 95th, 97th Regiments, etc.

* "Turres Veteres" of the Romans.

"The Officer commanding the Royal Artillery reports having received fourteen guns, two ammunition waggons, twenty-one carts, forty horses, and about 20,000 rounds of musket ammunition, taken in the action of 21st."

And on the night of August 22nd, in General Orders, appeared the full terms of the Convention.

Now how were these two events voiced in camp ? The conduct of Burrard in refusing to Wellesley the right to crown his victory at Vimiera, by securing Torres Vedras and placing his Army between the enemy and Lisbon, was condemned by all, for the enemy was routed and in complete disorder. Besides, the popularity of Wellesley was at its height. His active solicitude for the welfare of his troops, his soldierlike disregard of himself, and the experience of his repeated and uninterrupted success, had gained the confidence of all, so there was nothing he could order which would not have been most cheerfully carried out. Morale is to the physical as three to one.

As a proof of the popularity of Wellington, Field Officers of the Army addressed him thus :—

"The Commanding Officers of Corps and Field Officers who have had the honour of serving in the Army under your command, anxiously desirous of expressing the high opinion they entertain of the order, activity and judgment with which the whole of that force was so ably and successfully directed, from the time of landing to the termination of your command in the action of Vimiera, request you will accept from them a piece of plate as a testimony of that sincere esteem and respect which your talents and conduct have so justly inspired."

The presentation was made by Colonel Kemmis, 40th Regiment, the senior Colonel.

A French officer, a prisoner, gave his opinion freely, and to the effect that if we had had two Wellesleys in the Army there would neither have been armistice or Convention.

This cessation from hostilities enabled officers to walk and visit places as they wished, and whilst the Army rested, they spent some days at Cintra ; some were given residence in the Palace, and became impressed by the story told of Alfonso IV., who was passionately fond of hunting and other sports. During his absence the administration was carried on anyhow.

The nobility, indignant at such abuse, called a Council to which they invited the King. He attended, and immediately began by describing the sport he had enjoyed, quite regardless of any other business. When he had concluded, one of the noblemen stood up and addressed him in these words :—

"Sire, we came here not to hear of adventures of the chase, which are intelligible only to grooms and falconers, but to consult the welfare of the people. Your Majesty will find sufficient employment in attending to their wants, and if you will remove the grievances with which they are oppressed, you will find them dutiful and obedient subjects. If not—" and here the King interrupted him in a rage.

"If not! What then?"

"If not," replied the nobleman, in a firm tone, "they will look for a better king!"

His Majesty left the room in great indignation, but soon returned with a placid countenance and addressed them thus :

"I now perceive the truth of what you have just advanced. Remember that from this day you have not to do with Alfonso the Sportsman, but with Alfonso IV., King of Portugal." His Majesty was faithful to his engagement, and became one of the best monarchs his kingdom ever saw.

The next evening Sir Harry Burrard gave a dinner. The Convention was not much the subject of conversation, but when he did speak of it he seemed to enjoy the inestimable happiness of "Mens conscia recti." But whatever may be the feelings of the principal movers in the measure of the Convention, the Army entertained the most decided abhorrence of it. They condemned it in principle and detested it in detail, and would not believe their hero, Sir Arthur Wellesley, had had any share in the transaction, by which they thought their character was compromised and their glory tarnished.

The expected was at last announced by the recall of Sir Hew Dalrymple, for it was not possible to stem the tide of prejudice which the Convention of Cintra had raised against the Chief.

On arrival home a Court of Enquiry duly met, which ruled that "The Convention was an error, but the decision of General Burrard in not following in pursuit was, under the circumstances, sound;" and so the matter ended.

The doggerel ran, at home :—

> " Sir Arthur and Sir Harry—Sir Arthur and Sir Hew—
> Cock-a-doodle! Cock-a-doodle! Cock-a-doodle doo!
> Sir Arthur's a brave soldier; but of the other two
> Sing cock-a-doodle! Cock-a-doodle! Cock-a-doodle doo!"

Sir Hew remained at home, and became a full General in 1812 and a Baronet in 1814.

CHAPTER XXII.

THE command in Portugal now devolved upon Sir Harry Burrard. His first Order issued was :—

"Lisbon, October 8th, 1808.

"Lieutenant-General Sir Harry Burrard having received His Majesty's command to place a large proportion of the Army in Portugal under the orders of Lieutenant-General Sir John Moore, to be employed in a separate service, the following Regiments are to compose the corps to be so employed :—18th Light Dragoons, 3rd Light Dragoons (King's German Legion), 2nd, 4th, 5th, 6th, 9th, 20th, 28th, 32nd, 36th, 40th, 42nd, 43rd, 52nd, 60th, 71st, 79th, 91st, 92nd and 95th Foot, and one Battalion King's German Legion; with Generals Hon. C. Stewart, Lieutenant-General Frazer, Lord W. Bentinck, Beresford, Hon. John Hope, Hon. E. Paget" —20,000 men in all. Another force of 20,000 were to sail from Falmouth to Corunna, under Sir D. Baird.

The general instructions were for Sir John Moore to march through Spain with his face towards Burgos, which was to be the general rendezvous of British troops (including the force under Baird); and he was to combine his operations with those of the Commander-in-Chief of the Spanish Armies.

On the 22nd October, the force commenced its march from Lisbon—the objective being Madrid. Moore, with the main infantry, took the road *via* Abrantes; Hope, with the guns, cavalry and four regiments of infantry—6000 in all—*via* Badajoz; whilst orders were sent to await Baird on his arrival at Corunna, October 13th*, to march to Astorga, with a view of joining in at Valladolid or Burgos. On the 27th October, the advance column reached Abrantes†, and had their head-quarters at the house belonging to a nobleman—one of the fourteen sent to France by order of Napoleon as a hostage for the security or *bona fides* of Portugal—whilst regiments and other units were billeted about elsewhere.

The 32nd Regiment were located in a very fine old church. A silver chalice, and a crucifix said to be a sacred fragment of the original Cross, suddenly disappeared. Suspicion rested on the soldiers; and the inhabitants were tumultuous and enraged. The Colonel had every man's knapsack searched in vain.

The uproar increased; the clergy remonstrated, and besought a restoration by the most pious and energetic appeals; when, lo! the cup was found under the bed of a native priest, who declared he had removed and secreted it for safety!—the same with the crucifix.

The march was continued to Garvao—a wretched village, but acceptable—for the roads along the Tagus were all so bad. Next day the river was crossed at Villa Velha, and Castello Branco was reached, where the scenery was perfectly beautiful and romantic.

Most of these villages had been occupied by the French within the past year, and it was at Garvao that a French regiment almost succumbed to the heavy wet, the fatigues, and the intense cold at night. Even in our own force, six privates of the 5th Fusiliers had perished from the rain and cold within three days.

---

* Here Baird was astonished to find that the Junta of Gallicia refused him permission to land; and when at last he landed, he was disposed to wonder whether the Spaniards wished for his help or not.

The same impression was made on Sir John Moore when he arrived at Salamanca on November 13th. He found so little preparation made for his army, that he wrote to the Minister at Madrid, desiring him to tell the Spanish Government that if they expected his army to advance, they must pay more attention to its wants.

† Lisbon to Abrantes, 84 miles.

The last halt in Portugal was at Villa Formosa, and the first halt in Spain was at Ciudad Rodrigo; and what a change in men and manners marked the frontier—all to the advantage of Spain, for the Spaniards seemed full of enthusiam, while the Portuguese were replete with apathy.

Sir John Moore occupied the same house as Junot, Duc d'Abrantes, General Loison and other French Generals had lived in exactly a year ago. Here Sir John decided to concentrate his force, and orders to that effect were sent to General Hope at Elvas, to come straight north through Alcantara to Ciudad Rodrigo (November 15th).

The instructions to Moore were to combine his operations with those of the Commander-in-Chief of the Spanish Armies—for then the Spaniards, to the number of 140,000 men, with Castanos, Palafox and Blake, were around Vittoria, the rendezvous of King Joseph. Moore and Baird had commenced their march to join to co-operate with them in October, 1808.

Napoleon arrived early in November at Vittoria, prepared to cut in two the extended Spanish line; and at once dispersed from before Burgos the army of Estramadura, which he called "une infâme canaille fanfaronne." He remained in Burgos some days awaiting news from the two wings of his Grand Army—when the unexpected nearly happened to the Great Man, for in the words of his Military Secretary, M. de Ségur:—

"I had not yet placed the first posts when the Emperor himself arrived, with only his mameluke and Savary. He had travelled all the night like myself; he arrived post haste, covered with mud and half dead with hunger, cold and fatigue.

"This palace had not been spared much more than the rest of the city. The apartment destined for the Emperor was still in the greatest confusion, strewed with pieces of broken glass, overturned bottles of wine, and broken articles of furniture. We did our best to restore some order; then, Savary having gone to prepare some provisions with Rustan, I was left alone with the Emperor, who assisted me to light his fire.

"I had completed this duty by the help of a candle, when Napoleon, whose fine sense of smell was offended by the rank odour of the place, called to me to open a window near which he happened to be seated.

"I hastened up, and we began by drawing the curtains; but what a surprise! Behind these curtains were three Spaniards—armed to the teeth, upright, motionless, with their backs pressed against the shutters—who had taken refuge there to escape our plunderers, or had come with plundering intent, of which their army was accused like ours.

"During more than ten minutes, whilst Napoleon, alone with me, was there without distrust—one while seated, one while bent over the fire, and with his back to them—they might ten times over, by a single blow, have terminated the war. But, fortunately, they were soldiers of the line, not insurgents.

"These wretches, seeing themselves discovered, remained frozen with fear. The Emperor did not even think of laying his hand upon his sword; he smiled with a gesture of pity.

"I disarmed them, and delivered them to our soldiers; and after making sure that there was no other hidden enemy in the room, I hastened to reconnoitre the rest of this immense building."

On his left, Lannes defeated Castanos and Palafox on the Tudela; and on his right Lefébre and Victor had dispersed the army of Blake, the *debris* of which was crushed by Soult, whom the Emperor himself had detached from Burgos.

Signs of these reverses were now only too visible to Moore. News also reached him of Burgos being held by 20,000 French, and of Napoleon being there in person, with no other occupation than to crush the British. Thus free for a while, Napoleon crossed the Lomo Sierra on November 29th, with a corps of 13,000 *en route* for Madrid—there to instal his brother Joseph as King.

At first, Sir John Moore thought that Napoleon would advance on him straight from Burgos, and before Hope and Baird could reach him; but soon he learnt, by an intercepted despatch, that Napoleon in person had left Madrid with his army, in order to get in the rear of the British; that the army which had been stationed at Talavera had moved forward to Salamanca; and that Soult had been strongly reinforced at Saldana on the Carrion.

Retreat for Moore therefore became imperative. The only difficulty lay in the route that ought to be pursued on this momentous occasion—for the force now sent in pursuit of the British must be reckoned at 60,000, whilst Sir John Moore could barely muster 27,000, even with the forces of Baird and Hope, who had joined him at Sahagun on December 20th, 1808.

So rapid was the march of the main body of the French under Napoleon, and so closely did they pursue Sir John Moore, that at one time there were scarcely thirty miles between them—near enough for the cavalry to ward off many a nasty blow, as evinced in the following:—

"GENERAL ORDER.

"Headquarters, Sahagun, December 22nd, 1808.

"The different attacks made by the cavalry upon those of the enemy, during the march, have given them an opportunity to display their address and spirit, and to assume a superiority which does them credit, and which the commander of the forces trusts, and will be supported upon more important occasions.

"The attacks conducted by Brigadier-General the Hon. C. Stewart, of the 18th Light Dragoons, when upon the Douro; and that by Lieutenant-General Lord Paget upon the enemy's cavalry at this place; are honourable to the British Cavalry.

"The commander of the forces begs that the Lieutenant-General and Brigadier-General will accept his thanks for these services; and that they will convey them to Brigadier-General Slade, and the officers, non-commissioned officers and men of the cavalry under their command, for their conduct in the different affairs which have taken place."

Again, on the 29th, at Benevente, the cavalry worked right well—for Napoleon, in the act of passing his Imperial Guard (Cavalry) across the bridge, was at once repelled; the picquets of the 18th Hussars formed up under Otway; whilst the Brigade under Stewart (Castlereagh), with the 10th Hussars in front, swept around their flanks, entailing a loss to the enemy of sixty killed and wounded and seventy prisoners. Among the latter were General Lefébre and two Captains of the Imperial Guard.

Napoleon, by this time finding he could not come up with Sir John Moore, gave up the pursuit, and committed it to three Marshals of France, who—with as many divisions—were commanded to follow the British to destruction.

Napoleon was much disconcerted by the activity and ability of Moore. He wrote to his brother Joseph that his presence was required in Paris, giving as his reasons that Austria was on the point of declaring war, and that England would not hear of peace unless the Bourbons were given back their thrones—she was annoyed at the Convention of Cintra comparing the situation of the French then to that at Baylen, and had thrown in her lot with the 5th Coalition. The letter concluded by requesting his brother "de montrer plus de vigueur contre la canaille. Il faut vous faire craindre d'abord et aimer ensuite."

One reason for this sudden departure Napoleon kept to himself —which was that Moore had outwitted him.

The British soldiers in retreat are not good. They had been spoiling for a fight, and their behaviour had displeased Moore, for he told them straight that: "When it is proper *to fight a battle* he will do it; and he will choose the time and place he thinks *most fit*; in the meantime, he begs the officers and men of the army to attend diligently to discharge *their* parts, and leave to him, with the General Officers, their discretion on measures which belong to them alone."

The time and place which the great Commander chose to fight his battle was January 16th, at Corunna. He had made assurance doubly sure, and the soldiers had cheered him to the echo. Now no gibes were heard. All were in tears when they laid their great Commander, uncoffined and undressed, in the ramparts of Corunna.

At a dinner, when Shelley and Trelawney were the guests of Lord Byron, the conversation turned on poetry. "I will show you an Ode," said Lord Byron, "which compares with the very best;" and from the adjoining room brought forth a magazine which contained the newly-published poem (1810) by the Rev. C. Wolfe. He read it to his friends; they all agreed; but ere Lord Byron closed the book, he said, and said again, the two delightful lines:—

"He lay like a warrior taking his rest,
With his martial cloak around him."

For that is how his soldiers laid him. Hardinge (Lord Hardinge) had just returned from delivering a message for the Guards to prolong the left flank of the Highlanders, when a cannon ball struck Sir John and beat him to the ground. He was spared just to hear, like our hero Wolfe at Quebec, the British cheer and the shout of victory.

The command now passed to Hope, for Baird had received a ghastly wound, destroying his right arm. The despatch from Admiral de Courcy must tell the rest:—

"January 17th, 1809.

"Sir,—Having it in design to detach the 'Cossack' to England as soon as her boats shall cease to be essential to the embarkation of troops, I seize a moment to acquaint you, for the information of the Lords Commissioners of the Admiralty, that ships of war, as per margin,* and transports, under the orders of Rear-Admiral Sir Samuel Hood and Commissioner Bowen, arrived at this anchorage from Vigo on the 14th and 15th inst.

"The 'Alfred' and 'Hindostan,' with some transports, were left at Vigo, to receive a Brigade of 3500 men that had taken that route, under Generals Allen and Crawford.

"In the vicinity of Corunna, the enemy have pressed upon the British in great force. The embarkation of the sick, the cavalry, and the stores went on.

* "Ville de Paris," "Victory," "Barfleur," "Zealous," "Implacable," "Elizabeté," "Norge," "Plantaganet," "Resolution," "Audacious," "Endymion," and "Mediator."

"The night of the 16th was appointed for the general embarkation of the infantry, and meantime the enemy prepared for attack.

"At 3 p.m. an action commenced; the enemy, which had been posted on a lofty hill, endeavouring to force the British, on another hill of inferior height, and nearer the town.

"The enemy were driven back with great slaughter; but very sorry am I to add that the British, though triumphant, have suffered severe losses.

"I am unable to communicate further particulars than that Sir John Moore received a mortal wound, of which he died at night; that Sir David Baird lost an arm; that several officers and many men have been killed and wounded; and that the ships of war have received all such of the latter as they could accommodate, the remainder being sent to transports.

"The weather is now tempestuous, and the difficulties of embarkation are great. All except the rear guard are embarked, consisting perhaps at this moment of 2600 men.

"The enemy having brought cannon to a hill overlooking the beach, have forced a majority of the transports to cut or slip.

"Embarkation being no longer practicable at the town, the boats have been ordered to Sandy Beach, near the lighthouse; and it is hoped that the greater part, if not all, will be embarked, the ships of war having dropped out to facilitate embarkation.

"January 18th.

"The embarkation of the troops having occupied the greater part of last night, it has not been in my power to detach the 'Cossack' before this day; and it is with satisfaction I am able to add that, in consequence of the good order maintained by the troops, and the unwearied exertions of Commissioner Bowen, the Captains and other officers of the Navy, the agents as well as the boats' crews—many of whom were for two days without food and without repose—the Army have been embarked to the last man, and the ships are now in the offing, preparatory to steering to England.

"The great body of the transports, having lost their anchors, ran to sea without the troops they were ordered to receive, in consequence of which there are some thousands on board the ships of war.

"Several transports, through mis-management, ran on shore.

"The seamen appeared to have abandoned them; two being brought out by the boats' crews of the men-of-war, two were burned, and five were bilged.

"I cannot conclude this hasty statement without expressing my great obligation to Rear-Admiral Sir Samuel Hood, whose eye was everywhere, and whose exertions were unremitted.

"I have the honour to be, etc.,

"M. DE COURCY.

"Hazy weather rendering the 'Cossack' obscure, I detach the 'Gleaner' with this despatch."

Honours and mention for this campaign were freely given to Major-General Lord William Bentinck and the gallant 42nd; Major-General Paget with the 95th and 52nd; and to Lieutenant-Colonel Nicholls and his gallant 14th Regiment; for in a village situated in front of the line of battle which had to be taken, some companies of the 14th soon expelled the foe, to their war tune of "Ça Ira."

Great praise was given to the gunners, and to the Royal Naval officers—Hon. H. Curzon, Gosselin, Boys, Rainier, Serrel, Hawkins, Digby, Carden, Mackenzie; while equal honour was given to Brigadiers Manningham and Leith, and the Brigade of Guards under Warde, and to the light troops under Crawford—and last, but not least, on this occasion were the services of Colonel Murray (son of James Murray, of Quebec), the Quartermaster-General, and all other officers of the Staff, Graham, Gomm, Hill, Hardinge. Clinton, the Adjutant-General, was sick.

The brunt of the action fell upon the 4th, 42nd, 50th and 81st Regiments, with parts of the Brigade of Guards and the 26th Regiments.

The 50th was a Regiment which gave backbone to any force. At Vimiera it achieved great honour, for in that battle the 70th French Regiment, in addition to other severe losses, lost all their plumes and flags. After the fight, the band of the 50th appeared in all these captured decorations, and during the days of the Convention, fifteen of the French 70th petitioned to be, and were, enlisted into the 50th Queen's.* "Seldom since the days of Falstaff have such soldiers been exhibited as the Portuguese Army. Their appointments are beyond description, ridiculously bad. But dress must count for naught, for one of our most distinguished Regiments, the 50th, is the worst clothed and appointed on the expedition. They are ludicrously called the 'Dirty Half Hundred,' but no reproach is meant to be conveyed, as their name is never mentioned without applause."

From the leading journal of February 6th, 1809 :—"Votes of thanks were given from both Houses to Baird and others. Canning, Castlereagh, and Arthur Wellesley, all spoke from their seats in the House." The name of Moore, of course, was mentioned, but with profoundest grief.

---

* Sir John Moore had a corps of guides which consisted of deserters from the French Army, which joined at Lisbon, at the same time as these men enlisted into the 50th.

CHAPTER XXIII.

THE despatches of this glorious, although disastrous, victory were brought home by General Sir Charles Stewart, and delivered to his brother, Lord Castlereagh, at the dinner hour, when a party, including Sir Arthur Wellesley, were assembled.

Excitement and grief soon became general, not to be toned down by the preparations that were in hand to celebrate the Jubilee of the King—for joy would have been better placed to have welcomed home Sir John.

However, so it was. If for a little leisure, we stroll into the House of Commons (February 1st, 1809), there we find Sir A. Wellesley, Secretary for Ireland, obtaining leave to bring in a Bill to allow volunteering from the Irish Militia.

The Speaker next conveyed to Brigadier-Generals Fane and Frazer and Rear-Admiral Sir S. Hood the thanks of the House for their various services. He further read from the Chair a letter from Sir David Baird, in answer to the Vote of Thanks of that House on Wednesday last.

It was then debated if the last despatch of Sir John Moore, marked "Private," should be read to the House. Mr. Abercrombie (son of the late Sir Ralph—afterwards Lord Dunfermline) intimated that the sense of the relations be taken.

Now there was a great flutter in the Ladies' and the Speaker's Gallery, for Mrs. Mary Anne Clarke was shortly to appear at the Bar; for Mr. Wardle moved that the House should resolve itself into Committee on the charges against the Duke of York. He then read an extract from the *London Gazette.*

Lieutenant-Colonel Brooke was appointed Major of the 5th Dragoon Guards, *vice* Knight, who exchanged to half pay of the 56th Foot.

The first witness called was Dr. Andrew Thynne, who said he had attended Mrs. Mary Anne Clarke in his professional capacity for seven or eight years, and had told Mr. Knight that he thought Mrs. Mary Anne Clarke possessed sufficient influence with the Commander-in-Chief to expedite this exchange between his brother and Lieutenant-Colonel Brooke—but she would require money for the exertion of that influence. £200 was sent to Mrs. Clarke the day the *Gazette* appeared.

Mrs. Mary Anne Clarke was next called, when all the ladies who had been admitted to the Speaker's Gallery were heard to say: "She was not a perfect beauty, but her person and address were captivating, and her dress was elegant."

We have another instance of a Commander-in-Chief being under the management of his wife:—

The Duke of Marlborough had noticed the behaviour of a young officer in an engagement in Flanders, and sent him over to England with some despatches, and with a letter to the Duchess, recommending him to her to procure a superior commission for him in the Army.

The Duchess read the letter and approved of it, but asked the young man where the thousand pounds were for his increase of rank.

The young man blushed and said that he was really "master of no such sum." "Well, then," said she, "you may return to the Duke." This he did very soon afterwards, and told him how he had been received by the Duchess.

The Duke laughingly said: "Well, I thought that it would be so; you shall, however, do better another time"—and, presenting him with a thousand pounds, sent him over to England. The last expedition proved a successful one.

Mr. Whitbread later on, seeing Sir Arthur in the House, asked if, whilst he had been so gloriously fighting the battles of his country abroad, he had retained the emoluments of his Civil situation at home. Sir Arthur replied that he was absent two months, and as he did not receive any military emoluments, whilst heavy expenses attended the situation—which he confessed he was not in a condition to sustain—he conceived himself justified in taking his salary for these two months. So practically, whilst Sir Arthur gave us those two grand victories of Roliça and Vimiera, he was Secretary for Ireland on temporary leave of absence.

A heavy gale visited England at the time—and our poor troops, especially the sick and wounded, suffered terribly in the short voyage from Corunna to Portsmouth. "On Monday a transport with troops went ashore under Mount Wise. The 'Hindostan,' of 54 guns, arrived with 600 sick soldiers. Major Campbell (42nd), Ensign Hall, and several other officers, arrived from Corunna, have died here of their wounds. During the late tempestuous weather, seven transports were driven on shore on Southsea beach—five were got off.

These facts are only mentioned to give a faint idea of what a naval fight in our home waters would produce in the way of sick and dying; and how assistance on shore by the Red Cross would be of untold value.

Whilst awaiting plans for the next campaign, we will return to India with Captain Bayly, of the 12th, whose leave and Staff appointment at home had expired, and is now ordered to rejoin at Cannanore, accompanied by his wife and family, and a draft of 100 men.

On arrival their welcome was proclaimed by the band, and a refreshing breakfast at mess. Let us hope that the meeting of old friends (alas! how many had gone) gave warmth to the aching hearts of poor Captain and Mrs. Bayly, for during their short halt at Seringapatam, a cruel fever overtook them and carried off their three young children—a fever from which the Regiment had suffered terribly just a year ago.

In early 1808, a mutiny of the Rajah's troops in Travancore gave great trouble. The mutineers, headed by the Dewan or Prime Minister, had already got the upper hand; for the small garrison of two Sepoy Regiments and two guns, under Colonel Chalmers, were practically imprisoned in Quilon, a place some 300 miles down the coast.

It was to their relief that the 12th Regiment were hurried off. On the first detachment arriving, there arose a general howl amongst the rebel soldiery, who disappeared into the interior with their horses, elephants, and palanquins, only to reappear at night in immense multitudes, keeping up a heavy fire on the picquets, whilst heavy columns menaced the encampments. The whole force then sought shelter in an old Dutch fort, four miles distant on the shore.

It was whilst busily engaged in getting order out of chaos, that a flag of truce arrived, with propositions for surrender; failing which they were to suffer complete annihilation, as the forces of Cochin had united with Travancore, augmenting the enemy thus to 60,000 men.

A prompt refusal was given to this demand, and the bearer of the message retired, apparently to carry out his threat of destruction.

Now every man set to, to do his best, and soon fresh courage was infused—for in the offing could be seen a fleet of twenty patamars, with the remaining companies of the 12th and the 18th Native Infantry, bringing up the strength to about 3000 men.

A sad mishap took place, however, to a boat load of thirty men, under the command of Sergeant-Major Tilsey. He had been induced by the Travancore ruffians to land at a village, Aleppi, near Cochin, on pretence that the Regiment was landed and waiting for them two miles up country. They landed, and were surrounded before they could make the smallest resistance; their arms and knapsacks were taken from them, they were tied back to back and thrown into a deep tank, and, of course, all drowned.

The landing of stores and preparations for defence kept the force fully occupied until daylight on January 15th, when a tumultuous noise in the wood proclaimed the approach of an immense number of men, and the Travancoreans commenced an attack along the whole front of the British line; at the same time heavy columns were seen among the trees threatening both flanks.

Intending the attack for a surprise, the enemy fired his artillery at the tents; but when sufficient light enabled the Travancoreans to see the British ranks, they immediately directed their guns on the 12th Regiment, as if desirous of annihilating the Europeans first. Thus perilously exposed to the enemy's numerous artillery, the British infantry advanced; the right wing of the 12th and two battalions of Sepoys being directed against the enemy's left, and the left wing of the 12th, with one battalion of Sepoys, against the right of the enemy's line. The whole force was instantly brought into close action; but the British had only five small field pieces to answer the fire of the forty guns brought into action by the enemy. The British musketry was, however, well directed, and the incessant peals which echoed in the woods announced a vigorous contest, which was continued for several hours, during which clouds of barbed arrows, from the enemy's local troops, inflicted painful wounds on the men. About mid-day the 12th were ordered to charge with bayonets, and capture the enemy's artillery; they rushed forward with distinguished bravery, the soldiers shouting, "Remember our comrades at Aleppi!" The Travancoreans made a resolute defence, many of them being bayonetted as they stood, whilst a discharge of grape shot from one field piece killed eleven grenadiers of the 12th.

Finally both wings of the Regiment were triumphant; they captured eighteen brass field guns.

The loss of these guns disheartened the enemy, who retired about 3 o'clock in the afternoon.

"Fort St. George, February 6th, 1809.

"The Honourable the Governor in Council has recently received a favourable account of the action that took place on January 15th between the British troops and the troops of Travancore, in which, after a long contest, the Travancore troops were defeated with heavy loss.

"From the extent of the combined force which was opposed to the British troops, this signal victory reflects the highest honour on their discipline and valour; and the Governor in Council accordingly conveys to Lieutenant-Colonel Chalmers, who commanded the British detachment at Quilon in this distinguished action, his public thanks, and Lieutenant-Colonel Chalmers is requested to convey the thanks of the Governor in Council to Lieutenant-Colonel Picton of His Majesty's 12th Regiment, Major Muirhead, Major Hamilton, Captain Newhall, Captain Pepper, Captain Mackintosh, Lieutenant Lindsay, Lieutenant Arthur of the Engineers, and the officers of the Staff, Captains Cranstoun and Achmuty, with the other officers and troops of the detachment who bravely signalised themselves on the occasion.

"The Honourable the Governor in Council also takes this opportunity of expressing his warm approbation of the conduct of a detachment of troops stationed at Cochin under the command of Major Hewitt*, who with great skill and bravery repulsed a numerous and united force of the troops of Travancore and Cochin in an attack which they made on the British detachment on the 29th inst.

"The Governor in Council has particular satisfaction in expressing to Major Hewitt and to the officers and troops under his command, his public thanks for their highly deserving conduct.

(Signed) "G. BUCHAN,

"Secretary to Government."

To be mentioned in Despatches is a soldier's pride, and with Major Hewitt was another officer, Lieut. Thompson of the 12th.

---

* A company of the 12th under Lieutenant Thompson, and a battalion of Sepoys under Major Hewitt, were left for the defence of Cochin, and had been attacked by the troops of the Rajah. During the fight the Sepoys (17th Native Infantry) gave way, leaving the 12th alone to struggle with vastly superior numbers.

The 17th Native Infantry had been but recently raised. This was their first trial in an action where the enemy was so vastly superior—hence in their depression, irresolution, and timidity, they threw away their arms and military clothing, appearing in the simple dress of common natives, in order to give an idea to the enemy that they were only coolies.

He, with his Company, had been the backbone of the grand defence at Cochin, and he had been severely wounded.

He was listening silently to the reading of the Order. The omission of his name had an immediate effect—the disappointment raised a fever, and he was a corpse the next day.

Further disappointments followed—for many other hard and distressing days were passed in quelling this great rebellion, and it was not until March 31st that the British Resident could announce terms of peace.

Then, naturally, the 12th Regiment, as well as the other troops, expected some recompense in the shape of prize money; but, to the disgust of all, the Madras Government decided that all pretensions to prize money were inadmissible, as the war was undertaken against the Dewan (the Prime Minister), and the English regiment had merely aided the Rajah to quell the rebellion.

Here I sympathise most strongly with the 12th and all others there concerned—for, roughly speaking, do not these Prime Ministers make Monarchs of themselves? The method is to clear the way by murder or intrigue, and then to reign as potentates, until some stronger man comes along, when they, in turn, are shot.

Besides, in Travancore, there is this annual ceremomy—traceable to the fourth century—called the "Tulabhara," whereby the Maharajah is weighed against a mass of pure gold, which is then dispensed in charity. The Maharajah that year weighed in over eleven stone, and the 12th would have been quite satisfied with half of this in gold.

During the encampment of the Quilon force at the village of Attingurry, Captain Bayly obtained leave to visit Aujenga, about five miles distant. Proceeding down a beautiful river, studded on both sides with luxuriant cocoanut trees, mere curiosity actuated him on this occasion to take a view of the birthplace of Sterne's "Eliza Draper."

Yorick's letters to Eliza, and the writings of Abbé Raynal, placed this lady in 1767 before the world as the one idol. Certainly, she was possessed of rare mental gifts, and great personal attractions. Born at Aujenga, our furthest settlement in Southern India, she married Mr. Draper, Commissioner of Bombay. It was whilst at home she became the "lioness" of the London season.

On her return passage to India, amongst the passengers was a Miss Light, who afterwards married George Stratton, Esq., a young soldier, whom Mrs. Draper (Eliza) described as susceptible of tender impressions. In a sort of playful jest, Eliza hinted that "before Miss Light had sailed a fortnight, he will be in love with her!"

"But," says Mr. Sterne, "five months with Eliza, and in the same ship, and with an amorous son of Mars! But thy discretion, thy wisdom, thy honour, thy own self."

Alas! she abandoned age and austerity in the shape of an old husband, and adopted the young officer, who appeared to be all her vivid fancy imagined.

In Bristol cathedral is a monument "Sacred to the memory of Mrs. Elizabeth Draper, in whom genius and benevolence were united; she died August 3rd, 1778, aged 35"—but on her gravestone is the simple inscription "Mrs. Draper, 1778," for some still surmise that her body lies beneath a stone near the Court House at Tellicherry, near to the humble cottage, close to the seashore, where she was born.

A glorious victory, a long list of killed and wounded, always attracts attention—it is the miseries, the sufferings, and privations which the Regiment here underwent, which are apt to pass unnoticed: for even now, there remained a homeward march—from Quilon to Seringapatam—which was beset with all imaginable *malaise*.

This march commenced on May 23rd, 1809. Soon the monsoon burst with extra fury, making every river impassable, for in attempting some the regimental baggage was swept away, and many men were drowned.

The Regiment was fairly swept off its course, and, instead of making Seringapatam, had to head for Trichinopoly (July 24th, 1809), to rest awhile ere they left Madras for good.

Lord Lake and Sir Arthur Wellesley, in the conduct of their Mahratta War, won the confidence, esteem and gratitude of all men, the native army especially. It is sad to think that soon their good influence waned, for with other men came other measures, with results not near the same. This must be further dwelt upon in the Chapters on the "Mutinies."

CHAPTER XXIV.

FRANCE was now most troublesome at sea; for as long as Bourbon and Mauritius belonged to her, she held the key and could work her wicked will. An expedition, in which the flank Companies of the 12th formed part, was quickly organised, and arrived off Rodriguez, an island just suitable to act as a barrack, depot or base for the troops now following *en route*; for Bourbon and Mauritius must be ours.

Captains Rowley and Willoughby were in command of the conducting squadrons, and bore away on July 7th, to the different points of debarkation. At first the weather was favourable, but soon began to change, when the surf became so high as to render a landing dangerous. An attempt was, however, made. The Light Infantry of the 12th in a small schooner, and about three hundred men of the 33rd and 69th Regiments in boats, approached the shore and effected a landing with the loss of a few men drowned; but the schooner and boats were dashed to pieces, the soldiers' ammunition damaged, and many of their arms lost.

As no more men could be landed, Lieutenant Foulkstone, of the 69th Regiment, volunteered to swim through the surf and convey orders to Lieutenant-Colonel Macleod to take possession of St. Mariè.

This order was instantly obeyed, and the Light Infantry of the 12th distinguished themselves in storming the batteries, in which service they had two private soldiers killed; Lieutenants John Spinks and John B. Whannell, with five rank and file, wounded.

The grenadier company of the 12th, and other corps, under Colonel Fraser, afterwards landed at Grand Chaloupe, and by their spirited conduct, particularly the gallant behaviour of the 86th Regiment, the reduction of the island was speedily accomplished. Lieutenant Monro of the 86th was killed, and Lieutenant McCreagh wounded.

Mr. Farquhar (grandfather of the present Lord) was now installed as Governor. He had accompanied the expedition from Madras, armed with the usual Proclamation of changing wrong to right, and promising future happiness to all who would not fight.

It was now determined to annoy the good people of the Mauritius. Accordingly the four frigates "Sirius" (Captain Pym), "Magicienne," "Iphigenia," and "Nereide" (Captain Willoughby) attacked and carried the entrance to Port Louis. The "Nereide" entered; the remainder cruised outside. During the temporary absence of the latter, three French frigates ("La Bellone," "Minerva," and "Victoire") crept inside, when the "Nereide" became exposed to their concentrated fire. The sequel of the fight we learn from Captain Pym's official despatch to Governor Farquhar:—

"La 'Nereide' me fit le signal qu'elle etait prête pour le combat; alors je m'approachai d'elle et ayant reconnu la position de l'ennemi je dccidai de l'attaquer"....

"Le 'Syrius' toucha sur un petit banc inconnu 'La Magicienne' qui etait près de 'L'Iphigenie' toucha sur un banc de maniere a la priver avec plus de six pièces de canon.

"La malheureuse 'Nereide' prit son poste a peu de chose près, et soutient de le maniere la plus courageuse le feu dirige sur elle et celui qui etait destiné pour le Syrius"....

"Il m'est penible de dire que le Capitainé, les officiers, et tout le monde à bord de cette frégate sont on tués on blesses."

The bravery of Captain Willoughby was extolled by the French in the following distich:—

"Au brave Willoughby, Commandant la 'Nereide' frigate de sa Majesté Britannique.

"A la grandeur du vrai courage
Tous les peuples rendent hommage:
Recois le nôtre O Willoughby!

> Du grand Nelson tu montras la vaillance
> Le ciel le fit semblable a lui
> Et sur les traits voulut aussi
> Du même sceau marquer la ressemblance."

Willoughby the "Immortal" (born 1777, died 1849), like Nelson, had lost an eye. He became Sir Nesbit J. Willoughby, K.C.B. His early life was chequered—always in hot water with his Captains, he was tried for insolence, and dismissed the Service. He joined again as a volunteer under Sir John Duckworth, in 1803, and for good conduct was re-instated Lieutenant. He was tried for losing his ship the "Nereide," but was acquitted most honourably.

Captain Pym also became a distinguished officer. He was born in 1778, and died in 1855 as Sir Samuel Pym, K.C.B.

The foregoing regrettable incident gave the French, for a time, a decided superiority at sea. They greatly harassed the approach of General John Abercrombie (son of Sir Ralph), who, in command of large re-inforcements (with the 12th Regiment, less the flank company who had preceded them), had left Madras on September 21st, 1810, escorted by the "Russell" (74), "Clorinde," "Cornwallis," "Cornelia," "Bucephalus," and "Hesper." It was so near a capture of Abercrombie that he consigned his official papers and other documents to the deep.

The expedition was enabled to land at Rodriguez, from whence, on November 29th, 1810, they again embarked—this time in battle order for the capture of Mauritius. The landing was unopposed, which made the resistance in the interior all the more stubborn. It was after two days' hard fighting that, at 10 o'clock on December 2nd, 1810, General Du Caen offered terms of capitulation, which were accepted by Sir John.

The troops were thanked in Orders, and special comments given to :—

1. The steadiness and gallantry displayed by the flank battalion of the 59th Foot.
2. Condolence with H.M. 59th Foot on the loss of a most valuable and excellent officer, Lieutenant-Colonel Campbell.

3. The great steadiness of H.M. 12th Foot.

4. Tenders of thanks for the most efficient services of the seamen under Captain Montagu, R.N.

"Now Colonel Keating commanded the flank battalion, and every allusion to his conduct was sedulously avoided in the foregoing Order.

"In leading this choice corps, he descended from his horse as the firing became brisker, and complained that the wound in his leg gave him great pain. All who heard him were much surprised, as previous to that moment he had never come in contact with the enemy; but on the surgeon inspecting the wound, it appeared like a small puncture by a small-sword through the calf of the leg, which appeared to have been perpetrated by his own sword.

"General Warde rode up and said, 'Colonel Keating, if you *will* not or *cannot* lead on the flank battalion, I will'; and then, placing himself at their head, moved on with Colonel Campbell.

"The latter was shot through the head five minutes after, and at the termination of the affair, the gallant Colonel Keating retired to hospital, where (on visiting some of our officers who had been severely wounded) I saw him in apparently great pain. This was the same Captain Keating of fighting notoriety, whom we met at the Cape in the year 1796."

Reinforcements from France, escorted by three frigates—to save Mauritius—were speedily sighted by Captain Schomberg and his squadron of three and a brig. One was taken, another escaped after she had struck her colours, and the third was overtaken at Tamatava Bay, in the Isle of Madagascar, and there captured with the fort and other vessels in the harbour.

After the surrender of Mauritius, the French troops were embarked for France, and the principal part of our army returned to India; whilst the 12th, 72nd and 82nd Regiments remained to garrison the newly-acquired possession, Captain Bayly being appointed Brigade-Major. But, alas!

> "Never morning wore
> To evening but some heart did break,"

for when his wife and children joined him, the wife seemed fearfully changed—and within a few days her noble self was borne, on the shoulders of the old 12th she loved so well, to her last home, leaving her husband and young family disconsolate and crushed.

News shortly followed of the reduction of Java—the last colonial possession of the French Empire. Lord Minto (Governor-General of India) was present, and his Commander-in-Chief was Sir Samuel Auchmuty, who had come out of the unfortunate affair at Buenos Ayres with good repute.

General Janssens (who surrendered to Baird at the Cape of Good Hope) made a gallant defence, but the works fell by general assault on August 6th, 1811. Lord Minto, in his despatch, stated that "The French flag was nowhere to be seen between Cape Comorin and Cape Horn."

At the same time—but in European waters—was England, with her wooden walls, adding fresh victories to her splendid and lengthy roll; for in this year of Corunna, 1809, she had promised Austria, now again embroiled with France, to make with her fleet a powerful diversion on the French coast.

The French fleet, consisting of eight sail of the line and two frigates, was blocked up in Brest by Lord Gambier* and his two Admirals, Stopford and Beresford. The English fleet were beaten off their watch and driven out to sea by a strong westerly gale—only on their return to find that the French fleet had escaped to the Basque Roads, and were joined by another squadron of four battleships and two frigates, the whole lying under the guns of their own forts, with the river Charente open in their rear.

To attack them meant a very hazardous proceeding. Lord Gambier wrote home for fire-ships, stating:—"The enemy's ships

---

* Lord Gambier—born 1756, died 1833—first to break the French Line off Brest on the glorious June 1st, 1794. Governor of Newfoundland for five years, 1802-7, then recalled to undertake the capture of the Danish Fleet. In four days Copenhagen capitulated, and the Danish Fleet was removed to England. He presided in 1814 at a Conference in Ghent, when peace was signed between England and America.

lay very much exposed to the operations of fire-ships ; it is a horrible mode of warfare, and the attempt very hazardous, if not desperate ; but we should have plenty of volunteers for the service."

In reply, Lord Gambier was informed on March 19th, 1809, that twelve transports were fitting as fire-ships ; that Mr. Congreve,* with a supply of rockets, with men to work same, and five bomb vessels, were under orders to fit for sea and proceed to Basque Roads. Lord Cochrane was appointed to work these fire and bomb ships, under the directions of Lord Gambier. This dual control meant trouble.

On the night of April 11th, the fire-ships (led by Captain Woodridge) and the explosion ship proceeded to the attack. A boom was stretched across for protection ; this, however, was soon broken, and the English advanced towards their prey, undisturbed by the heavy fire from ships and forts. The French fleet, dismayed and thrown into confusion, cut their cables and ran ashore ; six escaped up the Charente.

Here was an opportunity to follow up this success, but Lord Gambier hesitated to entangle his fleet amongst the shoals; so he rested content with a capture of fourteen ships, destroyed and burnt.

Lord Cochrane, on his return home, was created a K.C.B. He opposed, from his seat in the House, the passing of any vote of thanks to Lord Gambier for his conduct at Basque Roads.

Lord Gambier requested a Court Martial, the charge against him being :—"That on April 12th, the enemy's ships being then on shore, and the signal having been made that they could be destroyed, he did for a considerable time neglect or delay taking effectual measures for destroying them."

The Court most honourably acquitted him, stating :—"That Lord Gambier's general conduct as Commander-in-Chief of the Channel Fleet, employed in Basque Roads between March 17th and April 29th, 1809, was marked by zeal, judgment and ability."

* A distinguished Military Engineer, born 1772. His rockets were a great success at Basque Roads, Walcheren, Leipzig and Algiers. He died in 1828, a full General in the Army.

The thanks of Parliament were given to Lord Gambier, Sir Harry Neale, Stopford, Beresford, officers and men.

The French Captains were nearly all tried. One—Captain Lafon—was condemned, and suffered death.

Certainly, the French trials disclosed the defenceless state of the grounded ships; and perhaps it was well for Lord Gambier that these facts were not fully known at his trial. The French trials lasted from June 21st to September 8th—the English Court Martial on Gambier from July 26th to August 4th.

Napoleon said, in after years, at St. Helena: "The French Admiral was an imbecile, but yours was just as bad. I assure you that if Cochrane had been supported, he would have taken every one of the ships. They ought not to have been alarmed by your *brûlots;* but fear deprived them of their senses, and they knew no longer how to act."

A few months later—in 1809—the greatest armament of the century, consisting of 240 warships and 400 transports, left the shores of England on a bright day in July, with orders to destroy everything afloat or ashore in the Scheldt, Antwerp or Flushing.

Such orders would have delighted more than half our Admirals or Generals, but to the Earl of Chatham and Sir Richard Strachan, this venture gave no proper zest.

The British landed on Walcheren and South Beveland, and invested Flushing. The conquest of the latter place was no easy matter, for the investment was incomplete, and the enemy easily obtained reinforcements and provisions from opposite coasts. However, on August 14th, after heavy cannonading the town surrendered, with a garrison of 4000 men as prisoners. Antwerp could easily have been taken then by a *coup de main*, but Lord Chatham did everything possible to ensure the failure of the expedition.

A Council of War was held on August 26th, and England agreed to abandon the enterprise; for, as regards the military, everything seemed out of gear—the expedition was ill-planned, ill-timed, ill-executed.

The French said of the Earl of Chatham that he was the most temporizing General in the British Army. His countrymen reproached him with being occupied exclusively about his health and his turtle soup, and forgetting that the first of British interests is for the coasts and harbours within a few hours' stretch of her shores to be in the safe keeping of a friendly people, and as inviolate as our own territory. Thus, if Antwerp had been but kept, how many of our ships and frigates could have been freed for other things?

The Earl of Chatham, with the greater number of troops, returned to England; the remainder were left to die of fever in Walcheren. "C'était le comble de la bêtise et de l'inhumanite," said Napoleon.

The fleet under Collingwood,* in the Mediterannean was confronted by Admiral Gauteanne; but he blocked and watched the latter to such good effect, that the Spaniards were enabled to recover several of their towns on the coast. Barcelona had long been besieged by the patriots of Spain, and was in sad need of ammunition and provisions. This relief the French attempted to give by sea, but as soon as they were discovered, Collingwood was at them. He drove three large battleships ashore, and attacked and destroyed the transports. This, the taking of Barcelona, was Collingwood's last exploit; for the above harassing work of watching, watching, fighting, fighting, at last told upon his Lordship, and he died on board his flagship, the "Ville de Paris," March 5th, 1809.

The command of the Midland Sea now passed to Sir Charles Cotton. The same activity continued, and other fights ensued, having for their reward either praise or censure. It was in the sea and shore fighting, in the operations of Sir John Stuart from Sicily,

---

* Collingwood, from eleven years of age to his death, was nearly always at sea—first with Nelson on the Spanish Main and West Indies, then with Howe on June 1st, 1794, and Jervis off St. Vincent in 1797, and at Trafalgar, where he led one of the two lines. "See," said Nelson, "how that noble fellow Collingwood carries his ship into action." His services were so noble and so useful that the Government refused to relieve him, even when turned sixty.

Ischia and Naples, that a midshipman—George Rose Sartorius—made his mark for valour. His ship was then the "Success" (Captain Brenton, afterwards Sir Jahleel Brenton).

Sartorius first learned what fighting meant on board H.M.S. "Tonnant" (a vessel captured at the Nile) at Trafalgar (Captain Charles Tyler), where they lost twenty-six killed, fifty wounded, and all three topmasts.

He was now on board the "Success" on April 4th, 1810, off the coast of Calabria, where the Captain noticed three vessels on the beach, and the crew hard at work loading them with stores. This offered quite a tempting prize, so boats under Sartorius, Oliver, Coates and Pearce, were at once despatched to attempt a capture. The boats, on nearing shore, struck on a sunken reef and swamped. The men swam to shore with cutlasses in their mouths; but on the moment of landing, a masked battery of two 6 pounders and four wall pieces suddenly opened upon them. Regardless of this, Sartorius and his party rushed on; captured and spiked the guns; destroyed two of the three vessels; and, baling out the swamped boats, returned to the ships, having lost but two men drowned and two wounded.

Again, on April 25th, he performed similar valiant service in capturing one ship, three barques, and several feluccas, whilst at anchor under the castle of Terracina. The boats went in, commanded by Sartorius and Baumgardt, and supported by the fire of the men-of-war; and in spite of much resistance, these two young officers took and brought off their prizes.

On May 1st, Captain Brenton, with the "Spartan" and the "Success," chased the "Ceres" and her consort into Naples, and remained cruising off the port. The French, however, were wily; for in the night, Prince Murat crammed his frigate and corvette with 400 Swiss troops, and with an additional seven large gunboats, he sailed out next morning to re-commence the fight. The "Spartan" soon disposed of the gunboats, exposing herself at the same time to a very heavy fire; and here Captain Brenton, whilst standing on the capstan the better to view his various opponents, was badly wounded in the hip. Lieutenant George Willes then carried on the fight. A slight breeze enabled the "Spartan" to take up a better position, and to force the French squadron to seek cover under the shore batteries of Baia. The "Spartan" then wore, and captured the "Sparviere" and crippled the "Fama."

The relative force of the French squadron and the "Spartan" stood thus:—French, 95 guns and 1400 men; "Spartan," 46 guns and 258 men.

Great praise in this action was given to Captain Hoste, Royal Engineers, who—although only a passenger on board—took entire charge of the quarter-deck guns, whilst Captain Brenton (until wounded) and his First Lieutenant Willes manœuvred the ship.

The "Spartan," having repaired her principal damages, and being proud of her prize, stood directly across the Mole of Naples, to the great chagrin and mortification of Prince Murat, King of Naples. Young Lieutenant Willes, for this engagement, was deservedly promoted Commander, and all others were honoured in the shape of loud encomiums—and thus came an ending to these pretty fights at sea.

CHAPTER XXV.

TO resume our story in the Peninsula. Since Corunna, it still remained for England to fulfil her promise to free Spain and Portugal of the French; so she sent out at once Arthur Wellesley, with a well equipped and provided force. Marshal Beresford was also in the country raising levies and issuing proclamations; so when Wellesley arrived in April, 1809, Portugal and Spain were in full heart again.

The campaign commenced by the passage of the Douro, which came upon the French as a great surprise, and so complete that Wellesley that night sat down to the dinner prepared for Soult. Soult in his retreat had to abandon his artillery and baggage; and Wellesley now watched Victor, who was marching on to Lisbon.

In the meantime, the Spanish patriots were defending their mountain passes with alternate success and disaster; but so well did they defend the passes of Serra Morena, which barred the way to Madrid and the south of Spain, that Joseph Buonaparte, with Sebastiani as his Chief of Staff, had to abandon their project of marching thither, and proceeded to join Victor, who was stationed in the neighbourhood of Talavera, along the banks of the Alberche.

Here Joseph, Sebastiani,* Victor and Jourdan made a strong

---

* Sebastiani commanded at Talavera, and had the confidence of Napoleon. He was extremely pompous, and a great boaster. His mother used to say of him, *apropos* of this battle: "Mon fils ressemble à un de ses tambours—plus il est battu, plus il fait du bruit." Sebastiani was French Ambassador in 1834. On June 18th, after the Waterloo Banquet, the Duke attended a reception at Lady ———, when on entering the room, everybody rose, and the band played "Rule Britannia!" Sebastiani considered this a personal affront, took his wife's arm, and walked out.

attack on General Cuesta, who soon fell back upon Wellesley. They then attacked the united English and Spanish armies, but were repulsed with the heavy loss of 10,000 men. Generals Lapisse and Malot were killed, and Sebastiani and Boulet wounded. The English had heavy losses, amounting to 6000. Generals Mackenzie and Langworth were killed. Wellesley was made a Peer for Talavera on July 30th, 1809.

Napoleon early in 1810 divorced Josephine, and married Maria Louisa. This matrimonial alliance afforded a little breathing time and short Continental peace, which enabled Napoleon to give all thought to Wellington, who was far too successful against his Marshals in the Peninsula. His first measure was to recall his troops after Wagram (July, 1809) and to send, under Massena, the corps of Regnier, Ney, Junot and Drouet, to prosecute the war in Spain.

They crossed the Pyrenees together, and Wellington had to witness the fall of Badajoz, Ciudad Rodrigo and Almeida, and to retreat down the valley of the Mondego, with Massena in pursuit. Wellington turned, and gave battle to him at Busaço (September 27th, 1810), and completely shook him off. This only increased Napoleon's ire and eagerness to urge Massena (with freshened troops) in pursuit again; but Wellington safely slipped his army within the lines of Torres Vedras, which he had so secretly prepared. For two months did Massena gaze upon the English army so comfortably housed and fed from home, whilst he—outside—was suffering from the stress of cold and hunger.

Massena could stand it no longer, and retired to Santarem in hopes of meeting Soult. Soult failed him; so, left with Wellington on his heels, he was driven out of Portugal, leaving his only post—Almeida—to be captured by Sir W. Erskine and his cavalry.

Massena thus, for a time, had to remain on the defensive. However, reinforcements, especially in cavalry, soon reached him, with which he took the field again, and met Wellington at Fuentes d'Onoro on May 3rd, 1811. The village of that name was held throughout the day by Colonel Cadogan and the 71st Regiment, whilst Ramsey, with his troop of Horse Artillery, became the *point d'appui* which enabled Wellington to win the fight. Now we pause for a while, only to remember that on March 20th, 1811, to Napoleon was born a son and heir—the King of Rome—but where was the happiness? He would not rest on his laurels and nurse his dynasty. Mauritius and Bourbon had fallen. Battle after battle in the Peninsula went against him. In fact, the tide had turned; it was ebbing hard to exile and to death ten years later.

Beresford now laid siege to Badajoz, which brought Soult forward once again for its relief. Beresford, not wishing to raise the siege, effected a junction with the Spanish General Castanos and Blake, and took up a position at Albuera on May 16th, 1811, on ground between the enemy, and twelve miles from Badajoz. A battle quickly followed, and at one time all was touch and go. "Retreat!" was even whispered. Colonel Hardinge now found the last straw to break the camel's back, and sent up the Fusilier Brigade to carry on the fight. Napier says: "These battalions, struck by the iron tempest, reeled and staggered like sinking ships; suddenly and sternly recovering, they closed on their terrible enemies; and then was seen with what a strength and majesty the British soldier fights. . . . 1500 unwounded men, the remnant of 6000 unconquerable British soldiers, stood triumphant on the fatal hill."*

The border fortresses of Ciudad Rodrigo and Badajoz were won again by Wellington—the former on January 19th, 1812, the latter by assault on April 6th, 1812—amidst the grandest scenes in war. Shortly after, he defeated Marmont at Salamanca, and entered Madrid in triumph on August 12th, 1812. This battle produced from Napoleon a letter demanding from Marmont categorical answers to a multitude of questions, censuring him in advance. The letter is dated Ghiast, September 2nd, 1812, and is always quoted as a striking example of Napoleon's moral and intellectual strength;

---

* Here fell General Crawford, the leader of the Light Division.

for it was written on the eve of Borodino, from the heart of cold Russia.

The French, thus beaten and temporarily dispersed, were by no means powerless. They concentrated afresh around Burgos—a fortress and *place d'armes* on the high road from Bayonne to Madrid, and the key which Wellington must get ere he could enter France.

There were hopes for Wellington that the Spanish army under Ballasteros, and an English force from Sicily under Lord William Bentinck, might be of some assistance. But Ballasteros was not the man for Wellington—he was insubordinate and jealous to a degree; he was removed from his command and exiled. The force under Bentinck was too weak to create a diversion in his favour, so Wellington had to raise the siege of Burgos, and retreat as best he could. In men alone he lost 7000; as regards discipline, he wrote: "Never had he seen, heard, or read of such abandoned conduct." In the same month, October, 1812, was Napoleon in that terrible retreat from Moscow—his route being marked with a dark and sanguinary chain of corpses until his army was dead.*

However, Napoleon found time to write the following letter from Moscow to Clarke, M. le Duc de Feltre:—

"I have read with interest the journal of the siege of the Castle of Burgos; I think that it ought to be put into the *Moniteur*.

"Let me have a draft of a Decree for rewarding General Dubreton and the officers and soldiers who distinguished themselves in that defence.

"NAPOLEON."

He returned to Paris on December 18th, 1812. Plots were thick against him—(General Malet† and eleven others were shot at

---

* Of 600,000 human beings in the vigour of life who crossed the Muscovite frontier in June, 1812, 42,000 only returned.

Of the 110,000 Russians who commenced the pursuit, only 35,000 remained.

† "Avez vous des complices? Demanda le Président du Conseil de guerre. Oui repondit Malet La France l'Espagne et vous même si jávais reússi."

once on the plains of Grenelle)—and he found his country almost drained to the core of men and money. However, he had one ally left in the King of Saxony; so bidding adieu to his wife and son for ever, he hastened off to finish up his war with Russia.

To continue the retreat from Burgos. Soult, whilst in pursuit, was offered battle by Wellington. It commenced in a pretty cavalry skirmish, which, unfortunately, deprived Wellington for a time of his brilliant second in command, Sir Edward Paget, for he was taken prisoner. Soult then desisted, leaving Wellington free to enter Ciudad Rodrigo and replenish his army from the well-stored magazines.

Mortifying as this retreat was, it still left to Wellington the two strong fortresses of Ciudad and Badajoz, and the whole of the south of Spain. Soult was now required by Napoleon in Germany for a while. King Joseph took his place, and Marshal Jourdan became his Chief of Staff. Their headquarters were at Valladolid, when Wellington took the field and harassed them until they were forced to make a stand near Vittoria, where, to quote from an officer, "The French were beaten before the town, and in the town, and through the town, and out of the town, and behind the town, and all round about the town."

Europe now breathed again; and next year all the Powers joined in the War of Liberation against the despot. At Lutzen and Bautzen, Napoleon, with fresh levies, held his own, but at Dresden and Leipsic he was signally defeated. Refusing all peace proposals, he fought the Allies on the soil of France, and displayed the most consummate skill. He checked them at Brienne, and afterwards—with Blucher advancing on Paris from Chalons, and Schwartzenberg by the valley of the Seine—he beat them in detail on nine successive days. The tide, however, could not be stayed; and on the night of March 30th, Napoleon, with Berthier and Caulincourt, drove into Paris—the former to capitulate.

Napoleon had never displayed so much true greatness as during this last campaign—resisting inch by inch the onset of the allied invasion. The same can be said of Wellington in his battles of the Pyrenees, forging and fighting through the mountain passes into the plains of France.

Hope, Hill, Picton, Graham, Lord Edward Somerset, Lord William Bentinck, Earl Dalhousie, Colborne, Cotton, Paget, and scores of others, all pressed on to victory. It was at Orthez and Toulouse, in the rough severe wintry weather of February, 1814, that Wellington and Soult—like Wolfe and Montcalm—fought it out to the last.

> "Type of the sprites who wait before the Throne
> Of the great kingdom of the Great Unknown,
> To future ages winged messenger;
> Old as God's lightning, but to us, whose ken
> Sees but the distance of the deeds of men,
> Youthful as yesterday, a child new born,
> Just waking from its sleep, yet whose first stir
> Jars the old order from its groove outworn."—ROGERS.

If electricity could have flashed a message on that Easter Sunday (April 10th, 1814) that Napoleon was no longer Emperor, and that Louis XVIII. was King, what a blessed Angel of Peace she would have been. For the Easter Sunday that witnessed this useless battle and needless loss of 9000 men, also saw the Monarchs, Marshals and Generals of Eastern Europe *en grand tenue* at a review in Paris, at which a group of elderly men attended. They were the brothers of Louis XVI. The eldest was proclaimed Louis XVIII., Bourbon King of France. The others—Comte d'Artois, Duc de Berri and Duc d'Angouleme—were *en evidence* in hopes of something good.

It was on his return from this last battle of the war, the Marquis—now the Duke of Wellington—dined with Lord and Lady Castlereagh in Paris, and afterwards went to the Grand Opera in their company.

The Duke was in plain clothes without any decoration, and sat in the back of the box. Soon, however, he was recognised by some-one in the pit, and a voice cried out "Vellington!" The cry was taken up by others, and at last the whole pit rose, and, turning to the box, called out "Vive Vellington!" He then stood up and bowed, when he was cheered and cheered again.

At the finish of the performance, the exit being so crammed, the Duke was allowed to pass out quietly with a frightened lady on his arm; when a voice was overheard, "Mais pourquoi l'applaudissez-vous tant? il nous a toujours battus." "Oui," came the answer, "mais il nous a battus en gentilhomme."

Napoleon was, by order, soon deported to Elba, on board H.M.S. "Undaunted." On a ship with such a name he was bound to show himself the same—here to-day and gone to-morrow—for he seemed no sooner landed than the "Inconstant" took him back again to France, landing on March 7th, 1815. Just one incident *en voyage.*

"En se reconnaissant, les deux bricks se saluèrent selon l'usage et tout en continuant leur marche échangerent quelques paroles. Les deux capitaines se demandèrent le lieu de leur destination. Le Capitaine Andrieux répondit qu' il allait à Livourne; la reponse de 'L'Inconstant' fort qu' il allait à Gênes; Andrieux remercia et demanda comment se portait l'Empereur; a cette question, Napoléon ne put résister au desir de se meler à une conversation si interessante pour lui, il prit le porte-voix des mains du Capitaine Choulard et repondit: 'A merveille!' Puis ces politesses échangées les deux bricks continuèrent leur route, se perdant réciproquement dans la nuit."

This reads but an escapade *tres naturelle,* for the eyes of Europe were off him. The Powers were squabbling hard at the Congress of Vienna, each one trying to get more than his neighbour; and in some cases so high did these disputes rise, that parties prepared for war. It was at a great ball* in Vienna that the news arrived: "Napoleon had left Elba!" This saved the situation, for the diplomats fled as if a thunderbolt had fallen, and soldiers drew their swords again.

* "Le Congrès danse et ne marche pas" was a common expression at the time.

CHAPTER XXVI.

NAPOLEON landed at Cannes on March 7th, 1815, and at Grasse —a pretty hamlet above the town—he held his first little open-air Court, listening to all complaints, receiving petitions, and promising justice. He then proceeded by mountain route to Cerenon, Barème, Digne, and Gap.

"Soldats ne tirez pas, nous sommes Français" seemed a pass word well agreed upon, although Soult, as War Minister, had told the Army of whom they went to meet, and furnished Ney with the following instructions :—

"To assemble as great a disposable force as possible, for the purpose of seconding efficaciously the operations of his Royal Highness Monsieur, and to manœuvre in such a manner as to harass or destroy the enemy"—for Ney was to be in chief command, although the Duc de Berri, the King's nephew, claimed to be the figure head.

Ney left Paris on March 8th, and in bidding farewell to the Bourbon King, is reported to have said he would return with Napoleon, his old companion, in a cage. Thus all went well with Ney, until the night of the 13th, when perhaps Hortense (for she was a woman *remplie d'esprit*—her mother was Josephine, her father Beauharnais), or it may have been some other, called secretly to see the Marshal, and handed in a letter from his old master, with parcels containing Tricolors, Orders of the Day, Eagles, Routes, etc.

Ney, when asked, on his trial (November 9th) before the Peers of France, for the name of this person, answered, "I will implicate no one."

His Generals, le Comte de Bourmont, Le Courbe, and other Staff Officers, had all dined with the Marshal that evening, and the menu was strictly "Bourbon." Their surprise was naturally great the next morning, at *dejeuner*, to be told that the Division would assemble, as the Marshal had something to say; after he had spoken, there was but one shout from the men, and that was "Vive l'Empereur."

Soult declared for Napoleon the same day. The superior officers, however, showed much consternation, and said to their Marshal: "This is a horrible affair; had we known this, we would not have come hither." Marshal Ney gave leave to all who wished, to return.

The following is the narrative of Comte de Bourmont and Le Courbe, his two Divisional Commanders:—

"Up to the 14th of March, the orders given by Marshal Ney, and by me transmitted to the troops, appeared to me conformable to the interests of the King. On the morning of the 14th, Baron Capet, Prefect of the Ain, arrived at Lour-le-Saulnier. He told me that the town of Bourg was insurgent, and that the 68th Regiment had assumed the tri-coloured cockade, in spite of the General who commanded the Department, and of the superior officers of the regiment, who, up to the last moment, opposed it with all their force. We went to communicate this news to the Marshal. He said little to us.

"On the morning of the 14th, the 8th Regiment of Horse Chasseurs having arrived, the Marshal ordered me to draw them up in the Square, and to put all the troops under arms, because he wished to speak to them. I returned to the Marshal's lodgings, and being alone with him, he said to me: 'Very well! my dear General, you have seen these proclamations of the Emperor; they are well done—they must produce a great effect.'

"'Without doubt,' I replied, 'care must be taken that the troops be not allowed to read them; there are phrases in them which would have great influence on the soldiers. For instance, this one: "Victory marches at the charge step."' ('L'aigle avec les couleurs nationales, volera de clocher en clocher jusqu' aux tours de Notre Dame.')

"The Marshal said to me: 'But, comrade! were you not surprised at seeing yourself deprived of one-half the command of your Division? at seeing that the troops were made to march by two battalions and three squadrons? Very well! the case is precisely the same throughout all France. All this was arranged; we were agreed about it three months ago. Had you been at Paris, you would have known it as well as myself. All the troops are disposed in their march in such a way that they will successively arrive to escort the Emperor. The King must have quitted Paris; if he does not quit it, he will be carried off. However,' continued the Marshal, 'no harm shall be done to the King; woe to him who shall harm the King! He is a good Prince, who never injured any one; woe to him who should do him injury!'

"I said to the Marshal: 'Whatever happens, nothing shall induce me to fight against the King.'

"He replied: 'I insist upon your doing it; I am certain the Emperor will treat you well. However, you are at liberty to decide for yourself; if you will not, General Le Courbe will go with me.'

"General Le Courbe then came in. The Marshal said to him: 'I was telling the Count de Bourmont that all was prepared in such a manner that the troops might reach the Emperor; the King had quitted Paris—no harm was to be done to him. Woe to the man who should do any; he is a good Prince, but he will be sent on board a ship. What now remains for us to do? Join Buonaparte?'

"'What?' said Le Courbe. 'I have no reason to rally under that....The King never did me anything but good, and the other nothing but harm. Besides, I have honour, and therefore will not join Buonaparte.'

"'And I, too,' said the Marshal; 'and therefore I *will* join him. No more humiliation for me; I will not have my wife*

* Ney had married an intimate friend—in fact, a lady chosen for him by Buonaparte and Josephine; so, perhaps, the Bourbon ladies doubly gnashed their teeth at poor Maréchale Ney.

come every night, with tears in her eyes, on account of ill-treatment.' "

The crucial—Now—was soon to come, for on June 18th, Wellington met Napoleon for the first and last time. The fight commenced at nine a.m., and Napoleon hammered hard all day. He knew the English expected help—at five p.m. that help had not arrived; so he staked his all on one grand and desperate effort. It failed—then, slowly and doggedly, he fell back. The retreat soon became a rout; by six p.m. the Prussians were up, and joined in pursuit.

One letter of the battle :—

"Cateau, June 24th, 1815.

" I have not had time, my dear father, to write you a more circumstantial account of our terrific day at Waterloo, since my short note to say I was well. Buonaparte, on the 16th, attacked our left before the whole of the troops cantoned on the right of the line could be brought up.

" The day certainly was to his advantage. The French cavalry behaved dashingly, and succeeded in breaking one of our squares of infantry. The Guards behaved with great bravery, and covered themselves with glory.

" The same day Buonaparte attacked Blucher and the Prussians with part of his army. He beat Blucher during the day, but did not drive him from his position. At night Buonaparte made one of the most extraordinary and brilliant movements ever heard of. He collected the whole of his cavalry in one great mass, charged through the centre of the Prussian army, took eighteen pieces of artillery, the whole of the ammunition, and the Prussian loss was 15,000 men by Blucher's own account. The Prussians were in *deroute*.

"The following day (17th), in consequence of the retreat of the Prussians, we retired to the position of Waterloo—a position quite or nearly flat; wide open country, the whole one fine glacis. In the right centre a small wood with an old convent in it. This point was the most essential for the enemy to gain. Lord Wellington sent the Guards into it with orders to defend it *coute qui coute*.

"The evening of the 17th, the enemy rattled in our rearguard and gained some advantages over our cavalry.

"On the morning of the 18th at eleven o'clock, we perceived the enemy's columns collecting in our front for the attack; the dispositions were immediately made, and at twelve o'clock to a moment the first cannon shot was fired.

"An awful gun; for we perceived by the enemy's columns his attack was in earnest, and the annihilation of one army must be the consequence of the shock. Our guns, to the number of 100, were placed in the front line, the squares of infantry to support our artillery, and the cavalry at intervals to act as they could.

"The enemy began a cannonade from 250 pieces of artillery upon the main point of our position, endeavouring evidently to cannonade our columns into confusion, and then for their cavalry to take advantage of it. They attacked at the same moment the wood and convent, in order to debouch their columns from it when taken. The Guards foiled them in this attack. The cannonade continued and the butchery was terrible. But we obliged our columns to remain firm; luckily enough it was we did, for presently appeared their cavalry, coming forward in the most brilliant manner, and in a body charged our centre to the very muzzles of our guns. They rode through us, passed our artillery, galloped through our intervals of infantry, went through some of our cavalry, and then attempted to form in our rear. Nothing could be seen equal to the conduct of these troops. The whole of this first attack was made by Cuirassiers.

"A most extraordinary circumstance happened now. Our artillery (Donald Crauford amongst them), after having been rode over and passed by the French cavalry, ran up their guns, turned them, and actually fired at the enemy when in our lines. Our artillery performed wonders of steadiness. Notwithstanding this our squares of infantry remained untouched, our artillery was remanned, and again began the tremendous cannonade. A second body of the enemy's cavalry then repeated the bold attack, and did exactly the same. Then we saw their infantry advancing in one grand mass to attack us. They advanced, under terrible slaughter from our guns, and nearly gained our position. The day seemed to turn on a straw! Lord Hill threw forward one division of his corps, before in reserve; the enemy could make no impression; their cavalry charged our squares with the greatest determination. They could not break them. Our cavalry rode round but did not charge the enemy. There was a general row. The enemy was beaten back, or rather retired in very good order to his position.

"We now thought it all over; it was about four o'clock. The enemy had made great efforts, but our infantry had foiled him everywhere. We could not follow him; he had retired rather than been beaten back; his position was very strong. About six o'clock we perceived a formation of columns. Cavalry and infantry formed in a grand mass. The enemy's artillery was brought to a more forward position, and again he began to cannonade us. He opened a fire, the most tremendous ever known, I believe, in the annals of war—250 pieces, very close, throwing shells, round shot, grape and every instrument of destruction. It is really not exaggeration to say one could not ride quick over the ground for the bodies of men and horses.

"Under cover of this cannonade advanced Buonaparte at the head of his Imperial Guard; cavalry in a column on the left flank, the Grenadiers of the Guard on their right flank. They advanced quite steadily up to our line in one great mass. They halted and commenced firing. Our troops were literally mowed down. The fire was so great nothing could stand. Our guns were moved close up to the flank of this column, *foudroyant* with grape into it. Lord Hill moved a brigade (our *élite*) round the flank. I brought up six squadrons of cavalry, and we made a general charge. The Cuirassiers of the Imperial Guard had their ranks much thinned by our artillery. They went about—we with the cavalry pursued them, leaving the French infantry steady on our flank.

"Marshal Ney was with this cavalry, and I was within twenty paces of him. He was alone, with about six orderlies.

"The enemy ran down his guns (about twenty) and fired such a shower of grape, and the infantry then opened upon us. The French Cuirassiers came clean into us. I was on my old brown horse; a grape shot went through his body, and a round shot struck my hat at the same moment. He fell dead. I was a good deal stunned, and could not get from under him. The French Cuirassiers rode over me, knocked my hat off, but did not wound me.

"I lay there till the French were licked back; they again rode by me. One of their Cuirassiers was killed passing me; I seized his immense horse, and with some difficulty got upon him; I rode off, and hardly was I clear of them, before a round shot struck my horse on the head and killed him on the spot. An officer of the 13th Dragoons dismounted a man of his regiment and gave me his horse; this was shot in the leg about half an hour afterwards.

"The enemy were now beat back. Buonaparte had led his own Guard and been beaten! The Prussians now came up on the enemy's flank, and this obliged them to hurry their retreat. Our cavalry and artillery then advanced, the enemy were pursued, and began a precipitate retreat.

"The fruits of this victory were about 200 pieces of artillery taken in the pursuit, four Eagles, the baggage of Buonaparte and his army.

"It is needless to enlarge upon the conduct of the troops of the two armies. Lord Wellington called it "the battle of the giants," and he says there is no hell for any fellow that escaped that cannonade.

"I believe in the whole army Lord Wellington was the individual who fought hardest, and the only individual not touched. Currie was killed by a grape shot close to me. Lord Hill, in the grand *melee* with the Imperial Guard, had his horse killed and was rode over; we lost him for an hour, and I thought he must have been killed; I saw him at last, knocking along a French horse.

"Of Lord Wellington's staff were Colonels Canning and Gordon killed, Fitzroy Somerset lost an arm, Delancy badly wounded, etc.

"The admiration of the whole army was the gallantry and determination of the French cavalry and the steadiness of our infantry. Three times the cavalry rode clear through our lines, and hardly a man escaped; they would not surrender! At one moment the infantry of the two armies were all in squares. Never was such devotion witnessed as that of the French Cuirassiers. Our cavalry, with the exception of two brigades, never could face them. . . . .

"A grape shot passed over Lord Wellington's saddle and broke Lord Uxbridge's thigh. He is doing well.

"I could not help exclaiming (when the grand *melee* was going on) to Hill, 'By God, these fellows deserve Buonaparte: they fight so nobly for him;' in which everyone agreed. I had rather have fallen that day as a British infantryman or as a French Cuirassier, than die ten years hence in my bed! I did my best to be killed; Fortune protected me. I was struck by a ball on the side of my thigh, which did not bleed me; one also struck me on the back of the shoulder, which I did not know till after the action was over. Hill was not touched; his cloak was *crible de bals.* I rode over yesterday with Lord Wellington to see Blucher. We saw Buonaparte's carriage, his hat, cloak, coat and all his Orders taken in it. His hat fits me exactly; would that I had such a head under it!

"This is such an extraordinary day that that must be my excuse for being so tedious in my account.

"Lord Hill has begged Lord Wellington to give me my lieut.-colonelcy; he does not know whether he can; he is very well inclined. I am now not far from my company. What a slaughter in the Guards! You cannot read this; it is very late at night and I am much tired. Donald goes on very well.

"The French have retired to Laon, where we shall find them ready to give us another row. God bless you. I will send Crauford, if I can, the cuirass of one of those noble fellows.

"Louis le Desiré comes here to-day; he will meet with a bad reception I think."

> "On this ball revolving so rapidly in space,
> A grain of sand in midst of infinity—why this bloodshed?"

"Blessed be the Lord my strength, who teacheth my hands to war, and my fingers to fight,"

must be the answer. Thus the long war ended.

CHAPTER XXVII.

1815.

"He is fallen! We may now pause before that splendid prodigy which towered amongst us like some ancient ruin, whose frown terrified the glance its magnificence attracted.

"Grand, gloomy, and peculiar, he sat upon the throne a sceptred hermit, wrapt in the solitude of his own originality."

AH! what a change in the aspect of Paris since the days we lingered there to catch a glimpse of the great Napoleon, before and after his campaign in Egypt, 1799. Now, in 1815, there was an English army with the Duke and Duchess of Wellington established in a grand hotel.

Lord Hill,* with Prince Blucher and Sir Henry Hardinge, were in the Hotel Montesquieu; whilst Lord and Lady Combermere (Cotton) resided at Malmaison—(for poor Josephine had died there just a year before, when Paris—then as now—was filled with emperors and kings. Alexander of Russia, the Castlereaghs, and Madame la Maréchale Ney, were her frequent visitors. At last she

---

* Sir Robert Hill, wounded at Waterloo; Sir Noel; Colonel Clement; Lord Hill. This quartette—all brothers—sent a joint letter home to their father, Sir William Hill, after Vittoria, June 23rd, 1813: "We are all at this moment together in the same room, and in perfect health."

After Waterloo the four brothers met—Lord Hill and Sir Robert were in bed with wounds; Clement and Noel came to nurse them. Lord Hill had somewhat comfortable quarters, for he had been in the neighbourhood of Brussels for some time. The English Cabinet sent him out directly Napoleon was on the move, for fear of anything disastrous happening before the arrival of Wellington from Vienna.

died very suddenly from the effects of a cold and chill). Then the streets were filled with soldiers, who sauntered along as if in a garrison town at home. The 2nd 12th Regiment were then in Paris, for they had been sent from Ireland to join the army of Flanders, but arrived too late for Waterloo.

Of balls and dinners there were plenty, and the salons that were held were varied and eccentric. Amongst the latter may be mentioned Madame de Krudener, the prophetess of the Holy Alliance. She kept Alexander quite spellbound by her religious flattering ways—he was the White Angel and Napoleon the Black. To quote from Lady Morgan: "There were always two or three crowned heads in attendance: Alexander on one side, dressed to effect in black and diamonds; the King of Prussia, nowise remarkable except by contrast, on the other. On a low stool at the feet of the prophetess, sat her disciple Bergasse, and her high-priest Jung Stilling.

"'Ecoutez donc!' said Madame de H———, nudging me; 'est il artiste, notre Dénou, quelle groupe!'

"'Attendez, attendez!' said Dénou. In the midst of a solemn silence she rose, and extending her arms, exclaimed with a strange and penetrating tone, 'Prions!' Down on his knees went the Emperor of all the Russias, followed by everyone present—kings, aides-de-camp, and valets included.

"'And this,' said Ségur, starting up, 'was the grandson of my Great Catherine!'"

Then a Conference was held—a revival of the one so abruptly ended at Vienna, where Napoleon upset all their inkpots by setting foot in France again.

Next came the reviews, when in one stand alone were the Emperor of Russia and the three Grand Dukes, the Emperor of Austria, the Prince Royal and the Arch Duke, the King of Prussia and his two sons, Wellington, Hardinge, Schwartzenberg, and many others—and in front of all was massed a huge army, for it was a thanksgiving day.

At last came the giving back to Cæsar the things that were Cæsar's, which consisted of those lovely paintings which the French had pillaged when Napoleon was the conqueror. France had now to watch, with pangs of great regret, the English fatigue parties packing up these treasures, whilst our Royal Engineers removed the statues from their pedestals, and at last those splendid horses from the Place de Carrousel.

And whilst we are on "Art," it is pleasing to narrate that Napoleon had asked of Isabey to paint his little son. The abdication came before the work was done, but in the "Hundred Days," Isabey—to please the Emperor—went to Vienna, where the poor little Napoleon was residing with his grandfather, the Emperor of Austria. He took his unfinished drawing to complete if so able; but when admitted to the "Eaglet uncrowned," he found his fair locks cut off, his head powdered, pomatumed and covered with a cocked hat, and himself in the white uniform of an Austrian General, decorated with the Order of Maria Theresa. Isabey left. Next day the news came of Waterloo, and that Napoleon was a captive in England.

This poor boy, his son, died from life-weariness and consumption in 1832. His mother, Maria Louisa, quitted France for ever in 1813. The next we hear of her is as Countess Neipperg; for she never lived with Napoleon again, and died in 1847.

Then the day came for the trial of Marshal Ney, which for a time put an end to all gaieties.

What we have just read from Count Bourmont and Le Courbe (see page 238) was their evidence on the Court Martial held at the Palais de Justice, Paris, on November 9th, 1815. There in the crowded Court sat Lord and Lady Castlereagh, Prince Augustus of Prussia, and others of great distinction, to hear the trial of this great soldier. The President was Marshal Count Jourdan; Marshal Massena, Marshal Augereau, Duke Castiglione, the Duke of Trevise, Marshal Mortier, etc., were the members.

An incident at once occurred, for Massena* rose and said: "I cannot assist as Judge on the Prince of Moskwa. We have been soldiers together; when last we met our enmity was great—it arose on the battlefields of Spain. I must claim exemption." The Council of War deliberated, and decided that it was impossible that the slight resentment of a General of an Army could operate on the conscience of a Judge. The Marshal then took his seat, and the trial proceeded—and somewhat abruptly ended.

THE PRESIDENT: "Gentlemen, defenders of the accused, continue the defence by confining yourselves within the circle marked out for you."

MARSHAL NEY: "I forbid my counsel to speak any further; I forbid them to speak, unless they are permitted to employ all the means in their power."

[A profound silence reigned for a short time in the Chamber.]

M. BELLART (after a conference with the King's Ministers) rose: "We have a right, and it is our duty, to refute the captious means resorted to; but since the Marshal renounces all further defence, we renounce the right of reply. I now present the requisition upon which the Chamber shall retire to deliberate: 'The Commissioners of the King require of the Chamber of Peers, in consequence of the proofs furnished, to declare Marshal Ney guilty of having kept up with Buonaparte, intelligence calculated to second the progress of his arms on the French territory on the night of March 13th; of having furnished him succour in soldiers,'" etc., etc., etc.

After finishing the list of charges, the President asked the usual query: "Accused, have you anything to say on the application of the penalty?"

MARSHAL NEY (rising as if to bid "adieu!" to this august assembly, and with a firm voice): "Nothing at all, my Lord;" and withdrew.

---

* Massena died in 1817, buried at Pere La Chaise; only one word on his tomb—Massena.

It was now nine o'clock; the Marshal had dined, and was smoking a cigar. Whilst thus *en reverie* the Secretary of the Chamber arrived to read the sentence. He first expressed his pain, and offered condolence in a few pathetic words, when the Marshal interrupted and said: "Read! Read! To the fact, to the fact at once!"—and when his titles were being detailed, he observed: "What good can this do? Michael Ney—then a heap of dust—that is all."

On being told he was at liberty to take leave of his wife and children, he desired they should be written to come between six and seven in the morning. "I hope," he added, "that your letter will not announce to my wife that her husband is condemned. It is for me to inform her of my fate."

When this sad interview was over, the distressed Maréchale rushed off to the Tuileries to implore the King. Her interview took place at ten o'clock. Then only was she to learn that the poor Marshal no longer existed; for he, in the interval, had been warned that the Rector of St. Sulpice had arrived. To the priest he said: "It is always good to reconcile oneself to God. I have seen many battles, and every time I could, I confessed myself, and found myself always the better for it. We ought to die as honest men and good Christians."

At half-past eight, the Marshal having dressed himself in black breeches and stockings, blue frock coat and round hat—for he wished to die as simple Michael Ney—he was informed that all was ready. Then saying, "Get in first, M. le Curé; I shall be above sooner than you," he was driven to the Luxembourg Gardens, where, by the wall of the Observatory, was waiting for him a small picquet of veterans. Facing them, he removed his hat and said: "Comrades! straight to the heart, fire!" (December 7th, 1815).

Wellington had done his utmost to stay this execution; he approached the King not only once but thrice; it was the rudeness of the King that closed the last interview. Poor Marshal Ney, with Labedoyere and Lavalette (who met the same fate), remained in

Paris. Soult and others avoided France altogether for a time; they called themselves exiles. It was a pity Ney had not done the same.

The last ball given was one by Sir Edward Barnes, Adjutant-General of the Forces, and the first president or founder of the Army and Navy Club. Thus the stay in Paris ended.

The troops went home again, but to a rather chill and cold reception—for the country was tired of war, and now that the bands had ceased to play, the soldier was soon forgot. The 14th Regiment, on their return, went first to Dover Castle, then to Ramsgate to take shipping on to Ireland; then a sudden change of orders sent the 59th in their place, with the following result:—

"On the 26th of January of this year, the 'Seahorse' sailed from the Downs, having on board, instead of the 14th Regiment, the 59th, and a few days later was wrecked off Kinsale.

"The numbers on board, counting women and children, amounted to 394. Of these, 365 were drowned; among the saved was neither woman or child. The troops that relieved us at Deal met a like fate.

"The 'Lord Melville' and the 'Boadicea,' transports, sailed at the same time with the 'Seahorse.' Like their consort, they also were lost off Kinsale. The 'Lord Melville' saved all her crew but seven. Out of 280 in the 'Boadicea,' only 60 were saved."

In a short four years, the Centenary of this Waterloo battle will be kept. Let not all the money be wasted on an empty shout; but endow a University and give scholarships profusely, and let it be for all things military—like the Navy have Osborne, Dartmouth and then Greenwich—and make all War Ministers pass an examination in the principles of war.

From Waterloo backwards, England had practically been at war for sixty years; and it is interesting to note how this state of affairs had checked the flow of coinage. Practically very little silver or copper money was minted during those years; to remedy the inconvenience, gold pieces of the value of 5s. 3d. and 7s. were coined (an Ensign's pay was 5s. 3d. per diem). The scarcity in copper money was met by the issue of tokens—coined privately. The old Mint for same is now known as Tokenhouse Square, and used as an auction house for properties. The deficiency in silver money was met by a huge importation of Spanish dollars, which were earmarked by the King's head (the Goldsmiths' Hall plate mark) being placed across the neck or ear of Carolus IIII. This latter expedient caused some derision and confusion, and perhaps trouble; for in the *Morning Post* of June 3rd, 1797, is given the following:—

"A soldier is now under trial at the Horse Guards for giving as a toast—

'The only contrivance to make dollars pass:
The head of a king on the ear of an ass.'

The couplet, as above, is what the soldier gave in his defence; but the charge was substituting for 'king' the word 'fool.'"

In 1799, when the first Earl Spencer was First Lord, and "Under wise councils the British navy prospers," Britannia appeared for the first time with a Trident and a Man-of-War in the distance. Hitherto she had been represented with a Spear and seated on the trail of a gun.

CHAPTER XXVIII.

THE RETURN OF THE 12TH REGIMENT.

HOME.

" We leave
Our home in youth—no matter to what end—
Study, or strife, or pleasure, or what not;
And coming back in few short years we find
All as we left it outside: the old elms,
The house, the grass, gates, and latchet's self same click:
But lift that latchet—all is changed as doom."

—Bailey.

IN 1813, the 12th Regiment went from Mauritius to Bourbon, and there remained until the abdication of Napoleon in 1814—when Louis XVIII.* quitted his pleasant home at Hartwell, stayed a night or two in London, and crossed the Channel (escorted by an English squadron) to resume the stern realities of kingcraft.

The Isle of Bourbon was then restored to France by treaty (May 30th, at Paris). The ceremony of handing back the island partook more of a Fête to Bacchus than a military or political function. Four frigates, in the name of Louis XVIII., arrived with Bouvet de Lozier on board as Governor, and one Regiment. The ceremony of Relieving Guards took place; indiscriminate hospitality followed; and, thank goodness! the return of the 12th Regiment to Mauritius the next morning was performed without accident.

Time rolled on until early in 1817, when orders came for "Home." The Regiment embarked on three transports, touched at the Cape of Good Hope, and reached St. Helena (in October, 1817) —now the home of the greatest man the world had ever known.

---

* Louis XVI., Louis XVIII. and Charles X. were brothers.

A request was forwarded for the honour of an interview; and the Emperor Napoleon appointed the following day for the officers of the 12th Regiment to be presented to him. The Regiment, however, never had that honour—for when all were dressed in their best, and ready to land, a note from General Montholm was received, saying that the Emperor was too ill to receive them that morning. Captain Bayly took a walk to Longwood, and there saw the great man amusing himself at a billiard table.

Lord Amherst and Sir George Staunton (Bayly had spent some happy days with the latter at Canton in 1801) were also off the island, returning from a mission to China, which somehow had proved abortive—for the same reason as Lord Macartney gave: "They would not *kow-tow* to the Emperor."

The homeward voyage was now continued, and the 12th Regiment reached Portsmouth in November, 1817. Portsmouth, from July, 1796, to November, 1817, must have witnessed scenes beyond description. The "Victory," from having brought her Nelson home for burial, rode at anchor then as now. Wellington and other Generals, with troops fresh from Waterloo and Paris, had just passed through on their homeward trip. The thousand and one other goings and comings would also have cast their spell either for woe, interest or joy.

Even this wintry November day of 1817, which saw the return of the 12th Regiment after twenty-two years of active service in Hindostan, had also spanned the life from cradle to the grave of a bright young girl—the heir direct to England's throne. So Portsmouth looked most sore depressed; all flags were dipped half mast; and the wooden walls of England, with their yards *a cockatrice,** completed this woe-begone appearance.

* Our sailors, when they see a rope out of place, or showing a loose end, deem it to be only fit for the hanging of Judas; the canting of the yards gives an untidy appearance; hence the word *cockatrice* is used as a synonym, for "The *cock* shall not crow till thou hast denied Me *thrice*." The loose roping is imaginative of "He went out and hanged himself."

The Russians always put their yards *a cockatrice* on Good Friday. Other nations on shore cast adrift their broken china, in imitation of stoning Judas.

Included in this sorrow was the feeling that someone had blundered, and that the lives of Princess Charlotte and her babe might have been saved. But, alas! who can tell? Her dying words, perhaps, assert that maternity cases are not all treated thus: "'You make me drunk; pray leave me quiet; it affects my head.' So she spoke to her medical attendants, who had been administering brandy, hot wine and sal volatile."—Raikes to Wellington.

The now widowed Prince Leopold obtained what solace he could in the sympathy and companionship of his sister Victoria of Coburg, then a widow with two children. Kindly time, however, took these woes in hand and cured them, leaving just the scar to mark the wound; for Prince Edward, whom we have previously noticed at the capture of Martinique, St. Lucia and Guadaloupe, under Grey and Jarvis in 1794, shortly after wed the widow Victoria of Coburg, and became the father of our late Lady Queen Victoria—whilst Leopold remained to our Queen, throughout his life, a good and trusted uncle; even more, for at her father's death in 1820, he assumed a parent's care and love.

Prince Edward was quartered in Quebec (Louis Street) when the order came for him to join the above expedition to the West Indies in 1794; and Dr. Mountain, the newly-appointed Bishop to Quebec, was then seeking for a suitable residence; hence the opportunity came to secure 7, Louis Street, and to purchase what he wished from the outgoing tenant, Prince Edward.

This fact is only mentioned as an introduction to our roll of the Bishop's youngest son, Armine, destined in later years to become Adjutant-General in India, with higher prospects, when death cut short a most promising career. Young Mountain joined the 96th Foot at Newport, Isle of Wight, in November, 1815, at a time when most young officers thought they were out of it, and had arrived too late—for when would there be another Waterloo, with all its honours and promotions? But Mountain turned these quiet times to good account by obtaining leave to study. He passed one winter at Augsburg, where Queen Hortense, the wife of King Louis and mother of Napoleon III., held her Court or salon. He wrote in February, 1821:—

"On a bien le temps de s'appligner a Augsbourg c'est une belle ville, mais de-peuplée et par consequent un peu triste; cependant, je suis fort content de mon séjour ici. Le soir nous allons quelquefois au spectacle, et de temps en temps chez la Reine Hortense, à laguelle nous avons été présentés par Monsieur de Saxenhofen, qui nous a montré bien des amitiés. Sa suite est composée de deux dames d'honneur, du Gouveneur du jeune Prince, et de l'aumônier, et l'ou trouve chez elle le bon ton de la cour san etiquette ennuyeuse. C'est une dame remplie d'esprit que la reine, et jáime mieux encore aller chez elle quand il n'y a point d'etrangers, que les jendis, ou elle recoit les personnes distinguées de la ville."

As the above letter reads "February 21st, 1821," there is nothing to show that Queen Hortense was in grief or at all anxious about her brother—then bedridden at St. Helena, building castles in the air, dreaming of a new life in America, and at more lucid intervals drawing up his will and arranging all his letters. Then came the days of delirium until the morning of May 5th, when—a few incoherent words of "Armée France, tête d'armée," and

"Life's fitful fever's o'er."

Napoleon liked to dwell upon the recollections of his youth; and as May 5th approached, the ruling passion—strong in death—probably recalled the same anniversary, May 5th, 1789, when at Versailles he witnessed his King so virtuous, humane and so mild—but so powerless! He wanted the six Regiments of Cavalry or more —"L'armée France, tête d'armée," as he said—for to stay the execution.

"I enterèd and I saw him lie
Within the chamber all alone,
I drew near very solemnly
To dead Napoleon.
"But calm—most calm—was all his face,
A solemn smile was on his lips,
His eyes were closed in pensive grace—
A most serene eclipse!"

Travel and study had sufficiently engrafted in young Mountain the pleasure of self occupation, enabling him to say with Boileau

> " Je ne trouve point de fatigue si rude ;
> Que l'ennuyeux loisir d'un mortel sans étude ;
> Qui, jamais na sortant de sa stupidité,
> Soutiens, dans les langueurs de son oisiveté
> D'un lâche indolence esclave volontaire,
> Le penible fardeau de n'avoir rien à faire."

It is said of a Dr. Mountain—perhaps one of an earlier generation—that, when seeking for preferment in the Church, he took for his text, on preaching before his Sovereign: " If thou hadst faith as a grain of mustard seed, and said unto this Mountain, 'Be thou removed and cast into the *see*,' it should be done."

It now came to young Mountain, like his ancestor, to bestir himself and seek promotion. At last he obtained a Lieutenancy in the 52nd Light Infantry, a regiment then in Canada, and commanded by Sir John Colborne—of Peninsula and Waterloo renown, and destined for greater things in the Canadian Rebellion of 1837, when his firmness and promptness greatly saved the situation. For this service he was created a peer (Baron Seaton), and became Governor-General of Canada on the retirement of Lord Durham. Lord Dalhousie also held a command in Canada. We first hear of him as a young Colonel commanding the 26th at Vinegar Hill; then in the Peninsula; and now shortly to be Commander-in-Chief in India.

In the spring of 1826, Mountain (now a Captain) was posted to the 76th Regiment in Jersey, where Sir Colin Halkett* was Governor; his late Colonel, Sir John Colborne, was Governor of Guernsey. The former was also a distinguished Waterloo officer, and the next nominee for the appointment of Commander-in-Chief in Bombay. The 12th Regiment was also serving in the islands at the time. From arriving at Portsmouth in November, 1817, they lay for three weeks through stress of weather off the Mother Bank before proceeding to Cork; then after the usual tour of Irish service,

* He died Governor of Chelsea Hospital, 1856.

they changed quarters to Manchester and Macclesfield; then again to Portsmouth for embarkation to the lesser Isles of Jersey and Guernsey.

While stationed at these islands, the appearance of the Regiment, the conduct of the men, and the excellent system of interior economy which existed in the corps, elicited the highest commendations of General Sir Colin Halkett; and the conduct of the four Companies at Guernsey, under Major Bayly, was also specially commended by the Lieutenant-Governor, Colonel Sir John Colborne, K.C.B. (Inspection Reports).

It is now time to congratulate kind Major Bayly on his promotion to Lieutenant-Colonel, and as such he embarked with his Regiment for the Rock of Gibraltar, the scene of their former triumphs. This was to be his last station

"E'er he ceases from his labour, and takes his rest."

Gibraltar was then in the hands of a fine old soldier, Sir George Don—well into his eightieth year, but full of hard work and common sense, as we shall see when the terrible epidemic of plague and sleeping sickness overtook them in the following year. Shortly after the arrival of the Regiment, the opportunity came for them to receive new colours, which Sir George presented with the following address, speaking straight from the heart:—

"It appears by the record of the 12th Regiment, to which I have the honour of presenting these colours, that among the many valiant deeds of the corps, it achieved distinguished glory at the battle of Minden. In 1797 I attended the renowned Duke of Brunswick on the spot where this battle was fought; after His Serene Highness had shown me the position occupied by the British, he said, 'It was here the conflict was most obstinate, and it was here that the British Infantry gained immortal glory.' This Rock and Seringapatam were afterwards among the scenes where the 12th Regiment distinguished itself, and which are immortalised in the history of our country. Being myself a soldier of fifty-seven years' standing, I am alive to every instance of meritorious conduct in my brother soldiers, and it is extremely gratifying to me to reflect

The Mysterious Lady Doctor.

To face page 259.

that the 12th Regiment, which so early established its fame, has continued to augment it on every occasion; and I am confident that whenever these colours shall be displayed before an enemy, the regiment will, by its valiant conduct, add to the number of glorious records with which they are graced."

Now came this terrible year of sickness, when not a family escaped; and in six weeks 4000 bodies had been interred. Six regiments were then serving in the garrison, one of which numerically was swept away, *id est*, 800 men perished, including officers, from the different corps.

At the height of this distress, a transport arrived from England laden with medical men and comforts. Amongst the arrivals were Drs. Pym and Barry, who vaunted much of their abilities in arresting the progress of this disease.

Dr. Barry was of brusque manner, very reserved, and was known to have fought a duel. At this time he was thirty-four years of age, an M.D. and a Surgeon-Major in the Army, and most skilful as a physician. A certain effeminacy of manner he tried hard to overcome. This brusqueness of manner is best explained by the following incident:—It was at Malta, when accompanying the General round the hospital, that a young assistant surgeon told the General that the man he was speaking to had heart disease, and might die at any moment! Dr. Barry, fearing the man would die of fright, at once exclaimed: "It's a d——d lie! There's nothing serious the matter with him."

Dr. Barry was promoted Deputy-Inspector-General in 1851, Inspector-General in 1858, retired in 1859, and died at 14, Margaret Street on July 25th, 1865 (born 1795)—when an official report was sent to the War Office that Dr. James Barry was a woman. She joined the Army as a Hospital Assistant in 1813, and an early love attachment for a young military surgeon determined her to follow his career. This could only be done by adopting the same calling. This "lady" doctor was well known to the Governor of Malta, General Sir F. Ponsonby, and to his wife Lady Emily; it is from heir son, Arthur Vallette Ponsonby, that I have heard many of Dr. Barry's quaint sayings and doings.

The saying is that at forty years of age the spirit of venture dies out, and the love of comfort creeps in. Lieutenant-Colonel Richard Bayly was now over fifty—thirty-four years of which had been spent in hard and strenuous service—and his son, Edgar Bayly, is now well up the "Lieutenants."

The epidemic at Gibraltar was over, and the Governor, Sir George Don, had expressed himself as most highly pleased with Colonel Bayly and his battalion—so the 30th March, 1830, witnessed the Ensign Bayly of 1796 taking passage in a Dutch brig for home.

"The last link is broken
That bound me to thee;"

but he remembered unto his dying day the "Auld Lang Syne" from the band, and the farewell cheers of his men.

Colonel Bayly was a soldier, a savant, and of great humanity, and whilst *en rétraite*, no one could say he had *n'avoir rien à faire.* He died at Boulogne on October 8th, 1851.

Thus we sever for a time our connection with the 12th Regiment, only to resume it a little later—for our Empire extends to where the sun never sets, and wars never cease.

Sir George Don remained on as Governor, until his death in 1832. He could read the smallest print without spectacles, and a more active, indefatigable old gentleman never lived; in fact, Gibraltar owes all the beauties it possesses to his great ingenuity and zeal for improvement.

CHAPTER XXIX.

> "As in a theatre, the eyes of men,
> After a well-graced actor leaves the stage,
> Are idly bent on him that enters next,
> Thinking his prattle to be tedious."

THUS Napoleon had played his many parts—he made the world his stage—his exit now was final. But there still remained the other actors, who richly deserved applause, for they kept the properties and arranged the cues, whereby each one was warned of his approaching turn—in fact, "to ride the whirlwind and direct the storm" was what they had to do; and it is to this grand triumvirate that India owes so much—Malcolm, Elphinstone and Metcalfe.

Sir John Malcolm (1769 - 1833) landed in Madras aged 13 years—one of 17 children—studied hard at Persian, and became Interpreter to Lord Cornwallis, the Governor-General. He then went to the staff of General Harris (where we have already met him at Seringapatam, 1799). Afterwards, as Resident at Hyderabad, he induced the Nizam to part with all his French troops, and also did similar work at the courts of Scindhia and Holkar. He crowned his good work at Teheran and Herat by securing a Persian Alliance in case of trouble with France—a man of untiring energy. He was Governor of Bombay from 1827-30. (Lord William Bentinck, Governor-General; Earl Dalhousie, Commander-in-Chief.)

The Hon. Mountstuart Elphinstone (1779-1859) reached India aged 16, and at 20 was British Resident at Poona at the Court of the Mahratta Princes. He here assisted Arthur Wellesley in diplomacy; when diplomacy failed he became Political A.D.C. to Wellesley. He fought at Assaye. In 1808 he was made head of the most important Mission to Central Asia, becoming Envoy to the Court of Cabul, and securing their alliance in case of a French invasion; and with this good work he will always be remembered. From 1820-27 he was Lieutenant-Governor of Bombay (when Sir John Malcolm succeeded him), and was many times offered the blue ribbon of the Governor-Generalship. This he declined, being fond of travel and literature. He settled at Limpsfield in Surrey. His nephew, Lord Elphinstone, was also Governor of Bombay. (He died in 1860).

Lord Metcalfe (1785-1846) reached India aged 15, and at 25 was selected by Lord Minto, the Governor-General, as Envoy to Ranjit Singh at Lahore (died June 27th, 1839). He ran up the ladder quickly, for in 1835 he succeeded Lord William Bentinck as Governor-General *pro tem.* He left India finally in 1838, and shortly became Governor of Jamaica; then Governor-General of Canada; for which services he received a peerage. The Queen herself wrote on the application sent in by Sir Robert Peel: "He has shown such a desire to do his duty in the midst of so many difficulties, and such extreme disinterestedness, that he richly deserves this mark of the Queen's entire approbation and favour."

Clive died in 1776; so when young Malcolm first saw Madras in 1782, the name of Clive must have been as Lord Curzon, our late Viceroy, *now* wishes it to be, *i.e.*, well known to everyone. Thanks are due to Lord Curzon that the deeds of this great man are to be rescued from oblivion and given to posterity, by the erection of statues in Whitehall and Calcutta.

Malcolm would surely have met and enjoyed the society of Warren Hastings, Mr. Shore (Lord Teignmouth), Lord Macartney* and perhaps Sir William Jones, the High Court Judge at Calcutta (and the greatest linguist and scholar the world has ever known;

*Lord Macartney was Governor of Madras (1780); became our First Envoy to China (1792); afterwards Governor of Cape (1798).

one work alone will prove his great calibre—the translation from Persian into French of the Life of Nadir Shah).*

Warren Hastings and Mr. Shore left for England together in February, 1785, when Sir William Jones noted that "he felt their loss immensely." Mr. John Macpherson took up the reins *pro tem.* as Governor-General; he likewise was a man of letters.

In September, 1786, Mr. Shore returned, and with him arrived Lord Cornwallis, having in his pocket the appointment of Governor-General.

On his arrival in England, Warren Hastings was received at Court and everywhere else with immense favour; and it was thought that rewards and honours—a Peerage—would surely be conferred upon him. But bye-and-bye, instead of recompense, there arose a murmur about punishment; that celebrated indictment was prepared; and then began that historic trial, which afforded Burke, Fox and Sheridan an opportunity for the display of their eloquence, but which in every other respect was wholly fruitless. After seven years' trial, in April, 1795, his acquittal was pronounced by a huge majority. Admiral Lord Hood spoke on the occasion as follows:—

"His Lordship, in a solemn manner, called the serious attention of the House to the consequences of proceeding, with too scrupulous a nicety, to investigate the conduct of those who had filled stations of high difficulty and important trust. Certain actions, which appeared to those at a distance in a very criminal light, were yet on a nearer investigation perfectly justifiable on the grounds of absolute and indispensable necessity. Should the fear of an Impeachment by Parliament, said his Lordship, be hung out to every commander in whose hands was placed the defence of our national possessions, it must necessarily operate as a dangerous restraint to their exertions; when it was considered that no General or Admiral had

*Nadir Shah, a great conqueror and powerful monarch, was born in 1688, near Meshid—of no origin—and Persia being in want of an heir to the throne, he was elected king. After this, he made his celebrated expedition to India, conquered the Great Mogul, occupied Delhi, and returned home laden with countless treasures. Once adored by his soldiery, he lived to become detested; his generals rushed upon him, and killed him, 19th June, 1747.

scarcely ever been fortunate enough to conduct himself, in the performance of his duty, without occasionally falling into circumstances in which the public service compelled him to do things in themselves not pleasing to his feelings, nor strictly legal, but, from the indispensable necessities of their situation, perfectly justifiable. The example set by the House of Commons, in the present instance, would for ever stand before our future commanders, and create a great and dangerous clog to the public service."

Lord Cornwallis remained in power until 1793, and became known as a Governor of many virtues. Malcolm was his interpreter, and the services of Sir John Jones as a *vade mecum* in every tongue were always available—his digest of Hindoo and Mahommedan Law was a *chef d'œuvre.* Sir John parcelled out his day as under:

"Seven hours to law, to soothing slumber seven,
Ten to the world allot, and all to Heaven."

And when death overtook him in the following year, the Europeans, the natives, especially the pundits of the East, all deemed his loss irreparable. He was the first, by his tact and taste, to popularise in Europe the literature of the East, and to bridge over the chasm which separated the mind of England from that of her great Indian Dependency.

The following is his epitaph, as written by himself :—

An Epitaph.

Here was deposited,
the mortal part of a man,
who feared God, but not death ;
and maintained independence,
but sought not riches ;
who thought
none below him but the base and unjust,
none above him but the wise and virtuous ;
who loved
his parents, kindred, friends, country,
with an ardour
which was the chief source of
all his pleasures and all his pains ;

and who, having devoted
his life to their service,
and to
the improvement of his mind,
resigned it calmly,
Giving Glory to his Creator,
Wishing peace on Earth,
And with
Good-will to all creatures,
On the (Twenty-seventh) day of (April)
in the year of our blessed Redeemer,
One Thousand Seven Hundred (and Ninety-four).

and in his Bible was written: "I have regularly and attentively read these Holy Scriptures, and am of opinion that this volume, independently of its divine origin, contains more sublimity and beauty, more pure morality, more important history, and finer strains of poetry and eloquence, than can be collected from all other books, in whatever age or language they may have been composed."

Sir John Shore relieved Cornwallis, and after him came Wellesley in 1798; then, on the departure of the latter in 1805, Cornwallis again appeared for the third time as the great Pro-Consul—so soon, however, to surrender all into Higher Hands, for he died the same year at Ghazeepore.

This vacancy, so suddenly made, was filled by Sir George Barlow, pending the arrival of Lord Minto (1806-13).

Then came the Marquis of Hastings, who ruled for ten years, and acquired for England, after most stubborn resistance, Nepaul, and that splendid warrior race the Ghurkas; next, Lord Amherst, nephew of Geoffrey the first baron (fellow Brigadier with Wolfe in Canada); and here we pause for a time to welcome Lord William Bentinck as Governor-General in 1828, and Earl Dalhousie as Commander-in-Chief. Their past careers we know to have been brilliant—Bentinck, at twenty years of age, was Governor of Madras, then was with Wellington at Vimiera, with Moore at Corunna, and again with Wellington in the Peninsula; whilst Dalhousie, from a young colonel at Vinegar Hill, had been steadily climbing under Wellington up to his present post as Commander-in-Chief of India.

It was in September, 1830, that Major Mountain was invited by these two noble lords to visit them at Simla. He wrote: "I met with a most kind and warm reception from Lord and Lady Dalhousie. The latter looks quite well, and is as active as when I first saw her; the former is much recovered. . . . He wishes me to go down with him to the Plains."

It is worth seeing, the camp of the Commander-in-Chief—in all about 5000 souls—with elephants, camels, horses, oxen, cows, goats innumerable; then the tents vary in size from 50ft. long by 30ft. high by 20ft. wide to the simple pâl.

The Governor-General was to start about the same time to meet Runjeet Singh, the Sovereign or Mehemet Ali of the Punjaub, at some trysting-place on the Sutlej.

It was on this occasion that one of our officers had been showing to this astute old Lion of the Punjaub—Runjeet Singh—a map of India, and was explaining to him that the blue, the yellow, the green, and the brown represented the independent Native States, while the red was the British possessions.

After looking at it for a while, the far-seeing old king exclaimed, "Sub lal hojaega" ("All will become red.")

The officer tried to explain that it was not our wish to acquire more territory.

"That may be quite true," said old Runjeet Singh, "but a conqueror can never stop in his career. He is forced along by circumstances beyond his control, and cannot help himself—it is fate. No! no! Tumaum lal hojaega" ("The whole will become red.") [Certainly his own country the Punjaub, Oudh Sindh and Burmah *have* all become red.]

This remark requires a slight qualification, for Runjeet Singh also was a conqueror, and his boundary was the Sutlej; but when he commenced to spread, and wished the Jumna for his limit, Lord Minto sent Metcalfe to explain that, by the Treaty of 1809, he had very much exceeded, and back to the Sutlej he must go. He then became so impressed with the manners and discipline of the Englishmen, that he ever afterwards made a study of our ways, and remained at peace till he died in 1839. By his words "Sub lal hojaega," he doubtless meant that our good example paved the way to conquest, for there are victories in peace as well as victories in war.

Lord and Lady Dalhousie halted first at Meerut, where the Commander-in-Chief inspected his own regiment (the Cameronians). Here Mountain had the offer of the post of Military Secretary to Sir Colin Halkett, Commander-in-Chief of Bombay—which post he filled for two years, when a transfer was arranged by which his services passed to the Governor-General, Lord William Bentinck, then at Ootacamund, Madras (1834). From the latter place he wrote :—

"I look forward with pleasure to our return to the City of Palaces at the best season of the year, to Lady William – the kindest of the kind, a more mixed society, a more varied life. You may, however, be surprised that I have seen nothing of Lord William. He is, of course, so overwhelmed with business, that we see but little of him. He is indeed worth the whole lot of us ten times told. So much force has seldom been lodged in hands so pure. His views are all truly benevolent and philanthropic, and though he may have been sometimes mistaken (as who is not ?), and has incurred much unpopularity by measures of necessary economy, of which he has been the instrument, he is a man *comme il y en a peu*, and who deserves to be loved. The happiest moments I have spent here have been in his company—the most instructive hours in his employ."

CHAPTER XXX.

> " When I was young I said to sorrow
> ' Come, and I will play with thee.'
> He is near me now all day,
> And at night returns to say,
> ' I will come again to-morrow ;
> I will come and stay with thee.' "

LORD WILLIAM BENTINCK left India for England in the following year (1835), and Mountain travelled with the late Viceroy and Lady Bentinck home. To be in their society was delightful, more especially now that they were off duty.

Lord William died on June 17th, 1839. On the statue to his memory in Calcutta, are these words by Macaulay :—

" During seven years (1828-1835) he ruled India with eminent prudence, integrity, and benevolence, never forgetting that the end of government is the happiness of the governed. He abolished cruel rites ; he effaced humiliating distinctions ; he gave liberty to the expression of public opinion ; and his constant study was to elevate the intellectual and moral character of the nations committed to his charge."

A more amiable and excellent woman than Lady William never existed. She was overflowing with affection, sympathies, and kindness ; and when, on May 16th, 1843, this dear lady was committed to the grave, there was no hired pomp or business-like show, but in the plainest manner suitable to the simplicity and nobility of her character, she was laid to rest.

In more recent years (1857 to 1904), England possessed, in Sir Henry Norman (the young Captain and Adjutant-General of 1857), one as unassuming, agreeable, intelligent, and cultured, as perhaps was Lord William. Their careers seemed similar: Soldier, Statesman, Governor, and then the blue ribbon, viz., the Viceroyship of India—for at the age of sixty-seven, Norman was offered and accepted that important post. However, he was warned that the wear and strain, on what he had gone through before, might prove too much; and he had the courage to withdraw. He died a Field-Marshal in the British Army, aged seventy-six.

The senior Member of the Council, Sir Charles Metcalfe—the last of the Triumvirate—succeeded Lord William Bentinck provisionally, but not for the full term. Party exigencies led to the appointment of Lord Auckland, and from that date commenced a new era of war. One of his early acts was the upsetting of Dost Mahomed and the placing of Shah Shuja on the throne of Cabul. An act of such high politics required forethought, attention, and some gifted imagination, neither of which he had.

Dost Mahomed was a man of great ability and energy, and honesty of purpose—beloved by his people. Certainly to the English he behaved like a gentleman. So did his son Akbar Khan, for I have heard more than one hostage acknowledge this. Call them usurpers if you wish, but the descendants of Ahmed Shah, who founded the Empire of Afghan in 1773, had through inefficiency been dispersed. This accounts for Shah Shuja being in India *en recluse*. This is the man Lord Auckland now intended to re-instate, when all the people of Afghan hated him; besides, Lord Auckland's policy was forcibly opposed by the man on the spot, viz., Sir Alexander Burnes.

Russia undoubtedly was scheming, as the reports of Rawlinson will show, but this gave no excuse for ousting the Dost Mahomed who could be trusted, for a man who was an imbecile, or certainly a fool.

A PORTFOLIO.

(To explain the Cabul War of 1839-42.)

SEPTEMBER 20TH, 1837.—

Captain Alexander Burnes ("Bokhara Burnes," as he was known in London Society, from his travels in Central Asia) arrived at Cabul as the head of the Commercial Mission sent thither by the Governor-General, Lord Auckland. He was received with great honour by Akbar Khan, and conducted to the Court of his father Dost Mahomed.

At a Conference on the 24th, the Ameer said: "Instead of renewing the conflict with Runjeet Singh, it would be a source of real gratification to me if the British Government would counsel me how to act—in fact, to make peace—none of our other neighbours can avail me; and in return, I would pledge myself to forward its commercial and political views."

Burnes thereupon assured him that the British Government would exert itself to secure peace between the Punjaub and Afghanistan; and added that, although he could not hold out any promise of interference for the restoration of Peshawur, which had been won and preserved by the sword, he believed that the Maharajah intended to make some change in its management—but that it sprang from himself and not from the British Government.

OCTOBER 15TH, 1837.—

Major Rawlinson encountered the Russian agent, Vickovich, journeying with presents to the camp of Mahomed Shah, at Herat.

"I tracked them," he wrote, "for some distance along the high road, and then found that they had turned off to a gorge in the hills. Then at length I came upon the group, seated at breakfast by the side of a clear sparkling rivulet. The officer, for such he evidently was, was a young man of slight make, very fair complexion, with bright eyes, and a look of great animation. He rose and bowed to me as I rode up, but said nothing. I addressed him in French—the general language of communication between Europeans in the East—but he shook his head. I then spoke English, and he answered in Russian. When I tried Persian, he seemed not to understand a word; at length he expressed himself hesitatingly in Turcoman or Uzbeg Turkish. I knew just sufficient of this language to carry on a conversation, but not enough to be inquisitive. This was evidently what my friend wanted, for when he found I was not strong enough in Jaghatai to proceed very rapidly, he rattled on with his rough Turkish as glibly as possible. All I could find out was that he was a *bona fide* Russian officer, carrying presents from the Emperor to Mahomed Shah. More he would not admit; so, after smoking another pipe with him, I remounted, and reached the royal camp beyond Nishapoor before dark. I had an immediate audience of the Shah; and in the course of common conversation, mentioning to his Majesty my adventure of the morning, he replied, 'Bringing presents to me! Why, I have nothing to do with him; he is sent direct from the Emperor to Dost Mahomed of Cabul, and I am merely asked to help him on his journey.'"

Major Rawlinson at once returned to Teheran to acquaint Mr. M'Neill with the communication which was going on between St. Petersburg and Cabul.

OCTOBER 29TH, 1837.—

From Cabul, Burnes wrote to Captain Jacob: "With war come intrigues, and I have had the good fortune to find out all the doings of the Czar and his emissaries here, where they have sent letters and presents. After proving this, I plainly asked the Governor-General if such things were to be allowed, and I got a reply a week ago altering all my instructions, giving me power to go on to Herat, and anywhere, indeed, I could do good. The first exercise of this authority has been to despatch a messenger to Candahar, to tell them to discontinue their intercourse with Persia and Russia, on pain of displeasure; and not before it was time, for a son of the chief of the city, with presents for the Russian Ambassador, is ready to set out for Teheran."

Next day he wrote: "Here a hundred things are passing of the highest interest. Dost Mahomed Khan has fallen into all our views, and in so doing has either thought for himself or followed my counsel, but for doing the former I give him every credit; and things now stand so that I think we are on the threshold of a negotiation with King Runjeet, the basis of which will be his withdrawal from Peshawur, and a Barukzye receiving it as tributary from Lahore. What say you to this, after all that has been urged by Dost Mahomed Khan's putting forth extravagant pretensions? Runjeet will accede to the plan, I am certain. I have on behalf of Government agreed to stand as mediator with the parties, and Dost Mahomed has cut asunder all his connection with Russia and Persia, and refused to receive the Ambassador from the Shah now at Candahar."

NOVEMBER 23RD, 1837.—

The Persians, urged on, as was believed at the time, by Russia, renewed the siege of the Afghan city of Herat.

DECEMBER 19TH, 1837.—

The Russian agent, Vickovich, entered Cabul with credentials from Count Simonich at Teheran, and a letter (though its authenticity was disputed) purporting to be from the Emperor himself. He was at first coldly received by Dost Mahomed, but as the prospect of aid from the British Government gradually grew fainter, the Russian rose in favour, and was latterly paraded in public through the streets of Cabul.

DECEMBER 22ND, 1837.—

Involvement of the British Government with the Candahar chiefs. Writing to a friend, Burnes recorded that they had gone over to Persia. "I have detached them, and offered them British protection and cash if they would recede, and if Persia attacked them. I have no authority to do so; but am I to stand by and see us ruined at Candahar, when the Government tell me an attack on Herat would be most unpalatable? Herat has been besieged fifty days, and if the Persians move on Candahar, I am off there with the Ameer and his forces, and mean to pay the piper myself."

The Governor-General was at this time on his way to Simla, and caused Secretary Macnaghten to write to Burnes: "It is with great pain that his lordship must next proceed to advert to the subject of the promises which you have held out to the chiefs of Candahar. Those promises were entirely unauthorised by any part of your instructions. They are most unnecessarily made in unqualified terms, and they would, if supported, commit the Government upon the gravest questions of general policy. His lordship is compelled, therefore, decidedly to disapprove of them. He is only withheld from a direct disavowal of these engagements to the chiefs of Candahar, because such disavowal would carry with it the declaration of a difference between you and your Government, and might weaken your personal influence; and because events might in this instance have occurred which would render such a course unnecessary."

December 30th, 1837.—

From Cabul, Burnes wrote: "The present position of the British Government at this capital appears to me a most gratifying proof of the estimation in which it is held by the Afghan nation. Russia has come forward with offers which are certainly substantial. Persia has been lavish in her promises, and Bokhara and other States have not been backward. Yet in all that has passed, or is daily transpiring, the chief of Cabul declared that he prefers the sympathy and friendly offices of the British to all these offers, however alluring they may seem, from Persia or from the Emperor; which certainly places his good sense in a light more than prominent, and, in my humble judgment, proves that by an earlier attention to these countries, we might have escaped the whole of these intrigues, and held long since a stable influence in Cabul."

March 21st, 1838.—

The Ameer Dost Mahomed made a final and unsuccessful appeal to Lord Auckland to remedy the grievances of the Afghans, and to give them a little encouragement and power. Burnes soon after took his departure from the city. He was naturally disgusted at being unable to impress his superiors with his belief in the honesty of Dost Mahomed.

June 3rd, 1838.—

Rupture between Great Britain and Persia. For the purpose, apparently, of lessening British influence at Herat, the Ambassador at the Shah's camp was treated with studied disrespect, and some members of the Mission directly insulted.

To-day Mr. M'Neill addressed a letter to the Foreign Minister at the Persian camp, announcing his intention to depart for the frontier on the following day. "I feel myself called upon," he concluded, "to inform you that, until the reparation and satisfaction I have demanded for the indignities already offered shall have been fully given, the Queen of England cannot receive at her Court any Minister who may be sent thither by the Shah of Persia."

June 26th, 1838.—

Lord Auckland involved Great Britain in the politics of Afghanistan—a treaty of alliance and friendship being executed this day between Maharajah Runjeet Singh of Lahore, and the exiled ruler of Afghanistan, Shah Soojah-ool-Moolk, "with the approbation of, and in concert with, the British Government."

The main design of the treaty was the destruction of the power of the Barukzye Sirdars, as represented by Dost Mahomed, and the replacing of the old King on the throne. It was also provided that Shah Soojah's rights over Sindh and Shikarpoor should be arbitrated and adjusted by the British Government.

August 11th, 1838.—

Colonel Stoddart arrived in the Persian camp with an ultimatum that the Shah must quit Herat or fight Great Britain. The Shah undertook to raise the siege, and called off his troops in the direction of Teheran early in the following month.

Towards the close of the year, Stoddart was despatched by Sir J. M'Neill on a Mission to the Ameer of Bokhara, from which he never returned. He was at first received favourably ; but the tyrant Ameer, taking offence at his letter to the Queen being answered by the Governor-General of India, afterwards treated the Mission with disrespect and cruelty.

When the news of our fatal disasters at Cabul reached this demon of Bokhara, both Stoddart and Arthur Connolly were at once subjected to a fresh outbreak of villainy on the part of the Ameer, by being immured in a dungeon until they died.

Colonel Stoddart and Connolly were both able, patriotic and noble-minded men.

JUNE 27TH, 1839.—

Died at Lahore, in his sixtieth year—Maharajah Runjeet Singh, chief of Lahore. Four princesses (his wives) and seven slave girls were permitted to burn themselves on his funeral pyre. He left the celebrated diamond "Koh-i-noor,"* so long coveted by the Princes of India, as a legacy to be worn by the chief idol of Juggernaut.

On April 25th, 1839, the Bengal Army marched and took Candahar, and soon after entered Ghuznee, Jellalabad and Cabul. At this latter place Dost Mahomed made a grand resistance. Again, one year later, on November 2nd, 1840, Dost Mahomed reappeared and fairly beat the English and their allies at Purwandurrah. With this success he reappeared next day and tendered his submission. His sword was returned, and a residence and revenue provided for him in India. Exactly one year later (on November 2nd, 1841) the insurrection broke out.

---

* Only six very large diamonds, called "paragons," are known in the world: the "Koh-i-noor" and the "Pitt" are of most interest. The latter contributed much to the fortunes of the Pitt family—belonging to Governor Pitt of Madras, father of Chatham and W. Pitt.

The "Koh-i-noor" has a wonderful history, which if true ought to have told Auckland that Runjeet Singh was a bit of a bandit, for he obtained it in a most audacious way from the Khan of Cabul. In 1640, it was found in the mines of Golconda, and handed over to the Grand Mogul Shah Jehan. In 1850, the East India Company gave it to the Queen.

Honours to Sir John Keane and others were given for this first act of the drama. One of the Court of Directors objected, for he quoted the opinion of Lord Wellington, "That they should wait till they saw the troops safely out of Afghanistan." Another remark of Wellington's was: "To employ foreign troops (as the troops sent up with Shah Shujah were) to collect tribute and revenue, would mean, a little later, animosity and hostility." And so it did.

To fully understand the pluck, the endurance, and the terrible anguish of the ladies in this campaign, the personal narrative of one is partly given. From her own lips I have often heard it:—

"Miss Griffiths (so soon to be a prisoner and hostage) when at the age of fourteen, left England to join her parents, Major and Mrs. Griffiths, in India, and at sixteen married Lieutenant Robert Waller, Royal Horse Artillery. Shortly after his battery was ordered to Afghanistan; to join her husband, even if the countries were at peace, would have been no simple task. It was in the early days of the occupation of Cabul that a few ladies and their children were allowed to join their husbands. Mrs. Waller, then not eighteen, and with a daughter to her name, set out on this long journey in March, 1841.

"'You will know long ere this letter reaches you that I had started from Kurnaull to join my Robert, and you will, I am sure, be glad to hear I have got over my journey very well, and we met a few miles from this place (Jellalabad) after four long and tedious months of separation. Robert's brigade being ordered to remain here for the winter, he built a very snug little hut for me, which we now inhabit, and if it were not for the uncertainty of how long we may be together we should be very happy: for, according to my notions, it is far preferable to be with my husband even in a mud hut than to be in comfort without him. We expect to move towards Cabul about the middle of next month, and whether we remain there is still undecided; much depends on the state of the country round Cabul. Some people think there will be some work for the troops in the summer, but I trust my dear Robert will not be engaged, though my heart sometimes fails me when I think that ours, being the only troop of horse artillery in this part of the country, will certainly be called into action if there is anything to be done. My earnest prayers are offered to the Throne of Grace for his safety, and I trust our merciful Father will ever watch over and guard him from every danger.

"'The climate of this country is very fine and healthy, and far preferable to Hindoostan; the scenery very wild, and I should imagine beautiful when the trees are covered with foliage. At present all around wears a cold and dreary aspect—all the mountain tops covered with snow, though we have not had any in the valley where we are.

"'My poor mamma was very nervous at the idea of my coming up to Afghanistan alone with my baby, particularly, too, as I had no experience in managing a little one in case of sickness; but we enjoyed perfect health the whole time, and reached this after a journey of three months. I shall now soon have the heartfelt pleasure and gratification of making Robert and my dear father known to each other, for although they always correspond, still they have never been together to know each other's characters and dispositions. When this is accomplished I shall feel perfectly happy and contented.

"'Now I must tell you of my sweet little girl—as I know anything belonging to Robert will be interesting to you. She is now four months old, and such a dear little thing; so good-tempered, and the very image of her father. Robert is dotingly fond of his dear little treasure.'"

Time now passes to January 6th, 1842, when Mrs. Waller, with the other ladies, who had not tasted a meal or been under shelter for some time, set out on this death march. Their servants had nearly all deserted or been killed. Some had infants a few days old at the breast, and were unable to stand without assistance; others were advanced in pregnancy and could scarce walk across a room—yet these helpless women, with their young families, were compelled to rough it on the backs of camels or on the tops of the baggage yaboos (ponies).

A second child was born to Mrs. Waller in this distressing time, and her life would most assuredly have been taken, for she was left behind on the day of her confinement, hidden in a mud enclosure, amidst the wild Afridis. At last some friendly native persuaded Akbar Khan to arrange for her not to be deserted.

CHAPTER XXXI.

THE GREAT DISASTER.

GENERAL ELPHINSTONE, since April, 1841, had been in command at Cabul, and his health was shattered—no nerves, no decision, no method. The internal state of Afghanistan soon became appalling. The reports of able men, such as Major Pottinger (the hero of Herat, Political Agent in Kohistan) and Colin Mackenzie (Political Agent at Peshawar), predicting the coming storm, were unheeded and uncared for, until some Giljye chiefs suddenly gutted Cabul, and took up a strong position in the difficult defile of Khoord Cabul—thus blocking up the Pass, and cutting off communication with India.

In consequence of this news, Brigadier-General Sale—late Lieutenant, Captain and Major of the 12th; now Colonel of the 13th—marched his brigade to re-open and secure the road to Jellalabad. He left Cabul with the 13th Light Infantry, on October 11th, and very soon realised that the hydra of rebellion had raised its head full high—for every height was occupied, and every defile bristled with stone and earthern breastworks. Heavy fighting occurred on the 12th, but he cleared the Pass. It was then he sent his Adjutant, Henry Havelock, back to Cabul to report to Elphinstone. Havelock reached Cabul in safety, and then debated whether to stay with Elphinstone or return to Sale, for he was fairly "between the devil and the deep sea." However, he decided to return; and in his diary for that date is written that the Morning Lesson was Jeremiah xxxix. 16 to 18:—

"16. Thus saith the Lord of hosts, the God of Israel: Behold, I will bring my words upon this city for evil, and not for good; and they shall be accomplished in that day before thee.

"17. But I will deliver thee in that day, saith the Lord: and thou shalt not be given into the hand of the men of whom thou art afraid."

There was only one course for Sale to adopt, namely, to proceed to Jellalabad, sending back to Cabul all the troops he possibly could—for on November 2nd, 1841, Cabul, Charekar and Ghuznee, which are ninety miles from each other, broke into open revolt simultaneously. Everything had been pre-arranged.

At Cabul, Lieutenant Sturt, R.E., and Captain Lawrence were the first to be severely wounded. Then came the murder of Sir A. Barnes, Lieutenant Barnes (his brother) and Lieutenant Broadfoot. And all this within a mile of H.M. 44th Regiment and other troops.

This wretched supineness taught the enemy their strength, and they quickly set about to use it, for on the next day came further murders (Lieutenants Maule and Wheeler) and further indecision from our General. The commissariat fort was then attacked and lost.

Next came the conference between Sir W. Macnaghten (our Envoy) and the General. The former was at last forced, in consequence of the imbecility of the latter and the relaxed discipline of the troops, to consent to a treaty for withdrawal. Afterwards he tried to compromise with Akbar Khan—on terms not defensible.

On the way to this last meeting and his death, Macnaghten only had his sixteen troopers, although he warned the General that two regiments and two guns must be in readiness for secret service, for the interview was critical. He expressed his disappointment, saying: "It is all of a piece with the wretched military arrangements throughout the siege."

Sir William and Captain Trevor were soon barbarously shot down; Captain Lawrence and Colin Mackenzie, though wounded, miraculously escaped, but only to be hustled off to prison. The bodies of the murdered ones were hacked to pieces, their heads exposed to every possible insult; yet not a soldier stirred, not a gun was fired—and this on Christmas Eve!

For Christmas Day may we make a slight retrospect of how this terrible situation had arisen. When Lord Keane returned to India in September, 1839, he took with him a large portion of the Bengal force. The whole of the Bombay troops, under General Willshire, made a simultaneous homeward movement. Thus the army was reduced to a moiety. From Cabul to Ferozepore, the nearest Indian station, was 600 miles. From Cabul to Peshawur lāy all those dangerous defiles—Khoord Cabul, Tezeen, Jugdulluk, and Khyber. From Peshawur to Ferozepore was the country of the Sikhs, on whom then no reliance could be placed. And along this extended line of communications, there was but one small military post, Ali Musjid. So on Christmas Day, the last Council was held, when it was decided to evacuate on terms as follows [Major Pottinger, who succeeded McNaghten, would not agree to these terms; he suggested fighting *a l'outrance*, but he was over-ruled]:—

1. To surrender all guns but six.
2. To surrender all treasure.
3. To exchange present hostages—Captains Drummond, Walsh, Warburton, Webb, Airey, Connolly—for married men with wives and families. [To the wives and families, Elphinstone objected.]
4. To pay up fourteen lacs of rupees.

And what about Shah Shooja? He had been brought and installed by England, and by England was left to fight it out alone. His end was soon, for on March 18th, 1841, he was murdered by his godson, Shooja Dowla—a ruffian at whose christening Shah Shooja stood as sponsor, and gave to him his name.

January 6th, 1842, was the fatal morning on which an army of 4500 fighting men left the city of Cabul. A short week after, on January 13th, one man alone was left; he staggered into Jellalabad, maddened and near dead from the horrors of that direful march. It was Dr. Brydon, for he alone was left to tell the story of a great massacre. Dr. Brydon lived, however, to be again shut up with Sir Henry Lawrence at Lucknow, where he was severely wounded. He died on March 20th, 1873, at Westfield, Rossshire.

It was on January 8th that Major Pottinger and Captains Lawrence and Mackenzie were handed over as extra hostages, on promise that hostilities should cease, and the Khoord Cabul Pass be cleared of all the Giljyes. And in this belief, once more the living mass of men, women and children moved off.

But the Pass was filled with bloodthirsty barbarians. However, onward moved the crowd until a universal panic prevailed; and thousands, seeking refuge in flight, abandoned everything—baggage, arms, ammunition, women and children. Captain Anderson's eldest girl and Captain Boyd's youngest boy here fell into the hands of the Afghans. Lady Sale was slightly wounded. Lieutenant Sturt (her son-in-law) had nearly cleared the defile—which meant an outlet for the surging mass escaping from the slaughter in the Pass—when he was shot and speared. Lieutenant Mein, of the 13th, gave succour, and stood by him to the last, receiving in his turn a horrible wound on the head. Naught was heard but groans of misery and pain, as each one now met his doom: whilst the snow—how pitiless!—seemed to make a coverlet for all.

The whole scene was too heartrending; so much so that Akbar Khan insisted on the ladies and children, together with their husbands, being handed up to him for safe escort to India. The party — which consisted of Lady Macnaghten; Lady Sale*; Mrs. Trevor and eight children; Captain and Mrs. Anderson and three children; Mrs. Sturt† and one child; Mrs. Manwaring and one child; Captain Boyd, wife and three children; Lieutenant Eyre*, wife and one child; Lieutenant Waller*, wife and two children; Mr. Riley, wife and child; Captain Troup*, Lieutenant Mein*; Sergeant Wade, wife and family—were then shortly joined

---

Those marked thus * were wounded.

† Lieutenant Sturt was killed on the 8th. As Colonel and Mrs. Holmes were driving together in an open carriage at Meerut in May, 1857, one of his troopers rode up and shot them both dead. The wife was the widowed Mrs. Sturt,

by Major Pottinger and Captains Mackenzie and Lawrence, who had been given up as hostages at Bootkhak; and to the inexpressible delight of Captain and Mrs. Boyd,* there was with them their youngest boy, whom they feared had been lost for ever in the Khoord Cabul Pass.

"On January 11th they started for Tezeen, and as can be imagined, how terrible was the spectacle along the whole line of road!—the dead and dying victims of this Afghan treachery, amongst whom were recognised the bodies of many former friends.

"On January 13th, the whole day our path was strewed with the mangled victims of war. We reached Jugdulluk late in the evening, where the force had hopelessly sought for shelter; and there we beheld a spectacle more terrible than any, the whole interior of one enclosure being a crowded mass of bloody corpses. Generals Elphinstone and Shelton, and Captain Johnson, were here brought in as prisoners and hostages for the promised evacuation of Jellalabad—and here they halted until April 11th."

During this halt at Jugdulluk, Akbar Khan became most desirous to smash or drive Sale from Jellalabad to India. Sale was a fine soldier, and resisted all attempts of Akbar on April 9th. A few days later, on hearing that General Pollock in relief was forcing the Khyber Pass, Sale sallied forth, forced a battle on Akbar Khan, and completely defeated him.

Pollock from the Khyber, Nott at Candahar, and Sale now free, meant three forces to march on Cabul—to be as an army of revenge. Mahomed Shah had replaced Akbar Khan in charge of the prisoners—and in justice to the latter it must be said that he took his defeat in quite good form, and attributed it to the bravery of our troops, led on by the gallant Sale.

---

daughter of Lady Sale. Lady Sale died at Cape Town in 1853. On Brighton beach is a cutter—the last of her good game.

* It was not until May 10th that Captain and Mrs. Anderson recovered, or even heard of, their little girl; she had been lost on that terrible January 8th in the Khoord Cabul Pass. Fortunately, she had been treated with the greatest kindness, but had been taught to say: "My father and mother are Infidels, but I am a Mussulman."

"January 29th was rendered a joyous and eventful day by the arrival, from Jellalabad, of letters and newspapers from our brother officers there quartered. Many conducted this correspondence by dotting off letters of the alphabet in the newspapers—an easy mode, and one not likely to be detected by an Asiatic. We learnt of Wild's failure in the Khyber Pass; of General Pollock's advance from India to our relief; and that Dr. Brydon was the only living soul that reached Jellalabad.

"February 19th.—The weather changed to heat and stillness, and at 11 a.m. all were suddenly alarmed by a violent rocking of the earth; so much so, that it was difficult to stand. All the walls around one fell in with a huge crash, and a rumbling sound below seemed as of a boiling sea of liquid lava. Lady Sale was nearly caught in her house as it fell; the others had time to reach the open. General Elphinstone, being bedridden, was rescued by his servant.

"April 15th.—The General now was dying. The Sirdar, Akbar Khan, had personally handed him back his sword a few days since. It was considerate, to say the least; and the act no doubt brought solace to the dying soldier—let us hope so. After all, General Elphinstone, worn out and ill, was given work he was unfit for: the victim less of his own faults than the errors of others. As a Colonel in the 33rd Regiment at Waterloo, he was taken prisoner, and was summoned by Napoleon to tell all he knew. Elphinstone refused, on which the Emperor grew furious and abused him somewhat. At 7 p.m. on April 23rd, 1842, he died. The Sirdar offered honourable interment at Jellalabad; but when the sad *cortege* set out, and the body appeared slung across a camel's back, escorted by a few Giljyes and his soldier servant, the bigotted savages cruelly set upon them, and severely wounded all. They stripped the body of the General, and pelted it with stones. Directly Akbar Khan heard of this he quickly interfered; he rescued and repacked the body, which was buried with full military honours and Christian rites at Jellalabad.

"August 7th.—Poor John Connolly died, and his body was sent to Jellalabad for burial. The brothers—Arthur, Edward and John—

accompanied Sir John Keane's army into Afghanistan in 1839. All three speedily rose in their special political work; but in the course of three years, one was shot through the heart in Kohistan, one died in this Afghan camp, and the eldest (Arthur, the celebrated traveller) was killed, with Colonel Stoddart, in a loathsome dungeon at Bokhara.

"On September 5th, General Nott recaptured the fortress of Ghuznee, and removed the much-talked-about sandalwood gates of Somnauth. All prisoners were released—Colonel Palmer, Captains Alston, Poett, Harris, Nicholson*, Williams, Burnet and Crawford.

"On September 20th, the many hostages and captives were rejoiced to hear the voice of a British officer saying that Sale's Brigade was but a few miles distant. A squadron of the 3rd Hussars suddenly burst upon their view, and General Sale was amongst them. The meeting with his wife and widowed daughter reflected in his countenance the index of his feelings."

Lord Auckland now went home to England, and Dost Mahomed returned as Ameer to Cabul—back to their own countries, which it is a great pity that either ever left.

"On December 19th, Ferozepore was *en fête*, for it was to be a grand reception by the Governor-General (Lord Ellenborough) of the Afghan Army. General Sir R. Sale crossed the Sutlej on the 17th, and was received with special honour, whilst his own regiment (the 13th Light Infantry) took the head of the column in their march amidst streets lined with elephants caparisoned in their best."

It was at this review that Ellenborough wished to arrange the army in the form of a star, with the artillery at the point of each ray, and a throne for himself in the centre. "And he ought to sit upon it in a straight waistcoat," said the Duke of Wellington.

One extract from the "Malmesbury Memoirs," and this chapter closes:—

"November 30th, 1844.—Mrs. Portman and her sister Lady Sale, and Mrs. Sturt (the widow of an officer killed at Cabul), came

---

* Of Delhi renown.

to luncheon. The latter is pretty and very interesting looking. I never saw a countenance that bore such evident traces of past sorrow. She does all she can to appear cheerful, but her gaiety seems forced; and I saw, once or twice, an expression of agony come over her face. She has a pleasant manner and is wholly free from affectation; has a very graceful figure, and a fascinating expression of countenance when she talks and smiles; and does not look above eighteen. But suffering of mind and body have deprived her of the freshness of youth, and she is deadly pale.

"I was agreeably surprised in Lady Sale, whom I imagined to myself to be a tall, masculine, overbearing sort of woman. Far from this: she is not above middle height, with quiet ladylike manners, very proud of her husband, and of the number of enemies he has killed in battle with his own hand. She is very fond of India, and delighted to return there: which Mrs. Sturt is not.

"We made Sir Robert Sale's acquaintance the same day at dinner, who, being ill with an attack of ague, could not come to Heron Court. He did not talk much about India, as his daughter was just opposite and would have heard what he said. He is a little man, with blue eyes and very white teeth—which a gentleman at table, who was suffering from toothache, said he envied him more than his victories."

Mrs. Waller, whose diary in a measure we have given, lived to the advanced age of eighty-four, and died at Brighton, surrounded by her numerous children. The adventures of her two eldest (then aged two and one years) we have read in connection with her own. All through this great disaster she played a noble part.

"O! let not woman's weapons, water drops, stain my man cheeks."

She, with her husband and her children, were but one.

CHAPTER XXXII.

LORD Auckland had, in addition to the Afghan War which ended so disastrously, to outfit an expedition for China. He selected Major Mountain as Adjutant-General, and a Colonel Oglander to command. They were to join in with General Burrel at Singapore. A hurricane of most unusual severity nearly caused their disappearance altogether; but at last a concentration was effected, and then commenced a business of half war, half diplomacy. The merchants had to be protected in their opium business, unpleasant as it was to China; whilst diplomacy tried its best to stay the import of this odious drug.

At last an atrocity or two rendered war imperative, and Sir Hugh Gough was given the command, with the 18th (Royal Irish), 49th and 55th as his fighting units. Hope Grant was the Brigade-Major and Lord Saltoun the Brigadier.

Soon the heights above Canton and Shanghai were stormed, and the towns of Amoy, Chusan, Chapoo,* Woosung and Shanghai were captured.

Finally Sir Hugh undertook a bold and great enterprise, which he carried out with great success. Seeing that the great Canal (1200 miles in length) which led to the Imperial City, was the channel through which the whole internal commerce of the country flowed, he, with his gallant comrade Sir W. Parker, took his fleet and army 200 miles up an unknown river to the intersection of the

---

*In the fight at Chapoo, Colonel Tomlinson and Captain Campbell (of the 18th) fell mortally wounded, and Colonel Mountain received six bullet wounds —a marvellous escape.

Canal, and attacked the town of Ching-Kiang-Foo, which, after a gallant resistance, was taken. In the arsenal 60,000 dollars were taken, and paid over to the Military Chest. The matchlocks, gingals, powder, etc., were destroyed.

Mountain writes from Ching-Kiang-Foo :—

"July 26th, 1842.

". . . . John Bull proportions merit by loss, but I can tell you this, that if we have done a good deal with a few men and trifling loss, it is to be attributed in a great degree to the nerve and confidence with which Sir Hugh Gough has led on his men.

"The Chinese are robust, muscular fellows, and no cowards; the Tartars desperate; but neither are well commanded or acquainted with European warfare.

"Ching-Kiang-Foo fell on the 21st. Contrary to the usual custom of the Chinese, who generally make a vast show, the Tartar garrison kept quiet and concealed; and so bad was the information derived through our interpreters, that up to the moment of attack, Sir Hugh expected no opposition. The Tartars, however, had made up their minds to fight desperately for their homes, and did so."

Here Captain Collinson (of the 18th) and Captain Gibbons (of the 49th) were killed.

The Tartar General, when he saw all was lost, retired to his house, which he commanded his servant to set on fire. He then sat in his chair until he was burned to death.

The first man to reach the top of the battlement was Lieutenant Cuddy, of the 55th, who planted the English Ensign on the walls under the fire of the enemy, and then coolly assisted the foremost of his party up the scaling ladders. This officer was urgently wanted by the police in 1847 on a charge of manslaughter, for he had acted as second in the duel between Lieutenant-Colonel Fawcett and his brother-in-law, Lieutenant Munro. The former was killed, and Munro was tried for murder. This duel created a shudder at the time, and the law was exercised with great severity on all concerned.

Peace was signed and sealed on board H.M.S "Cornwallis" on August 29th, 1842—a date which also gave release to our prisoners and hostages in Afghanistan.

Sir Hugh Gough then sailed for India—to which country he had been appointed Commander-in-Chief *vice* Sir Jasper Nicholls.

To pick up our thread with England during these fearful days of bloodshed and bewilderment, we find that King William IV. had died on June 20th, 1837. Everybody regretted him, for he was a good man, though not a great king. On the previous day, the Duke of Wellington had sent him, according to custom, the banner by which he held his estates. It was handed him by Lord Munster; and the last words of the dying monarch were: "Ah! that was a glorious day for England."

Princess Victoria was at Kensington Palace when told of her uncle's death. The world generally is acquainted with the portrait of our youthful Queen, in her girlish garb, as she received her Prime Minister and Archbishop; and from that hour one and all became impressed with Her Majesty's sense, discretion, and kindly heart, her caution and her prudence.

The Duke of Wellington, at the time, was busily engaged in writing military maxims for the Government, conferring with Lord Hill, the Commander-in-Chief, on the Canadian Question, the State of the Army, and other important measures—and also on the *tapis* were outbreaks in Turkey and in China.

Sultan Mahmoud of Turkey declared war against Mehemet Ali, the Khedive of Egypt, for driving his troops out of Syria. The Turkish Admiral was ordered to take the fleet to Syria, but treacherously handed it over to the Khedive at Alexandria. Mehemet Ali was then promptly told by the Powers (France excepted) to clear out of Syria and restore the Turkish Fleet—which eventually was done. Great tension existed for a time between England and France, owing to the latter siding with the Khedive.

And in China we were at war on the much vexed Opium Question.

These were the questions of moment when Queen Victoria came to the throne on June 20th, 1837. In the same month—perhaps to herald and spread her good name far and wide—was born a child of science*; and to her must be given a foremost place of welcome, in that the two railway stations, Euston Square and Camden Town, were connected by a simple wire, along which orders and messages were rapidly sent by means of an electric spark. This infant has grown up beyond recognition, for she soon discarded her leading strings, and took a wireless form, talking to us from the realms above, as well as from the deep, deep sea.

Lord Hill resigned the command of the army in August, 1842. His health had failed. The Duke of Wellington relieved his old comrade and became Commander-in-Chief again.

Lord Hill died on December 10th, 1842, in deep humiliation to his God. "I believe I have not an enemy in the world," were his last words.

His wish was for a private funeral, but most of the 53rd (his old Regiment) found their way there unobserved, and mixed themselves with the village mourners, out of real affection for their Chief.

---

* Electricity.

CHAPTER XXXIII.

IN February, 1842, when Lord Auckland left India, affairs in Afghanistan were at their worst—the Army of Occupation under Elphinstone had been effaced, and the avenging forces of Nott and Pollock had yet to mobilise—so Lord Ellenborough, on arrival, seemed beset with difficulties and dangers, requiring a comprehensive turn of mind to grasp them. To "swop horses in the middle of a stream" is never considered desirable; hence our sympathies must go out to Ellenborough, who had to stay the bleeding and stitch up the wounds, upon which his predecessor had so badly commenced.

In December, 1842, he welcomed back to India the avenging armies of Nott and Pollock, and the band of heroes—the prisoners and the hostages—under Sale, of Afghan fame. This jubilation and display, however, was not sufficient to keep in check the various Ameers in Scinde, who, encouraged by our disasters in Afghanistan, refused to observe, much less be bound by, treaties framed at Simla. They refused an honest peace, and declared for open war, which Lord Ellenborough (through Sir Charles Napier) quickly waged upon them.

Sir Charles Napier, the Major of the 50th at Vimiera and Corunna, gave them their *quietus* at the battle of Miani, in a manner always quoted as a most brilliant feat of arms—for it was 3000 British against 20,000 Baluchis. This is referred to later.

Next followed trouble in the Court of Gwalior. There were pretenders to the throne, and a succession had to be established. The army had to be pacified and settled with, which was done by the victories of Sir. Hugh Gough at Maharajpore and General Sir J. Grey at Punniar—both gained on the same day, December 29th, 1843. Hope Grant, Harry Tombs, Nicholson, and other rising men, first saw fighting on that day.

This strong, determined policy of Lord Ellenborough carried full approval with the Government of the day; but the Directors of the East India Company saw things in a different hue. They censured him for language perhaps too strong, 'tis true, and gave him his recall. This crisis was unexpected, and to meet it was not easy—for Sir C. Metcalfe was in Canada, Lord Elgin in Jamaica; whilst Lord Seaton (Colbourne of the 43rd), Fitzroy Somerset (Raglan), Gladstone and others would not go. The next choice fell upon Hardinge, who accepted, and arrived in India on July 24th, 1844.

"January, 1845.—'The movement of the Sikh Army towards the Sutlej, and its mutinous, ungovernable state,' will be your answer to any question which may arise as to British reinforcements arriving on the Sutlej or its immediate neighbourhood." So wrote Lord Hardinge to his Agent-General at Lahore. In Loodiana, Ferozepore, and Umballa was a complete Division of English troops at this time.

A long interval occurred for *pourparlers*, viz., January to December 14th—on which latter date, the British forces moved up from Umballa by double marches on alternate days; and on the 18th reached Moodkee (20 miles from Ferozepore), after a march of 21 miles, in a very exhausted condition. The British force was 16,700 men and 69 guns, chiefly horse artillery; while the Sikh forces were 60,000 men and 103 guns in fixed batteries. On December 7th, the Sikh Army crossed the Sutlej, and lay in readiness to make a sudden attack on the reinforcements coming up from Umballa. The Sikhs were, however, repulsed with a loss of 17 guns, and retired to the formidable entrenchments of Ferozeshah, defending them with 80 guns and 60,000 men. It was now for the Governor-General and Lord Gough to effect a junction with the Ferozepore Division under General Littler. This was done, and on the night of December 21st—the right under the Commander-in-Chief, the centre under the Governor-General, and the left under Sir John Littler—the attack commenced. The time was not sufficient to complete the victory, and the troops bivouacked where they stood. Thus wrote Lord Hardinge, the Governor-General:—

"The night of the 21st was the most extraordinary of my life. I bivouacked with the men, without food or covering, and our nights are bitter cold. A burning camp in our front, our brave fellows lying down under a heavy cannonade, which continued during the whole night, mixed with the wild cries of the Sikhs, our English hurrah, the tramp of men, and the groans of the dying. In this state, with a handful of men who had carried the batteries the night before, I remained till morning, taking very short intervals of rest by lying down with various regiments in succession, to ascertain their temper and revive their spirits. I found myself again with my old friends of the 29th, 31st, 50th and 9th, all in good heart—and with them that Regiment which has earned immortal fame in the annals of the British Army, H.M. 80th Regiment. My answer to everyone was that we must fight it out, attack the enemy vigorously at daybreak, beat him, or die honourably in the field The gallant old General, kind-hearted and heroically brave, entirely coincided with me.

"When morning broke, we went at it in true English style. Gough was on the right; I placed myself and son in the centre, about thirty yards in front of the men, to prevent their firing; and we drove the enemy, without a halt, from one extremity of the camp to the other, capturing 30 or 40 guns as we went along. The brave men drew up in an excellent line, and cheered Gough and myself as we rode up the line, the regimental colours lowering to me as on parade. The mournful part is the heavy loss in my officers."

Such was the battle of Moodkee and Ferozeshah.

"The remains of the Khalsa* army are said to be in full retreat across the Sutlej, at Nuggurputhur and Tilla, in the greatest confusion

* Khalsa—Commonwealth, the army of the Sirdars.

and dismay: but are breaking off in small parties to occupy the many villages on the river bank." Such was the report brought in. Loodiana (under General Wheeler) still remained a threatened point, and there Sir Harry Smith was sent to its relief as well as to keep open the communications with Bussean and Delhi in the rear. This he accomplished with the loss of his baggage. Being joined by Wheeler, he moved off to attack the new position which the enemy had entrenched near the river Aliwal. "And, my Lords," said Wellington in the House, "I will say upon this, I have read the account of many a battle, but I have never read the account of one in which more ability, energy, and experience have been manifested than in this." The victory was complete.

The following is a copy of a pencil "Express" written on the field of Aliwal by Sir Harry Smith to the Commander-in-Chief:—

"Bank of the Sutlej, 28th January.

"Hearing the enemy had received a reinforcement yesterday of 12 guns, and 4000 men last night, I moved my troops at daybreak this morning to attack. I think I have taken every gun he had, and driven him over the river. My guns are now battering him from the opposite bank. He came out to fight me. I expect fifty guns are on the field at least. My loss, I hope, not great. The cavalry charged several times, both black and white, like soldiers—and infantry vied with each other in bravery. To the God of Victory we are all indebted. God bless you, dear Sir Hugh! My staff all right. Mackeson and Cunningham, of the Political Department, bore heavily on some villages.

(Sd.) "H. G. Smith, M.G."

The immediate result of the victory of Aliwal was the evacuation by the Sikh garrisons of all the forts occupied by the Lahore soldiers on that side of the Sutlej. They all fell back across the river to an entrenched position at Sobraon. This was on January 29th and 30th, 1846.

The Commander-in-Chief waited a few days to rest the victorious army of Sir Harry Smith, and for the arrival of the siege train and ammunition from Delhi.* The latter arrived on February 8th, and within forty-eight hours from that time, the enemy's entrenched camp was carried by storm, his army almost annihilated, and sixty-seven guns captured; and by February 12th the British Army lay encamped at Kussoor in the Punjaub, thirty-two miles from Lahore.

The son of the late Maharajah Runjeet Singh† was then installed and peace proclaimed, Major Henry Lawrence being appointed Resident. Sir Charles Napier strongly condemned the terms of this peace. He predicted another war, which came in 1848.

This is what Lord Gough says :—

"In the battles of Moodkee, Ferozeshah and Aliwal, I reported upon the noble bearing of Her Majesty's Army, both cavalry and infantry. At Sobraon, two new regiments contended for the prize so nobly won by their comrades in former actions; and I gave them the opportunity by placing them in the attacking columns. Nobly did they show that they would not be outdone.

"Her Majesty's 10th and 53rd have, by their steadiness and brilliant conduct, under Lieutenant-Colonels Franks and Phillips, established a name second to none. The former regiment never fired a shot until within the enemy's works, when they sent a withering volley into the densely-manned trenches."

The following story is related by Mr. F. A. Vincent, an Indian civilian, who knew the 10th Regiment during the Mutiny, and is a

* The transport of the military stores from Delhi put the energy and local knowledge of John Lawrence to the test—and with what grand results! A train of four thousand carts, each driven by its owner, was speedily organised. It made its journey of over two hundred miles to the front, and arrived without any straggling or deserting, without the failure of a man, a wheel, or a bullock, in time for Sobraon.

† The iron hand of this Marahajah kept his country quiet and on good terms with England. Our after experiences with this independent State were sad; for at his death anarchy and military insolence, and an encroachment on India, led to the first Sikh War. After Gc .gh's victories, the throne was kept warm for a young Maharajah under the protection of Lawrence, the Resident. This plan ailed, and led to the second Sikh War; then nothing was left but annexation.

voluntary appreciation of the fine bearing of the 10th Regiment at Sobraon :—

"It was at the height of the Indian Mutiny, when Sir James Outram was on his way up country travelling by steamer to relieve Lucknow, and inspecting all the stations on the river Ganges from Calcutta to Allahabad, such as Bhangulpore, Monyghyr, Barh, Patna and Dinapore, that I happened to be walking with Hookum Singh, a Sikh subadar, and a very gallant officer who had lately distinguished himself. We were in camp at the time, European and native soldiers all round us, when I asked Hookum Singh what he thought of the state of affairs—would we hold India and crush the mutiny and rebellion, or were the odds of a hundred to one of fighting men against us, more than we could subdue? He hesitated a little before answering, and then began: 'Sir, you know that I myself am loyal and a well-wisher to the Government, but I fear the worst; and my reason for thinking so is that you have exhausted all your fighting men and are now obliged to enlist mechanics in your ranks. Why, it was only last week that I was ordered down to Calcutta to receive these playthings (pointing to the medals he was decorated with), and I was taken on board one of the troopships which had just arrived there. I saw some newly-arrived recruits, such poor, puny creatures, tailors and shoemakers, some mending their shoes and others their breeches; then it was that I gave up all hope for the Sircar (Government). For when it came to hand-fighting I knew how poorly tailors and shoemakers will fight.'

"I tried to explain to him how in England there was no especial fighting caste, and that the whole nation were fighters; but I soon saw he could not take it in, and he added, 'The recruits, too, were poor, weak boys, whom I could fling on my shoulder with one hand.' All I could say to this was, 'Wait till these boys have been well fed and drilled for six months, and you will see a great change in their appearance.'

"Whilst we were talking, a big, burly, red-haired soldier approached us, whom I could tell at a glance was an Irishman. 'Could you throw that Irishman over your shoulder, Hookum

Singh ?' 'What number is on his cap, sir ? Please tell me quickly,' was his answer. 'The 10th,' I replied. Then, to my astonishment, the brave old subadar literally grovelled before the soldier, who seemed much surprised, but passed on with some laughing remark. 'What does this mean, Hookum Singh ?' I could not help saying, as I saw there was a tale attached. He then told the following story. I wish I could tell it in his own simple words, full of earnest feeling. It is many years since I heard it, but I can give the pith of it, as it made a great impression on me :—

" 'I began life as a gunner in the Sikh Army, and was in charge of a gun, when the Sikhs determined to fight the English for the possession of India. We had a large and powerful army, and we knew that we were, man for man, far superior to the native soldiers of the Indian Army, whereas the English soldiers were few, and unable to stand work in a hot sun. We soon found out our mistake. I was with the battery I was attached to, holding a strong entrenched position at Sobraon. It was on the afternoon of a very hot day that we were told that a British regiment was advancing to the attack, and we soon saw them marching steadily towards us. We were preparing to open fire when they got within range, but our Sirdar ordered us not to do so till he gave the word.

" 'Nearer and nearer they came as steadily as if they were on their own parade ground, in *perfect silence.* A creeping feeling came over me, this silence seemed so unnatural. We Sikhs are, as you know, brave, but when we attack we begin firing our muskets and shouting our famous war cry ; but these men, saying never a word, advanced in perfect silence. They appeared to me as demons, evil spirits, bent on our destruction, and I could hardly refrain from firing. At last the order came "Fire !" and our whole battery, as if from one gun, fired into the advancing mass. The smoke was so great that for a few minutes I could not see the effect of our fire, but fully expected that we had destroyed the demons ; so what was my astonishment, when the smoke cleared away, to see them advancing steadily, still in *perfect silence,* but their numbers reduced to about one half. Loading my cannon quickly, I fired again and again into them, making a gap or a lane in their ranks each time, but on they came in that awful silence till they were within a short distance of our guns, when their Colonel ordered them to halt and take breath, which they did under a heavy fire.

"'Then with a shout such as only angry demons could send forth, and which is still ringing in my ears, they made a rush for our guns, led by their Colonel. In ten minutes it was all over; they leapt into the deep ditch, or moat, in our front, soon filling it, and then, swarming up the opposite side on the shoulders of their comrades, dashed for the guns, which were still defended by a strong body of our infantry, who fought bravely. But who could withstand such fierce demons, with those awful bayonets which they preferred to their guns—for not a shot did they fire the whole time—and then, with a ringing cheer, which was heard for miles, they announced their victory. That's why I honour the 10th, as Gods or Demons, but not men.'

"'And you, Hookum Singh, what became of you?' 'By God's mercy I am alive, and my name is Hookum Singh.'"

All honour to the 10th for their day at Sobraon. The majesty of their fight was intensified from the fact that they stood 742 bayonets full fighting strength, eager to grapple with the foe—it was their first fight.

Alongside of them stood the 50th, the "Dirty Half Hundred." Their strength, after fighting at Punniar was—

675	
115	killed at Moodkee
560	
113	killed at Ferozeshah
447	
17	killed and wounded at Budeewal
430	
67	killed and wounded at Aliwal
363	
221	killed and wounded at Sobraon
142	

Thus the 50th at Sobraon stood but 363, and out of that noble band only 142 survived, and a subaltern brought the Regiment out of action.

The 1st Bengal Fusiliers were also splendid in their fight at Sobraon, leaving 209 officers and men dead on the ground. The Governor-General, Lord Hardinge, gave them much praise, both at Moodkee and Ferozeshah.

"It was in the midst of this battle, the 50th 'formed fours deep' under a terrific fire, to allow the 31st, who had occasion to retire, to pass through their ranks. This is without detriment to the 31st, who were mentioned in despatches for their gallantry. Sometimes it is impossible to advance against a fire." The above is given by one Colour-Sergeant Thompson, who afterwards rose to command the regiment and died a Major-General.

The field state of these battles will show how splendidly the British infantry—the 9th, 10th, 29th, 31st, 50th, 53rd, 62nd and 80th—fought and died between December 17th, 1845, and February 10th, 1846.

The 62nd won special praise—for by chance of war they had to face the strongest part of the entrenched position on their night march of the 21st. Then because the walls, like the walls of Jericho, did not fall at once—calumny tried to speak. The site of this midnight attack was surveyed afterwards by Lord Hardinge and Lord Gough; and there, to their wonder and amazement, they counted 17 officers and 185 men, mowed down by grape and canister from guns of various calibre. Special thanks appeared in Army Orders dated January 28th, 1846, for their noble conduct on the night of the 21st. The Artillery, as usual, fought admirably :—

		Officers.	Men.	Horses.	
At Moodkee	...	6	20	45	killed
		6	30	25	wounded
At Ferozeshah	...	4	30	120	killed
		6	71	80	wounded
At Aliwal ...	...	1	3	30	killed
		0	15	21	wounded
At Sobraon	...	1	6	17	killed
		1	43	23	wounded

Alas! death had made a harvest. **Men** of the highest distinction and promise were now no more :—

Sir Robert Sale, G.C.B., from the day he joined the 12th Regiment in 1799, on their march to Seringapatam, must have seen a hundred fights.

Sir John McCaskill, K.C.B., K.H., an old and much valued officer.

Major A. Fitzroy Somerset, Military Secretary to Lord Hardinge (son of Lord Raglan).

Major Broadfoot, the last of three brothers—an irreparable loss. He was as brave as he was able in every branch of the political and military service.

Lieutenant F. P. Haines*, aide-de-camp to Lord Gough, was very severely wounded. He lived to become Field-Marshal and Commander-in-Chief of India.

Sir Robert Dick, K.C.B., a most distinguished General, winning laurels at Egypt, Alexandria, Fuentes d'Onoro, Ciudad Rodrigo, Salamanca, Burgos, Waterloo.

Colonel Taylor, of the 29th Regiment, a heroic soldier (the father fell at Vimiera, where Lord Hardinge received his first wound).

And so the roll continues.

---

* An Impressive Spectacle.—Pall Mall to-day (June 11th, 1909) was the scene of an impressive spectacle. Outside one of the houses this morning were drawn up a gun carriage, two squadrons of the Life Guards, and a detachment of the 2nd Scots Guards. The assemblage of the troops caused a large crowd to gather, and presently heads were uncovered as from the doorway emerged the coffin with all that remained of one of England's most gallant servants, Field-Marshal Sir Frederick Paul Haines. The *cortege* attracted much attention on its way to Brompton Cemetery. A great many military officers were present, most of them men with Indian service. Over the Union Jack with which the coffin was covered, was spread a banner presented to the Field-Marshal by Queen Victoria. His age was eighty-nine.

CHAPTER XXXIV.

1848.

THIS was a year of great unrest — all Europe seemed upset. Venice had proclaimed a Republic; all North Italy was in open revolt; there were disturbances in Prussia; whilst in France, Louis Philippe had fled the country; and England was preoccupied with her Chartists, and an Irish *emeute* under Smith O'Brien.

Lord Hardinge, who had been succeeded by Dalhousie (the son of the Commander-in-Chief of a decade past), was posted to the Irish Command; whilst Louis Napoleon, whom we remember as residing with his mother Queen Hortense at Augsburg in 1821 (when Colonel Mountain was a *persona grata* at her salon), was now invited or rather proclaimed President of the French Republic—the first step to Empire.

After Sobraon, which meant the conclusion of the first Sikh War, a youthful Maharajah was seated on the throne, with Colonel Henry Lawrence as Resident or President at Lahore.

The rule of Henry Lawrence was further strengthened by others, such as his brother John Lawrence, Herbert Edwardes, Lumsden, Taylor, Nicholson, etc. It might have been assisted by the eight Sirdars, or the "Khalsa" (Commonwealth), but treachery again took ugly form—for whilst Lawrence was home on leave, two young officers, Vans Agnew (of the Civil Service) and Lieutenant Anderson (of the 1st Europeans), both in political employ, were murdered in May, 1848, and the Khalsa Army again revolted.

The first warning of this treachery was sent by express letter from Agnew to Edwardes on April 19th, 1848; the worst was only feared—it had not happened; so in hopes of saving life, Edwardes with a force (1st Native Infantry, two guns, 300 horse and 20 zumboorucks) hurried to Mooltan, which place he reached on the 27th, too late to save his comrades.

The personality of Edwardes soon brought many stray Sikhs, Pathans, Beloochis and Afghans to his standard, bringing up his force to 6000 men, many of whom, perhaps, had never fired a gun or been taught the first duty of a soldier, viz., to obey. On June 18th, a day of grand memory, for it came to Edwardes as a Waterloo, he fought two pitched battles against the whole army of Moolraj, from whom he captured more than half his guns. Lieutenant Lake and General Courtlandt were his two Centurions. The Moolraj now sought shelter within the walls of Mooltan, a fortress of considerable strength; whilst under General Whish, from Umballa and Ferozepore, two Brigades were being prepared to take the field—1st Brigade: H.M. 10th Regiment, 8th and 52nd Native Infantry, six guns, four squadrons. 2nd Brigade: H.M. 32nd and 49th Regiments, 51st and 72nd Native Infantry, six guns, six squadrons. Major Robert Napier (Magdala) was the Engineer in charge.

This concentration of troops before Mooltan was not effected before August 25th, although many of the troops had been on the march since early April, dragging along siege guns, etc., as well as their frenzied bodies, and over ground bleached with the bones of comrades killed at Moodkee, Aliwal and Sobraon only two years before. We have read of the horrors of Moscow, but were they worse than the sufferings of these poor soldiers marching to Mooltan in the scorching heat? Take one note from the diary of the 32nd: "The number of sick this day, 175; fourteen dead; heat in tents, 130°;"—then when the fiery sun had set, instead of sleep for these jaded men, it was tramp, tramp, tramp.

About the same time the 32nd lost their Colonel Potton, with the Quarter-master and many others, in the capture of a village; the command then fell to Major Inglis, who in 1857 so bravely held Lucknow.

The siege of Mooltan was long and tedious; it surrendered on January 3rd, 1849. On the tablet of those two murdered men—Peter Vans Agnew (B.L.S.) and William Anderson (Lieutenant 1st Bombay Fusiliers)—is given the rest:

"After two separate sieges,
the Fort of Mooltan was surrendered to the British troops,
and the bodies of two murdered Officers
(which had been treated with the most savage indignities)
were in all righteous vengeance
carried through the breach
made by the British guns
and buried with military honors on the summit
of the Citadel.
Thousands of Englishmen
stood round the grave.
Dewan Moolraj
was brought to trial at Lahore, convicted of
murder and sentenced to be hanged;
but was recommended to mercy, and
finally ordered to be transported for life.
His Rebellion
was followed by an insurrection of the
Sikh people, and brought on
the Second Sikh War;
which resulted in the annexation of the
Punjaub to British India,
and the restoration of peace,
after many years of anarchy
(with a brief interval),
to the Countries between the Sutlej
and the Indus."

Colonel Mountain (now Military Secretary to Earl Dalhousie) received permission to join his regiment, and overtook Lord Gough, his former chief, now forming a camp at Sobraon on the Sutlej. That night at dinner, the Chief looked grave, for he had just heard that the Sikh Army and Afghans were in full march towards them. The Punjaub had to be reconquered. Colonel Mountain was at once appointed Brigadier (24th Queen's, 13th and 30th Native Infantry), his Divisional General being Sir Walter Gilbert.

The first move of importance was made by Cureton, who on November 2nd—with his Horse Artillery, 3rd and 14th Hussars, 5th and 8th Light Cavalry, and 12th Irregulars, supported by infantry—advanced to drive the Sikhs beyond the river that they had so lately crossed. The Sikhs had laid a trap; for they lured poor Cureton on into a deep and heavy sandy swamp, when they opened on him, from the river bank, a withering and unexpected fire. General Cureton, Colonel Havelock (brother to Sir Henry), and about 70 all told, were killed in this affray at Ramnuggur.

The Commander-in-Chief, after this somewhat untoward event, rested at Ramnuggur, and waited for his heavy guns. The enemy, however, emboldened by their recent success, made sundry little ventures, all tempting Gough to fight. At last came Chillianwallah (January 13th, 1849); for the Sikhs were now in our front, about twelve miles off, with their right entrenched and their left protected by a belt of thick jungle. They mustered 35,000 men and 60 guns. We had 15,000, but a part of this force was at Ramnuggur. Herewith is a letter from Mountain to his friend the Governor-General, the day after:—

"My dear Lord,—We marched from Lussoorie on the 12th, and reached ground the name of which I forget, after a march of six-and-a-half hours. We marched again yesterday (13th) in order of battle, and about noon had a scrimmage with the advanced posts of the Sikhs, who were soon induced to abandon them after a salute from our heavy guns.

"We then formed up, and a Head-quarter officer told me in passing, 'Major Mackeson has persuaded the Chief not to attack to-day;' and our baggage was ordered up from the rear. About half-past one the Sikhs opened the ball with artillery; our heavy guns were then thrown forward and replied.

"My brigade was lying down in line, the round shot toddling spent were picked up, and only two men hurt. After a time the Chief passed down and said, 'Advance;' so up and forward was the word. We had what is the severest trial for infantry, to charge against grape through jungle. The Sikh had brought his field-guns into the jungle, dug trenches—which were evidently fresh—for his matchlock men, and supported them by cavalry.

"I had not gone 100 yards before I lost sight of any superior officer, as well as of any support; but we pushed on till we had taken the last gun in our front, on the skirt of the jungle. The Sikh cavalry were on the open to my right, and if I had had cavalry, I might have swept them before me; but as the enemy were all about the jungle, and on my flanks, I could not advance further, and after a time I got an order to move to my left to support General Campbell; thus the guns that we had taken were left to be carried off by others. We brought away two, however; and the rest I believe were brought away by spare horses from the artillery.

"I can give no account of the whole, as in such a jungle, each brigade, and in some cases each regiment, had to act for itself.

"My loss has been heavy: it has pleased God to spare me; but I grieve for officers of my brigade, and for men too, though I do not yet know the number.

"The 24th Queen's suffered severely—Brigadier Pennycuick and his son, Colonel Brooke, and a major killed, General Campbell wounded, Major Ekins, D.A.G., killed."

H.M. 24th Regiment advanced precipitately, and carried a battery with the point of the bayonet, but were attacked by a fresh Sikh division, and dreadfully cut up and driven back. [A wing of this same regiment was annihilated later on at Isandhlana, South Africa, in 1881].

The British force engaged was 13,000. 89 British officers, 43 Native officers, and 2200 men were killed and wounded. On the 18th Mountain again wrote:—"I went to see my wounded again last night. The sight of fine young fellows without legs or arms, or otherwise fearfully maimed, gave me a wretched feeling. It was I who led them on, who upbraided when they shrank, and cheered when they rushed on; but it would have been worse for them had I not so urged them, as if we had failed, we should have been cut to pieces. I had 800 killed or wounded. A man must have a heart of stone not to feel it .... for in each

> 'A mighty spirit was eclipsed—a power
> Had passed from day to darkness, to whose hour
> Of light no likeness was bequeathed—no name.'"

The result of and the heavy losses attending the battle of Chillianwallah were received in England as distinctly bad news. Lord Gough was recalled. "Indeed, I had not intended to attack to-day, but the impudent rascals fired on me—they put my Irish blood up, and I attacked them" was not considered sufficient excuse; and Sir Charles Napier was appointed to succeed.

Malmesbury Memoirs.—"On March 14th we dined with the Colchesters, and were introduced to Sir Charles Napier. He is a little man, with grey hair brushed back from his face, with an immense hooked and pointed nose, small eyes, and wears spectacles, very like the conventional face of a Jew. He is appointed to retrieve our affairs in India, and when the Duke of Wellington named him to the post, he at first hesitated, until the Duke told him, 'If you do *not* go, I must.' Great notice was taken of Sir Charles, and it was a curious scene altogether; and for a few minutes those two great men were engaged in close conversation in the centre of the room, and the company formed a circle round them, though at a respectful distance."

The division from Mooltan now joined Lord Gough, who complimented them most heartily on their marching so soon after their hard work at Mooltan. This gave a strength of British troops for the coming fight at Goojerat as under :—

H.M. 10th, 24th, 29th, 32nd, 60th, 61st and six companies of 53rd Regiments; H.M. Cavalry: 3rd Light Dragoons, 9th Lancers, 14th Light Dragoons; one regiment Bengal Europeans.

"It was a glorious day. The sun shone bright, but not hot. The snowy range was distinct and in great beauty, the country rich and green as England, interspersed with occasional trees. I never saw such a sheet of luxuriant crops; I felt as if I were treading down gold from the time we came within eight or ten miles of Goojerat.

"When we paraded on the morning of February 21st (Mountain writes) General Gilbert rode up to me and said: 'You are to open the ball; the enemy hold a post in front of you, with four battalions and six guns, and you are to knock them out of it.' Of course, I was delighted, but the ruffians did not hold it; and the second post, which they *did* hold, about half-a-mile further on, was in front of Penny, and he got the order for that. I was thus cut out of any prominent duty, though I was with the right wing, which, in fact, won the battle. . . . . You observed in your last letter, that no doubt the lesson given by the British infantry at Chillianwallah was not lost upon the enemy. One of their Sirdars, who was in the battle of Goojerat, being asked why the Sikhs ran, replied, 'What could we do, after three hours' cannonade, supported by that tremendous wall of infantry?' When their artillery, which stood manfully, was knocked over, they had no mind to try again the taste of the white man's steel and musketry. . . . .

"I had a few men knocked over, and my field battery lost a great many horses. I had one narrow escape myself. I had a light company out with my battery, and as the officer was much more under fire than was necessary, and I could not make him hear, and did not like to send any one, I went myself. A round shot struck the ground close to my horse's hoof. He started at the splutter of earth, so that I thought he was hit, but we were all safe. . . . . Goojerat, in its results, the utter rout of an army of 60,000 men, with 70 guns, the prostration of the Sikh nation, and the conquest of a whole kingdom, is certainly a great victory."

To follow up this victory, it was decided on the evening of the 21st to send a force under Sir Walter Gilbert in pursuit of the Sikhs now across the Jhelum.

Mountain writes that: "The last shots fired this campaign were the six rounds fired at General Gilbert and me as we attacked Sheer Sing's position on the other side the Jhelum. . . . . I think that the surrender of 38 guns and 18,000 stand of arms to General Gilbert, and of all the Sirdars and remaining troops beyond the Jhelum, is a much more grand result, politically considered, of the battle of

Goojerat, than a dozen more victories. Nothing could prove more triumphantly the utter prostration of the Sikh army and nation;" or as Gough expressed it in his farewell address: "That which Alexander attempted, the British army have accomplished." Gough on his return was created a viscount and given a pension of £4000 a year. He had commanded in more general actions than any officer of the age, the Duke alone excepted. When in residence at his Irish seat (St. Helens), it was a favourite route march of the 12th and 87th Royal Irish Fusiliers, then wintering at the Curragh (1862-63), to serenade this fine old General, to halt and talk awhile. He died on March 2nd, 1869.

The mail of February 2nd from England brought intelligence of Colonel Mountain's appointment as Adjutant-General; and on this occasion Lord Dalhousie wrote to him as follows:—

"March 16th, 1849.

"My dear Mountain, I wish you joy with all my heart of being Adjutant-General to Her Majesty's forces in India.

(Sd.) "DALHOUSIE."

The Adjutant-General was all one with Sir Charles Napier, the now Commander-in-Chief, in his plans for improving the barracks* in India, and promoting the discipline and healthy amusements of the troops. His next chief was Sir William Gomm (November, 1850). Within a short four years, the Commander-in-Chief had to halt his camp at Futtyghur, for his Adjutant-General was ill with fever; and on February 8th, 1854, this high-spirited soldier, Colonel Armine Mountain, was borne from his tent to his grave. "He was a distinguished soldier, the conspicuous man of worth, the fervent and consistent Christian, exemplary in all the relations of life." These were the claims of the departed to universal regard and esteem, and the enduring respect and affection of the Commander-in-Chief and the Head-quarter Staff.

* The barracks were unhealthy and insecurely built. Take the 50th Regiment: After its four glorious fights ending at Sobraon, it was quartered in Loodiana, when in a gale of wind the barracks fell, killing in men, women and children just three hundred.

CHAPTER XXXV.

GOLF.

IT may be pleasant now to break off awhile to play some rounds of golf. For the same mail which carried home General Pollock's account of the forcing of the Khyber Pass, Sir Robert Sale's victory at Jellalabad, the destruction of Ghuznee and fall of Cabul, the close of the China War, and the glorious pacification of India, also conveyed the answer from the Bombay Golf Club to an invitation from the members of the Royal Blackheath Golf Club to link up an alliance with them, and to interchange at times a game or two.

The reply was somewhat jocose: "That from Cairo one might drive to Alexandria, and then approach to Greece, clearing some good hazards at Thermopylæ, and hole out on the green at Corinth." Another suggestion was to tee off from the Pyramids, but then the Sahara is rather dry and sandy.

Again, as we approach 1852, the writer distinctly remembers playing golf at Blackheath, and listening to the guns at Woolwich as they boomed off each minute in sad homage for the Duke (his funeral). And from the betting book of the Royal Blackheath Golf Club, the secretary kindly allows him to reproduce a few conundrums and other scraps which the reader may ponder over at leisure:—

Extracts from Betting Book of the Royal Blackheath Golf Club.

"September 1st, 1792.—Mr. Duff lays Mr. Turner a hogshead of claret that Monsieur Dumourier, now Commander of the Northern Army of France, was not advanced to the rank of Colonel in the

French service previous to the present revolution. Mr. Turner lays the contrary. Lost and paid honorably by Mr. Duff."

"June 14th, 1794.—Mr. Hamilton bets Mr. Easterby a gallon of claret that the 'Ardent' man-of-war, taken by the combined fleet in the Channel, was retaken from the French in Rodney's action on April 12th, or in the following up of that victory. Mr. Easterby lost this bet."

"April 13th, 1799.—The following bet was this day laid: 'That the fleet of transports blocked up in the harbour of Alexandria are either burnt or destroyed by Nelson before this day.' The captain, Mr. Longlands, bets the company a gallon that they are not destroyed. The company say they are. Six on either side took this bet."

"Mr. Black lays a gallon against Mr. James Stein, that he will break a broomstick suspended betwixt two bumpers of wine without spilling the wine or breaking the glasses. Lost by Mr. Stein, and the £1 spent in this day's bill."

"July 12th, 1806.—Mr. Ruperti bets Mr. Callender a gallon that Jerome Buonaparte* is brought a prisoner to England before he returns to France. If he is killed in any engagement Mr. Callender agrees to pay."

"July 13th, 1805.—Mr. James Walker bets Mr. Broughton a gallon (one guinea) that Lord Nelson comes up with the French fleet before they reach port, either in America or Europe. Mr. Broughton says he will not."

"Mr. Callender, a worthy captain of the club, died in 1808, and bequeathed £50 to the club—which purchased £100 3 per cent. stock."

---

*Jerome was made King of Westphalia in 1806. His marriage with Miss Patterson had been annulled by the Pope—on the ground of his being a minor and his wife a heretic—also by order of Napoleon. Jerome brought her as far as Lisbon, but the lady was not allowed to land; so with her friends she proceeded to London, and took residence at the Grove in Camberwell.

"August 15th, 1812 (anniversary).—Sixteen members and twenty visitors dined at 2 p.m., among the latter being His Serene Highness the Duke of Brunswick and the Duke de Bouillon. Venison and turtles were often the gifts of members."

"August 14th, 1813 (anniversary).—Thirteen members and twenty-eight visitors present at 2 p.m. Sir John Eames's regimental band attended. The Duke of Brunswick present."

"October 15th, 1814.—The health of Professor Wilson was drunk with 'three times three' as a memorandum of his extraordinary powers, at the age of four score and four, in playing a match this day and winning it confessedly by the skill exhibited in the game." [He died May 11th, 1816.]

"June 24th, 1815.—That the club be put into mourning at the lamented death of the Duke of Brunswick—who had so often honored them by his presence whilst in England, and remembered them when abroad—where he had so gloriously fallen in the field of Waterloo."

"August 7th, 1824.—Mr. Croft paid a gallon on the birth of another young golfer, and the health of Mrs. Croft was drunk with 'three times three'—this being the fifteenth child."

"June 12th, 1830.—An account of the formation of a Golf Club at Dum Dum was read from the Calcutta papers, and 'Prosperity to it' given from the chair with all honors."

"June 26th, 1830.—His Majesty George IV. having departed this life this morning, 'His Memory' was drank in solemn silence."

"March 21st, 1831.—A bet to decide the age of John Carn Hobhouse, M.P. for Westminster—he being a contemporary of the immortal Byron.* The registrar of the club, Mr. John Masson, wrote an amusing letter, to solicit an answer from Mr. Hobhouse himself.

---

* Byron and Hobhouse were taking their morning swim at Brighton, and had great difficulty in making the land, for the wind was blowing off the shore and the tide setting out. Crowds of people collected on the beach to see them. yron, being the better swimmer of the two, had difficulty in saving Hobhouse.

[Reply]

"March 25th, 1831,

"21, Charles Street, Berkeley Square.

"Sir—In reply to your note, I have to inform you that on the 27th of next June I shall no longer be liable to serve in the Militia. This is the only advantage I shall derive from having been born so very long ago. Your enquiry needs no apology, and I am

"Very much your servant,

"John C. Hobhouse."

"March 21st, 1835.—The company this day was not numerous, but officially select. 'The healths of those who survive, and the immortal memory of those who fell at Alexandria, 1801, on this memorable day,' was drunk with truly patriotic sincerity and adoration."

"March 20th, 1840.—Vivat Regina! Prince Albert having sent no gallon on the occasion of his marriage: no record."

Domestic occurrences were a strong feature in the annals and funds of this club; for the dinner and wine bills were mainly met out of the fines and presentation gallons or guineas, and also from bets. Public breakfasts on the Heath were given on stated occasions, when ladies were invited. In the evening was a ball.

The carriers or caddies in the olden days at Blackheath were taken from the pensioners of the Royal Hospital of Greenwich, and looked most picturesque on the Green in their blue uniform and three-cornered hats. One, "Old Alick," passes down into posterity as "born in 1756; died in 1840; first sent to sea in 1769; fought in sixteen men-of-war during the war, under Sir Charles Cotton, Lord Gardiner, and Lord Gambier."

Further, it has been my good fortune to play golf in Europe, Asia, Africa, and America.

And in later years (1882), when the illness, and soon the death of my wife, gave me reason to appreciate a quieter life, I became Adjutant of the Cambridge Volunteers. An ex-Sergeant-Major

Adams, of the 12th Regiment, was my Chief of Staff. He was a man of great integrity and worth, and a soldier to the core.

A brother officer, named Major Harris, held a similar position to mine in the University Battalion. We knew each other well. He was a man "de bon augure" for any work on hand; this was entirely proved when he was Colonel of the 12th in 1886, for all things went right cheerily whilst he was in command.

He was a Cambridge "Blue" for cricket, and was quite in the swim. So for myself, not to feel quite out of it, I purchased a cap and gown, with a "Paley" thrown in, and thus became an undergraduate, specified as a "Non Coll." The society was delightful, both with the Dons and lesser men. The cricket and the rowing gave no spare time.

In a match against the Leys School, a ball most inadvertently shot straight into my face; for weeks and weeks I looked a ruffian. However, as our drills were held at night, perhaps my black eyes passed unnoticed in the dark—the scar I still retain. I loved the A.D.C., and remain still a member of the "Pitt." I left with my blue for golf, as the following will show:—

"OXFORD v. CAMBRIDGE (1882). *Times.*

"This match took place on Thursday, March 9, over Wimbledon links, by the kind permission of Mr. Henry A. Lamb, the secretary of the London Scottish Golf Club. Each University was represented by six players a side. A. N. Cumming, Balliol College, was captain of the Oxford team, and W. T. Linskill, Jesus College, of the Cambridge team.

"The players arrived at Wimbledon about twelve o'clock. The common was in splendid order, and the day everything that golfers could wish. At the end of the match it was found that Cambridge had won a closely-contested match by one hole only. The following is the result:—

"W. Welsh (Jesus) lost by 3 to Horace Hutchinson (C.C.C.).

"R. C. Faithfull (Trinity) halved with A. N. Cumming (Balliol).

"W. T. Linskill (Jesus) lost by 4 to Ludovic Grant (Balliol).

" E. B. Lehmann (Trinity) beat by 3 A. C. M. Mackenzie (B.N.C.)

" J. L. Cusson (Trinity) beat by 5 F. J. C. Mackenzie (Keble).

" Major Gardiner (12th Regt.) halved with W. B. Nicholls (Oriel).
" All played very well."

The crowd at the last hole worried me a bit, for it was on my last "putt" that the win or lose would rest. Thank goodness! when it was over, Cambridge was just one hole to the good.

CHAPTER XXXVI.

REPORT to Committee to the House of Commons on the Army and Navy and Ordnance Estimates, March 31st, 1848 :—

"You will scarcely believe, Gentlemen, that the economy of the House of Commons had at that time reduced the Field Artillery of the British Army to 34 guns, inadequately horsed, unskilfully manned, and even without the means of moving with their ammunition waggons. As for guns of position in the field, there was actually not a 9-pounder equipped for service in the whole United Kingdom."

The Duke of Wellington and Sir John Burgoyne were most urgent, throughout their lives, for the defensive power of Great Britain to be upheld; and about the last political official act the Duke performed, was to call upon the Foreign Minister, Lord Malmesbury (when the new Cabinet was formed in 1852), to press upon Lord Derby, the Premier, to increase our armaments both by land and sea.

England as a country is indifferent to its defence, and little understands the requirements of the Army or the Navy. It has no imagination, as the following little anecdote will show :—

The Duke of Wellington, when a guest of the Earl of Abingdon, was one day out shooting. In the course of the shoot, the head gamekeeper approached his lordship with a very grave and depressed look, and asked if His Grace could be the gentleman who had won the battle of Waterloo. "Ah! true, it is! and why not?" replied Lord Abingdon. "Oh, my lord! he has gone and missed a partridge."

Many of our Members of Parliament still think that "start a soldier with a red coat and a gun, and their task is done." How few understand what is meant by "the sinews of war;" how that the soldier has to be fed—if killed to be replaced, or if wounded to be nursed; or even how his gun is to be reloaded, after the few rounds he carries have been expended; and then how to make Wellingtons.

After reading of some of the Duke's work in India, a friend said to him: "It seems to me, Duke, that your chief business in India was to procure rice and bullocks." "And so it was," said the Duke; "for if I had rice and bullocks I had men, and if I had men I knew I could beat the enemy."

And to a later query, when asked on whom the command of his army should fall in the event of anything happening to him, the Duke replied, "Beresford;" qualifying his remark by saying: "Some of you fellows in handling troops might do as well, perhaps better than he—but what we want is some one who will feed our troops."

Even with Sir John Moore in the Corunna campaign, how often he had to complain that the troops sent to him were no more equipped for war than if the campaign had been in Hyde Park.

The century is now half over, when surely friends and foes, after these years of incessant bloodshed, would wish to cry "Halt!" and seek for peace. And the world thought so too, for the Great Exhibition of 1851 sent out its invitations, hoping that the great era of goodwill to all mankind had arrived.

But it seemeth Peace with her olive branch had not been so invited. For our hero of Aliwal, Sir Harry Smith, was still unable to sheath his sword; and as Governor and Commander-in-Chief in South Africa, he was once more engaged in active operations. There old comrades met again—the 12th and 74th (with memories of Seringapatam), the Royal Artillery, the 43rd, 60th, 73rd, 45th, 91st and 12th Lancers—for in early 1850 it was well known that the Kaffirs were preparing for war; and at an interview between Sir Harry Smith and the Gaika chiefs, all swore obedience and kissed the stick of peace except Sandili, the head, who refused to attend.

On December 24th, Colonel Mackinnon, Commandant and Chief Commissioner of British Kaffraria, was sent with a small force either to conciliate or to capture Sandili. He failed in both, for the Kaffirs attacked him in a narrow defile, and drove him back with great loss. This was a signal for a general rising of the Gaikas. Next occurred a horrible massacre of the military settlers in the village of Chunbie, on Christmas Day. The Kaffir police deserted, and the Hottentots (former allies of the British) rose in rebellion.

Flushed with this success, the Kaffirs attacked Fort Cox, where Sir Harry was; but his dauntless spirit was equal to the occasion, for with 150 men he cut his way through, and amidst very heavy fire reached King Williamstown, twelve miles off.

Reinforcements were now asked for and slowly arrived: The 12th Regiment from Mauritius, July 29th; the 60th Rifles, part of 12th Lancers, and drafts, in August; the 1st Rifle Brigade and remainder of 12th Lancers, in October, 1851. His Excellency, Sir Harry, then decided on a general attack upon the various camps and strongholds along the Eastern frontier, and the Amatola range of mountains. Columns under Colonel Percival (12th Regiment) and Colonel Eyre (73rd) were fully equipped for this service, which was accomplished with great success.

However, on November 6th, was fought the battle of the Waterkloof. It was from the bush that Colonel Fordyce* (74th), Lieutenant Carey (74th), and many others were shot, and it took some hours of stubborn fighting before the 12th, 74th, 91st, and levies drove the enemy from the forest, and gained the honours of the day.

* The successor to Colonel Fordyce—Colonel Seton—was to arrive on board the "Birkenhead." We all know the fate of this ill-fated ship—she struck on Point Danger, near Simon's Bay, and went down, carrying with her 9 officers and 349 men; but every woman and child was saved, for the men in perfect discipline received and carried out their orders, as if they were embarking instead of going to the bottom.

An old bay horse swam ashore from the wreck of the "Birkenhead," and enlisted as a charger with one Captain Moody, R.E., and again ran the gauntlet well, avoiding all the bullets, as she had shied off all the sharks. As other ships arrived, there were sundry tales to tell.

The Army List for 1851 gave the postings of two young officers, viz., Ensign Garnet Wolseley* to the 12th Regiment, and Valentine Baker, from the Ceylon Rifles, to be Cornet in the 12th Lancers.

The career of the former is well known—the last Commander-in-Chief under Queen Victoria, and first under King Edward, Peer of the Realm, and Field Marshal. It is not so much the honours, but the man, for he is one the country could not have done without. We reaped the full benefit of his organising power in the African War of 1900. Then the latter, brilliant as a soldier, and his one thought patriotism.

Baker joined during the war, and soon knew his way about, scouting and patrolling. He was a good judge of a horse, and a perfect horse master. On arrival, he purchased from Sir Harry Smith a strong compact little brown horse named "Punch," of perfect temper, very steady under fire, and wonderfully enduring; he was with his master all through the Cape War, India, then to Crimea, across the Desert and through Egypt, and then from Crimea to England.

Sir Harry Smith was also a great horse lover. He rode at Aliwal an Arab he had trained to fall flat to his command. It was during the battle when Sir Harry, contending for the front along

---

* It was in 1857 that H.M.S. "Acteon" was sent in haste to a ship in distress. The "Transit" was suspended amidships on a rock, her bows pointing to the sky, and her stern submerged in the sea. She had on board the 90th Regiment, with some units from other corps. Having embarked 20 officers and 200 rank and file of them, we made sail for Singapore; the "Beaver" brought up the remainder. The monsoon was fair; we had our brothers in red as messmates for three days, and enjoyed their company very much. Amongst them was G. G. Wolseley, a fair-haired young fellow, the junior Captain of the Regiment, now the veteran Field-Marshal Lord Wolseley, late Commander-in-Chief.

with the rest, found himself confronted with a hostile volley. "Down! charger! down!" shouted the General, when flat fell his horse; the quick recovery of horse and rider, after so pitiless a discharge, greatly astonished the Sikhs.

Sir Harry was always supposed to have a charmed life. It was on the eve of Bloemplatz that Pretorius, the Boer leader, paid him a secret visit, and out of liking for the General, explained to him that it was no good forcing events, for Bloemplatz would, in a few days, fall to him like a ripe plum from a tree; but if the place was rushed as he intended, the Boers would make a stand, and Sir Harry would be their target.

The two commanders were somewhat agreed on this point; but Sir Harry could not gainsay his instructions, which were to fight. Pretorius, on leaving, said, "Farewell! remember, I have warned and advised you." The next day, during the battle, it is well known that a bullet cut away his bridle, another his stirrup leather, and a third tore his saddle; but there was no bullet with a billet for Sir Harry.

Sir Hope Grant was another ideal horseman and horsemaster. The nobility of the man, his craft as a cavalry man and General, will be told in the Chapter on the Mutiny.

This Kaffir war, as planned and commenced by Sir Harry Smith, was handed over to General Cathcart to complete. The reason why! I know not—but Sir Harry, a hero in twelve Peninsular battles, the valued Staff Officer and trusted friend of Wellington at Waterloo (where he was one of three brothers in the field); again with Gough and Hardinge at Maharajpur, Ferozeshah, Aliwal and Sobraon; was now told by orders from home to depart.

Sir Harry was welcomed home, and Corporations vied in offering him complimentary addresses, dinners, etc. He, however, soon had to sorrow, with the world at large, over the death of his old chief, the Duke of Wellington, on September 14th, 1852. What a noble example of duty! For up to the last he worked at the Horse Guards as Commander-in-Chief, bringing his weighty experience into every detail of all things which govern the activity, both in peace and war, of man and beast.

The war continued. The unceasing labours, the night marches, the burning sun, the torrents of rain, encountered by our soldiers in the desultory fighting of this arduous campaign, were best verified by the appearance of the men, and more especially so by their garments.

"It was held (referring to a small fort) by a company of the 73rd, under Captain R——, and never did I see a more extraordinary looking set of beings in the shape of soldiers. They had lately returned from a round of patrols, having been incessantly in the field since the commencement of the war; scarcely a vestige of uniform was to be seen amongst them. Their clothing consisted of a mass of rags of every colour, patched together with leather; and their beards and moustaches had been unshaven for many a long month. The guard were just mounting as we entered the fort, and they certainly presented a semblance of uniformity; for every man had an old worn blanket round him in place of a great coat, four or five had handkerchiefs tied round their heads, some old wideawake hats, some nothing; but their muskets and bayonets were as clean and bright as if they had been in quarters, and their bearing as they marched off to their posts told that they were indeed real British soldiers, and notwithstanding their rags, beyond disguise."

Then the appearance of the 74th: they were dressed in their sea kits, dyed brown with mimosa, still wearing tartans, however, and forage caps with broad peaks like the French *kepi*. The broad peaks mentioned here had been removed from the old shakos, which became the property of the soldier after two years' use. The peaks were cleverly sewn on to the forage caps to give extra protection from the sun. The story goes that the Adjutant (a fine soldier who had risen from the ranks and was a splendid "drill," though not much of a scholar) was sorely puzzled how to put the Colonel's verbal instructions into plain unmistakable language. Eventually, however, it is said that he drafted the following order for the Colonel's approval: "Them shakos as hasn't been two years in use isn't to have their peaks cut off, but them as is is."

Another instance of Queen's English as it was written in the Navy occurred when Louis XVIII. was rowed ashore by a boat's crew of H.M.S. "Majestic" on his escape from France. The honest tars, in obedience to an order not to accept money, and being offered £15 as a present from the King to drink his health, transmitted to the Admiral the following letter:—

"'Majestic,' November 6th, 1807.

"Please your Honour,—We holded a talk about that there £15 that was sent us, and hope no offence, your honour. We don't like to take it, because as how we knows fast enuff that it was the true King of France that went with your honour in the boat, and that he and our noble King, God bless him both, and give everyone is right, is good friends now; and besides that, your honour gived an order long ago not to take any money from nobody, and we never did take none; and Mr. Leneve, that steered your honour and that there King, says he wont have no hand in it, and so does Andrew Young, the proper coxen; and we hopes no offence—so we all, one and all, begs not to take it at all. So no more at present.

"From your honour's dutiful servants."

And even since the creation of the School Boards, the following conclusion to a letter from a personal servant somewhat perplexed me:—

"I have sent you the account of the ten pounds you sent me. I paid Silvester two pounds the 9th of May that was the day you sent it this is enough for our bord wages on Monday next 4 shillings left to keep dogs and to pay for stamps (etc.)

"I remain your

"Obscene sivint,

"M. J. MILLETT."

The war finished by a march into Basutoland, to punish one Moshesh for trespass in the Transvaal. The headquarters of this monarch were reached on December 13th, and there the force halted until the cattle fine was paid. In this campaign Colonel Eyre (73rd) shone as a most able leader, to gain a higher reputation in the Crimea.

"Thus the concentration of troops under Cathcart enabled him to engage Moshesh in many minor operations without any decisive results.

"Moshesh was a jocular warrior, for although he held his own he satisfied the honour of General Cathcart by writing him a letter in which he said: 'As the object for which you have come is to have a compensation for Boers, I beg you will be satisfied with what you have taken. You have shown your power. You have chastised. I will try, all I can, to keep my people in order in the future.'

"In this instance, Moshesh repeated his cunning tactics of twenty years earlier, when he defeated Moselekatze by sending a letter after his retreating enemy, couched in the following terms: 'Moshesh salutes you. He sends you these cattle as a recognition of your bravery. He desires to live at peace with you.' Moselekatze never attacked him again."

Lieutenant-General Viscount Hardinge, the late Governor-General, now succeeded the Iron Duke as Commander-in-Chief.

CHAPTER XXXVII.

CRIMEA.

"To conduct a war successfully, there must be an adequate and certain supply of men."

WITHIN three days of the Peace closing the Kaffir war, there appeared in the *Times* of March 2nd, 1853, an ominous despatch from Lord Stratford de Redcliffe to Lord Clarendon, that at an interview with the Grand Vizier and Minister of Foreign Affairs, they had informed him: "That since the arrival of Prince Menschikoff, the language held by the Russian Embassy to them had been a mixture of angry complaints and friendly assurances, accompanied with positive requisitions as to the Holy Places in Palestine, indications of some ulterior views, and a general tone of resistance at times bordering on intimidation."

Thus appeared on the political horizon a small cloud that was destined, two years later, to burst in a storm of bloodshed and devastation around Sebastopol.

A most important Council was held on Midsummer Day, 1854, when the Duke of Newcastle read to the Cabinet a despatch placing the English Generals at Varna under compulsion to invade the Crimea. This meeting is supposed to have ended in a "Midsummer Night's Dream," for it is very certain that before the reading of the document was finished, the Cabinet—with one or two exceptions—were overcome with sleep.

After this sleep followed stupor. At this time the offices of the Colonial Secretary and War Minister were under one man—the Duke of Newcastle. On the outbreak of war it was deemed necessary to separate them, and it rested with the Duke to decide which to keep. The Duke elected for "War," stating that it was the unanimous wish of all his colleagues that he should do so; and as no one contradicted him, the matter ended. "I myself," said Gladstone, "and everyone else I talked to, were of opposite opinions, viz., that we all thought Palmerston was best suited for War, and the Duke for the Colonies."

A rude awakening followed. On March 13th, 1854, Mr. John Bright said in the House of Commons, he had read the proceedings relative to the dinner given at the Reform Club to Admiral Sir Charles Napier (with Lord Palmerston in the chair), with pain and humiliation; the reckless levity displayed being, in his opinion, discreditable to a responsible statesman of a civilized and Christian nation. Then quickly came the sickness at Varna—cholera claiming its thousands. This inglorious mortality greatly reduced the *morale* of the force; so when the order came to move from Varna to the Crimea, it was greatly welcomed.

On September 14th, the British force (24,000), the French (22,000) and Turks (8000) landed in the Crimea. On September 20th was fought the battle of Alma, our troops being victorious throughout—showing the same majesty in fight as carried them victorious through the Peninsula; and on September 28th, a brilliant flank march was made for Balaclava, to secure a base for future operations.

On October 25th, the cavalry battles of Balaclava* were fought—the charges of the Heavy and Light Brigades have been set to fame by the Poet Laureate of the day. And on November 5th was

* Hamley writes from Crimea: "I am disappointed in old Palmerston—he seems to meet every difficulty with a joke, replying like Jack Falstaff 'with a fool-born jest,' as if Grimaldi had taken Kemble's place in a tragedy." Further, "It has always struck me as not quite honourable for the Cabinet of England to allow the despatches of Burnes, relative to Cabul, to be presented to the House of Commons in a garbled and mutilated form—trimmed in such a horrid way as to aid, perhaps, their misjudged policy. Palmerston had to admit they were garbled, but contended, as the advice of Burnes was not to be taken, his policy therefore seemed to them mistaken. Queen Victoria always disliked Palmerston, and no wonder!" See McCarthy's "History of our own Times."

fought the "soldier's battle" of Inkermann.* A doggerel which I remember from boyhood best describes the *melee* :—

> "Please to remember the 5th of November:
> Sebastopol powdered with shot,
> When General Liprandi charged John, Pat and Sandy—
> And a jolly good licking he got."

The day was dark with fog and rain, thereby rendering it almost impossible to discover anything beyond the flash and smoke of the artillery and musketry fire. Heavy were our losses that day—2612—including Lieutenant-General Sir George Cathcart, the late Commander-in-Chief at the Cape.

"All the world's a stage," and from the late war drama in Africa, poor Cathcart had come and gone. Eyre, the fighting Colonel of the 73rd, was now a Brigadier, and destined for further good work, earning at the Redan his K.C.B., also Arthur Ponsonby of the 43rd (later Colonel of the 12th), who was Aide-de-Camp to Sir G. Brown; whilst the 10th Hussars and 12th Lancers (amongst the latter being Valentine Baker), with many others, were now *en route* from India. Last, but not least, we have the brilliant author and soldier E. Hamley; "Chinese" Gordon—later known as "Gordon of Khartoum;" his brother (he and Hamley were the two Aides-de-Camp to Sir Richard Dacres); and then young Garnet Wolseley.

The fair autumn weather had now entirely gone, and the winter commenced with a terrible hurricane. The wind prostrated all the camp, the hospital tents (full of sick and wounded) were swept away,

* The day after Inkermann was of great moment, for a lengthened consultation as to the further conduct of the campaign took place between the allied Generals. The result of this consultation was a determination to persevere in holding the ground then occupied by the Allies, to fortify our position on the Inkermann heights, to defend the advanced trenches with firmness, and even if possible to carry forward the approaches. Pelissier described Lord Raglan as being the last to speak, in unusual agitation moving the stump of his lost arm convulsively; and when Pelissier voted for remaining, rushing up to him to shake hands, declaring that nothing would make him stir.

and the ships in harbour—many laden with creature comforts for the troops—were all destroyed.

In February, 1855, John Bright once more made an eloquent appeal to Lord Palmerston, when it was hoped that a prospect of peace might be near. "The Angel of Death," he said, "has been abroad throughout the land; you may almost hear the very beating of his wings. There is no one to sprinkle with blood the lintel and the side-posts of our doors, that he may spare and pass on: but he calls at the castle of the noble, and the mansion of the wealthy, equally as at the cottage of the humble; and it is on behalf of all these classes that I make this solemn appeal."

If John Bright, in his prayer that the sword might be returned to its scabbard, read to the bellicose as "Peace at any price," what censure can be too severe for the Government who, day by day, witnessed the approach of war, and furnished no sinews to resist it?

The army had, at Varna, shrivelled to a skeleton, and as a skeleton was sent to fight. Baker noted: "That although all the available men and horses were sent out, the strength of the regiments was so small that some had almost ceased to exist before the campaign commenced; and at Varna, before even landing in the Crimea, the 5th Dragoon Guards (or rather, a small remnant of some sixty men and horses belonging to that corps) had to be amalgamated with the 4th Dragoon Guards, the 5th having disappeared as a regiment."

The headlines of Chapter VIII in Sir Edward Hamley's "War in the Crimea" give some idea of the breakdown: "Privations of the troops—Want of transport—Transport done by the men—The cavalry horses starved—Sufferings of the sick—The hospitals—Indignation in England—Resignation of the Ministry." What an indictment!

"Scutari,* the longed-for haven for the sick, was for weeks the very climax and headquarters of suffering—crammed with misery, overflowing with despair. In those large chambers and long corridors,

* Hamley.

lay thousands of the bravest and most miserable of men. Standing at the end of any of the galleries that traversed the four sides of the extensive building, one looked along a deep perspective, a long diminishing vista of woe. Ranged in two rows lay the patients, feet to feet; the tenant of each bed saw his pains reflected in the face of his comrade opposite; fronting each was another victim of war or cold, starvation or pestilence. Or, frequently, the sick man read in the face before him not the progress of fever, nor the leaden weight of exhaustion, but the tokens of the final rest to which he was himself hastening. With each round of the sun, nearly a hundred gallant soldiers raved or languished out their lives; as the jaws of the grave closed on the prey of to-day, they opened as widely for that of to-morrow. It might be thought that, at this rate, the grave, so greedy, so improvident, would exhaust its victims—that some day it would gape in vain. But no! The sick flocked in faster than the dead were carried out, and still the dismal stream augmented, till the hospitals overflowed; while still faster poured the misery-laden ships down the Black Sea, feeding as they went the fishes with their dead."

And the poor animals! There they stood and suffered, fetlock deep in gore and mud.

Thus was Scutari, with thousands of sick and wounded on their pain-stricken beds, and alas! and with burials sufficient to remind one of Defoe's account of the Great Plague—when appeared, as "Angels of Light," Miss Nightingale and her band of nurses:

> "Lo! In this house of misery
> A lady with a lamp we see
> Pass through the glimmering gloom,
> And flit from room to room:
> And slow, as in a dream of bliss,
> The speechless sufferer turns to kiss
> The shadow as it falls
> Upon the darkening walls,
> As if a door in heaven should be
> Opened, and then closed suddenly."
>
> —"*Santa Filomena*" (LONGFELLOW).

England, in her forlorn state, was forced to apply the principle, which never answers, of "robbing Peter to pay Paul," and sent to India for the 10th Hussars and 12th Lancers. This "Peter-Paul" policy of taking these two regiments from India, on the eve of the great Mutiny, is commented on by Baker: "Great events have occurred, painfully proving the defects in our military system which I have urged.

"The Indian Mutiny, which might have been crushed in the bud had any cavalry force been at hand, still drags heavily on; whilst the successive Generals of successive armies raise the same cry, and lament their weakness in this most important arm.

"Havelock has sunk beneath the weight of nine victories that he could not improve. Nearly the whole of his gallant band has died away, consumed by its own glory.

It was well into the New Year (1855) before the 12th Lancers left Mangalore with 528 horses (entires) for Suez—where commenced the march across the Desert to Cairo, and on to Alexandria. A further shipment to Balaclava landed the regiment in the Crimea, with 509 horses (geldings). The march across the Desert cost the regiment nineteen horses, seven of which strolled away and were lost—a very small casualty list considering (April 30th, 1855).

After a few days' rest, the men and their little Arabs were all in good condition, and able to take their share in outposts, convoys, and other duties; an equally good marching-in state of the 10th Hussars was handed in on their arrival at Balaclava.

Baker relates a reconnaissance and outpost day under General d'Allonville as offering great instruction, which clearly proves that with reliable outposts, night attacks are best left to the enemy.

A military machine is merely a conglomeration of men, *i.e.*, units: when the temperature of a man departs from normal, he becomes helpless, weak, unreliable, and has to be kept quiet and nourished if he ever wishes to resume his daily work; but the military machine is allowed to go miserably below par, and is denied all rest and nourishment to tone it up again.

On the Queen's birthday, May 24th, 1855, the five regiments of the Light Brigade could only turn out one weak squadron amongst them; Lancers, Light Dragoons, Hussars, all being mixed together to form it. And when these attenuated regiments *were* fed up, it was by young, weakly, half-drilled recruits, who could barely sit on —much less manage—a horse. The same with the infantry.

How paramount it is, then, that in order to conduct a war successfully, there must be an adequate and certain supply of men. If it is the policy in these days for Germany to wrestle with us in the shipbuilding process, who can say but what our fall may come not in having the money to build the ships, but in not having the men to man them? This fatal insufficiency of men is the radical defect in our Army, and who can tell it may not become a radical defect in our Navy?

Let the country be converted to what is a just view of national defence, *all* being taught how to aid and assist in the hour of peril.

The Crimean War plainly told the country that this must be done. Let us not forget the Council of War that met after Inkermann, and even now—fifty-six years after that date—if England will but be true to herself and wake up, the awful torments, tortures, and horrors of the winter of 1854 in the Crimea may not have been endured in vain.

"Florence Nightingale" is now a household name, held in all praise and merit. She was noted from her youth for her love of doing good; and this love was fully exercised until her death (August, 1910). At a party in 1854 the Duke of Newcastle, then Secretary for War, became the object of the hour, when one observed:—

"I, who have been accustomed at home to see distinguished officers and men of grave experience at the head of the War Office, could not at first believe that this smart young fellow, with his free and easy manners and his happy smile, looking as if he had not a care in the world, could be the Minister into whose hands was committed so grave a charge. A veil, as it were, fell from my eyes, and

I then understood what had hitherto been incomprehensible to me: how it was that England had been so unsuccessful in this (Crimean) campaign.

"It is not that the Duke of Newcastle is alone responsible for the melancholy results, but I denounce the whole system, which must be utterly false and pernicious if it can lead to a young man without any military experience being given such a post. . . . When there is a question in England of making someone a Minister —they only think of the number of votes which he has behind him in the Lower House, and he is given first one portfolio and then another, as occasion requires, no matter if he is fitted for the post or not."

Sir George Cathcart sprang from a talented military race. His grandfather was an Aide-de-Camp at Fontenoy; his father was politically employed in Russia; and his son (Sir George), as Aide-de-Camp, was at Lutzen, Bautzen, Dresden, Leipsic, Quatre Bras, and Waterloo. For the last two fights, he was Aide-de-Camp to Wellington.

Lord Raglan was wounded at Busaco; and whilst riding slowly with the Duke from the field of Waterloo, a stray shot shattered his right elbow, which caused the loss of his arm. June 18th, 1855, killed him. The failure of Pelissier to advance to the attack of the Redan at the hour agreed upon, caused him great depression; and the most cheerful man in the whole camp became an easy prey to dysentery. He died ten days after the Redan—June 28th, 1855—aged 68 years.

CHAPTER XXXVIII.

THE late Duke had never minced matters. His opinion, so recently given, regarding the unpreparedness of this country for war and against invasion, must have been fresh in the memory of all. And on this very day (November 18th, 1852), when his body was buried in peace, there were strong apprehensions of war with France. The President had styled himself Emperor Napoleon III., and thorny questions regarding his recognition as such had arisen. A hurried Vote for 5000 seamen, 1500 marines, and 3000 artillery was passed. Fortunately, however, the clouds rolled by.

The Prince Consort was one who watched the Army and the Navy most carefully. It was he who greatly favoured the idea of extended manœuvres, and the formation of large military encampments. At Chobham, in the summer of 1853, a relatively small force of 4 regiments of cavalry, 3 battalions of Guards, 2 brigades of infantry, 1 troop of R.H.A., and 3 field batteries, with engineers and a pontoon train, were assembled—as if prepared for war—and thus gave the only little preparedness the nation could boast of, when in the autumn of the following year, the Crimean War burst upon us.

At the close of the Crimean campaign, many huts and other useful structures were brought home and re-erected in the same pretty district, between the Chobham ridges and Aldershot town. It was here that most of the troops repaired on their return from war, and formed an Aldershot Division, with Lieutenant-General Knollys in command.

On July 8th, 1856, Queen Victoria held her first parade at Aldershot, to welcome home her Army from the East; when Lord Hardinge, the Commander-in-Chief, on handing to Her Majesty the Field State, was seized with sudden illness, that left him but a few short months to live.

George, Duke of Cambridge, was appointed in succession to this late gallant soldier, Commander-in-Chief and ex-Viceroy of India.

Aldershot had come to stay; and as the periods for the summer drills approached, so like hardy annuals, for the next fifty years to come, appeared the Duke with his Headquarter Staff. "H.R.H.," as he was called, soon became honoured and trusted by all soldiers as the very best of friends.

Sir Richard (soon to be Lord) Airey was first Quartermaster, and afterwards Adjutant-General of the Army. He had a brother Talbot, and the two were soldiers of grand calibre. The latter had gone through the prison life in our Afghan struggles in 1837, and afterwards fought at Punniar and other battles, and finally in the Crimea.

A Council of Military Education came into power about this time, and threw open the portals of Sandhurst and Woolwich to public competition. Hitherto Sandhurst had been a school for youthlets of good family with proclivities for war—a certain number of marks on leaving qualified for a free commission; cadets below this standard were given the option of purchase (£450 as Ensign, £800 as Cornet), provided of course they were not atrociously dull or stupid.

The Senior Department, for Officer Students, was moved off about this time into a more palatial building called the Staff College,* with Colonel Napier (who had served as aide-de-camp to his uncle Sir Charles Napier in India and elsewhere) as first Commandant.

It is when the autumn of life has well set in that a fascination comes in revisiting the scenes of one's early youth. The "then" and "now," or the memories of Sandhurst in the late "fifties," return with great vividness: it is easy even *now* to trace the peg *then* allotted for your rifle, belt and bayonet, and your bed with birdcage overhead, which latter kept the requirements for a week's use. All other articles bearing any resemblance to a plain clothes description

---

* The Duke of Cambridge laid the foundation stone in the winter of 1859-60. The gentlemen cadets attended the parade.

were carefully stowed away in unknown regions, only to be resuscitated at the commencement of each vacation, or "vac." as it was called.

Then the Friday pay or pocket-money night, which meant a quick hand-over of all that came to Capner, Look or Grey—the tuck merchants of the day. The same names are now seen over most prosperous-looking shops, but the likeness of the owners betrays a younger generation.

The College boasted of a "Gym.," provided only with some very primitive poles and bars; but it had a second use as a Court of Final Appeal, where "Judge Fisticuffs" presided.

The Chapel which *was* is now used as an "overflow" for meals, where knives, forks, plates and spoons congregate in huge profusion. What a great change from what was then a rendezvous for daily prayer and Sunday worship. On Sunday the parade was a grand spectacle, when the whole armed force of Sandhurst turned out in *grand tenue*. First came the Cavalry with a nice man, Captain Brooke, in charge; he was named "the Murmuring Brook," for he always had a grievance.

The horses took their Sundays in, and kept it as a day of rest, or spent the time in conjuring up fresh tricks or plans of eviction on those daring to bestride them—they could be so nasty, no matter how hard one tried to please.

Captain Brooke, as well as the Sergeant-Instructors, repeatedly affirmed that if you felt the outward rein and pressed the inward leg, your horse would strike off true and united into a most delightful canter, in such like harmony as becomes a bride and bridegroom on leaving church. However, when this advice was tended, the result never came off, but the *rider* did. A horse or two would then immediately rear and squeal, which meant a general upheaval of all the rest; in this crisis the language of the "Murmuring Brook" was not exactly polite or comforting.

The Army also could muster some Engineers, and a gunner man or two, but the main phalanx meant Companies A and B of Gentlemen Cadets.

The band played well; it was led by one Sergeant Sullivan. Often a young boy, of early pinafore days, stood by and listened. This was little Arthur, whose talent we have all extolled in his later "H.M.S. Pinafore" days—afterwards honoured as Sir Arthur.

However, the whole pomp and panoply of the show was justly centred in our war-worn Governor, Sir Harry Jones, G.C.B., who had been in every battle of interest from 1808 to the Redan, 1855, where he was badly wounded.

There was also a Captain Fosbery, who had left his leg behind at Inkermann.

The cadets all felt proud as they marched behind these heroes into chapel, and wondered when their turn would come to lose their arms or legs.

Our Chaplain was one Dr. Chepmell. He had written a history full of dates; it was distinctly dry; neither was his preaching very thrilling. He gave to all, however, a good preparation for their confirmation, which was taken yearly by good Bishop Samuel Wilberforce, until the day came when on the Surrey Downs his horse fell, and Sam Wilberforce was no more.

There was a hospital, then as now, in the woods, for those cadets who needed isolation; and a sick bay or two within the College to meet the lighter cases.

Then came the lakes for boating, bathing, and for skating; for we had some *winters* then—and our good athletic sports, which took place on either Green. Sandhurst can boast of inaugurating athletic sports in 1812. The example was not followed by other schools until 1840 (the first Oxford and Cambridge Meeting was held for the first time in that year).

Aldershot, or rather Farnborough, in April, 1860, was the place selected for the last of the prize fights. Here the Benicia boy Heenan, an American (28 years, 6ft. 2in.) fought Tom Sayers (34 years, 5ft. 4in., champion) for the belt. I believe three counties (Berks, Hants and Surrey) meet within a small *confine* of Farnborough; so the police of each need not interfere until the pugilants crossed *their* line. The fight went on for thirty-seven rounds, the last five of which were fought amidst great confusion; for the police had interfered, and caused the referees to separate their men. Sayers then had a broken arm, and Heenan was almost blind. Both were presented with a silver belt—for the battle had to be considered drawn.

This trial of supposed "international pluck" gave rise to great excitement. Tinkers, tailors, peers, sailors, bishops, priests, plough-boys, Members of Parliament, and thieves, likewise the Gentlemen Cadets, all turned up at this great encounter. The scrimmage back to save a roll call—if they could—was a proof that in wind and limb they were already fit soldiers for their Queen.

The neighbourhood around Aldershot, Frimley, Sandhurst, and Bagshot, somewhat savoured of the rogueries of Dick Turpin and other masqueraders. The College grounds were full of caves and secret passages, which were supposed to have been the hiding-places of these "road kings." The Gentlemen Cadets somehow kept up the romance, and secreted themselves right well for a game of cards, a smoke, or a picnic off some forbidden fruit.

Frimley, a few years previously (1851), had been the scene of a horrid murder: the rector, the Rev. Mr. Hollest, was murdered in his bed, whilst his wife—who was below reading—escaped. Three of the ruffians were hung, and the fourth imprisoned for life.

The neighbourhood now has altered, and how populous it has become! The house which was formerly the property of Mr. Longman, the publisher, is now the home of the ex-Empress Eugenie, and the last resting-place of her husband Napoleon III., and her son the Prince Imperial.

Again, close by at Claremont and Weybridge, we have the homestead and the last resting-place of Napoleon's predecessor, Louis Philippe, and his Queen. God forbid that any member of *our* Royal House should ever have to seek asylum and burial on a foreign soil; but England and her Colonies must do their "Sentry go." "All's well!" must be the invoice of all on duty at their posts; then "God speed!" will become the pass-word for the unhindered progress of our lion-hearted race.

The memories of Sandhurst must now conclude with the great Board Day, when the Duke appeared with his Headquarter Staff—Major-General W. Forster, K.H., Military Secretary (his assistant was Colonel Greathed, C.B., R.E., of Indian repute); Sir James Scarlett, Adjutant-General; Sir Richard Airey, Quartermaster-General; and the Hon. Jim Macdonald, A.D.C. (of whom the Cadets have been told that when he had his second horse shot under him at Inkermann, he laughed and said: "This sort of thing should only happen to younger sons")—to give away the Commissions and the prizes—a proud day for many, for as they passed on to parade that day as Gentlemen Cadets, they left as Ensign or Cornet or So-and-So unattached. But not to be forgotten is our rendezvous in town that night, at the great "Commission Dinner," a menu no need to attach.

From the month of July, 1860, we date the opening meeting of the Rifle Association at Wimbledon, in which the first shot fired was by our good Queen Victoria, with result a bullseye. Mr. E. Ross became the first winner of the Queen's Prize (£250) and gold medal, and the Volunteers (of 1859 creation) took rank at once as first class shots. Lord Clyde reached home from India this month, and was greatly welcomed—and a fine old soldier passed away in Sir Harry Smith of Aliwal.

Now wars came on again—for China had imprisoned two Secretaries of our Embassy, Messrs. Parkes and Loch (the late Lord Loch), and Captain Brabazon, Lieutenant Anderson, and Boulby of the *Times*, whilst on mission to settle some weighty question. All were treated with the greatest indignity and insult. To abate all this arrogance, an army was placed in the hands of a cavalry officer, one only known for his Indian service—a "Sepoy General," as Napoleon would say. This officer, however, was Sir Hope Grant, which is sufficient to say that within four months the Allies (for the French were with us) had marched from the sea to Pekin, stormed the Taku forts, and occupied the Imperial Palace, where a treaty of peace was signed.

CHAPTER XXXIX.

MUTINIES.

**" There is some soul of goodness in things evil,
Would men observingly distil it out."**

ONE of the greatest dangers threatening India is Mutiny. Money or pay is safe to be one cause, and want of sympathy another. In the past century many mutinies occurred, the causes being varied.

The seeds of the mutiny of the Madras troops at Vellore in 1804, perhaps, were sown at home, for the Commander-in-Chief in England—whenever the Duke of York was not in disgrace—was one David Dundas. We have already met him at Bastia. Shortly after he brought out a book called "The Principles of Military Movements," based upon what he had seen at Potsdam when Frederick the Great was King. This drill became his hobby. Sir Walter Scott termed it "a confused hash of regulations, which for the matter of principle might be reduced to a dozen;" and relates how General Sir William Erskine, when all was in utter confusion at the retreat from Dunkirk, in passing Sir David Dundas, shouted out: "Davie, je donnert idiot, where's a' your peevioys (pivots) the day?"

Now imagine all the dry details of this confused hash being forced upon the elderly Subahdurs and other officers of the Indian Army. Surely they must have agreed with Erskine that Sir David was a "donnert idiot." A strong point is made of this in the following private letter written by Colonel Barry Close, Adjutant-General Madras Army, to Lord Wellesley, the Governor-General:—

" *Private.*

"In your letter to the Adjutant-General of the 16th of June, 1805, you have remarked 'that it is the prevailing practice of the

Service to withhold that respect and intercourse from the native commissioned officers to which their situation entitles them.' This remark appears to me to be so very important that I shall venture to enlarge a little on points connected with it. . . ."

On the point of drill the Adjutant-General commented thus:— "Previous to the epoch when the European discipline was taken as the standard for that to be enforced upon the coast, the Native Commissioned Officers were men generally of good descent and caste; and under the feeling of that time were certainly admitted by the European Officers to their level, and treated with the attention due to their situation in the military scale. But upon a change of system, when *minute accuracy* in points of exercise began to supersede all other objects, the European Officers found it difficult to exact that constant attention from the Native Officers which the performance of the new system of exercise required.

"Mistakes or negligence at drill, added to the disadvantages of infirmity, often brought reproachful language upon them.

"When soured by abuse or reproof, they felt the protection of their commission, and sometimes manifested disinclination, if not disaffection. This only further estranged the parties, making the European Officer reserved and austere—until, as your Excellency has observed, the intercourse ceased except on the parade or at field exercise."

Another reason given was: "During the late war, the prize money of a European Subaltern was about seven times more than that of a Subahdur; and I am acquainted with one instance in which a commissioned Native Officer spoke of this circumstance in terms which ought to excite the most serious reflection.

"As regards the turbans, the Hindoos of caste seem to regard leather as an object of disgust, or, at least, hold it to be too mean to be attached to the head, the seat of honour."

They also hated the fantastical shape and colour of the turbans. Then came the clipping of the Sepoys' whiskers, by order; and this was the last straw. The storm burst on July 9th, for at midnight, the Sepoys took indiscriminate vengeance on their officers, and thirty-three were slain.

How did the British Army at home overcome this minute accuracy required by these Potsdam lessons so haphazardly set by Dundas? "The soul of goodness" extracted from the above evil was arrived at by Sir John Moore, who begged of the Duke of York to allow him to select regiments, to train men and officers to act as fuglemen for other corps, and so create a system of his own out of the hazy compilations of Dundas.

Corunna soon robbed us of this grand master, so apt in training and in handling troops. However, his lessons had taken root sufficiently to enable Robert Crawford to carry on in a training camp at Shorncliffe. Our famous Light Division became a product of the above.

The above mutiny was no sooner over than Sir George Barlow, the Governor of Madras, upset the dignity of the whole Madras Native Army by his interference in matters purely military. He is supposed to have released a Colonel Monro from arrest, when placed there by the Commander-in-Chief, General Hay Macdowell. He also deprived the Commander-in-Chief of his seat in Council. This act caused officers and men to become jealous of their rights. Matters now went from bad to worse, until at last the discontents selected Seringapatam for their rendezvous, and Colonel Bell (of the Company's Artillery) as their Chief.

A stout resistance was made for a time. Colonel Bell was eventually dismissed the Service, but received a handsome provision for life from his brother officers.

Lord Minto, on assuming office as Governor-General (after Cornwallis, who died in 1805), made Madras his headquarters for a time, until matters went right again.

All through the "fifties" India was in mutiny. General Jacob had warned the country of what the climax would surely be:—

"Now all this was pointed out in 1851, and had been pointed out over and over again before; and we want to know, and the

English people will want to know, 'Who is to blame that these remonstrances were disregarded?' There is a heavy storm, if we mistake not, gathering over Lord Dalhousie's head, and justly so. He seems to have been absolutely and determinately blind to the state of the Bengal Army. It is well known that no officer at last dared to make any representation on the subject to the Government.

"The presage of this hurricane, which has swept over Bengal and the North-West, was distinctly recognised by all observant men acquainted with the other side of India, and its coming constantly predicted.

"The mutterings of the tempest were all unheeded, and no precautions whatever were adopted. Every warning was slighted, and the pilot even sought the shoals, while yet the clouds were gathering black on the horizon.

"Oude and Burmah were annexed, without a requisition for a single European regiment; and Delhi, with its arsenals, was left to the charge of three native regiments!

"Let Dalhousie make his defence to the English people. Too proud to take advice, and too blind to recognise the danger, we owe it to him that, at the close of an eight years' administration, he handed over the Army to his successor in a state of mutiny."

Sir Charles Napier gave the same like warning, perhaps more forcibly than Jacob. These two men had served together in India from 1841 to 1847, when operations in Upper and Lower Scinde gave scope for their great talents—one of which was the gift of conciliation. Then, if this good will is scorned, there must follow a prompt and vigorous blow. Sir Charles saw this with the Ameers of Scinde. They professed goodwill, but acted not so; so he was forced into the "Art of War." Jacob joined him with a supposed force of police, but he had so well trained and equipped them as to be the perfect model of a cavalry corps. This corps so loved him, that the place from which the men came was then named (and still so called) Jacobabad.

In 1842, Sir Charles, with 350 of these men, marched to attack a far-famed fortress, Emaum Ghur—quite in the wilderness. It took eight days to reach it. On his arrival, the enemy fled. Wellington always described this exploit as "one of the most curious military feats which he had ever known to be performed, or ever perused an account of, in his life." The enemy now tried to intercept him on his march to Hyderabad, and 25,000 infantry and 10,000 horse took up a position at Meeanee (February 17th, 1843), which he immediately attacked—his force being 1800 infantry and 800 horse (the 22nd Regiment behaved like steel)—and after three hours' fighting the Beloochis fled again, leaving 6000 slain. Napier lost 250.

Hyderabad, the capital of Scinde, now fell; and a victory at Dubba meant the Province had been won. Whether true or not, the one word "Peccavi"—"I have sinned" (Scinde)—was quite sufficient to proclaim the news. Jacob soon brought the Province into order, and interested many by his great experiments in his rifle gun and explosive shells—his commission was artillery. His health failed him in Persia, where he went in command of cavalry; but he returned to Jacobabad to die. He never once visited England during his service of thirty years. His whole life, and all his money, was given to his work.

Sir Charles Napier, with health impaired and a dying wife, left for England in 1847, only to return in 1849 as Commander-in-Chief; for whilst at home the news of the drawn battle of Chillianwallah created great panic, and Napier was begged to return at once. "If you don't go, I must," said Wellington. However, there lurked behind some veiled opposition. Napier begged hard that he might take on his staff Lieutenant Wood, of the Indian Navy—he knew the rivers, and could speak the language, and could easily raise a bridge train; for does not "Punjaub" mean "five rivers." But "No!" said the Directors. We remember in later years, when General Gordon was to rescue the Egyptian garrisons in the Soudan, he begged hard for the services of Zubehr. "No!" said the callous Government. Surely, is not this opposition calculated to shake the confidence of any great commander?

Napier left England on March 24th, 1849, and reached Calcutta on May 6th—forty-three days after leaving London. Then his suspicions were soon confirmed that some secret hostility was at work, for at his first interview with Lord Dalhousie, the latter said that in letters from England he had been warned against Napier endeavouring to encroach upon his power, and had answered, "He would take damned good care he should not."

Napier replied that Dalhousie was right in his determination, for there was no wish to infringe on his authority.

Goojerat had avenged Chillianwallah, so Napier on arrival found Peace—not War—but plenty of work for him to do, for he had to meet the discontent of thirty battalions either in or going to the Punjaub. The trouble arose from "Pay;" for before the country was annexed, service pay was drawn—but the Punjaub being now our own, this extra pay must cease. This transition required some tact, or certainly a little time, which Lord Dalhousie had not the statecraft to discern. He abused Sir Charles for the oil he used to quiet down the waves, and officially reprimanded him in a most offensive way.

"Myself have calmed their spleenful mutiny," Sir Charles could truly say; but not until he had fully thanked, for services received, his two distinguished Brigadiers—Sir Colin Campbell and Sir Walter Gilbert—for they, *tria juncta in una*, prevented without bloodshed what might have been a "Terrible Unknown," and only at the cost of £9 7s. 6d. per 1000 Sepoys, or £40 for the whole. And this was what Lord Dalhousie called "usurping the Civil Power." Alas! and the country in such a danger!

The reprimand in question, Sir Charles of course would not stand; neither would Sir Colin or Sir Gilbert. Sir Charles went home, to die in August, 1853, worn out at last with labour and his wounds. Sir Gilbert predeceased him. Sir Colin, thank God, was spared. But had this same trio only been in India in 1857, when Canning (so truly human) became the Governor-General and then first Viceroy, there would have been "Peace! Peace!" everywhere, and no horrors here to tell.

In 1855, Brigadier Colin Mackenzie (an Afghan hostage of 1842) had to quell an *emeute* at Bolarum, where the 3rd Regiment of

Cavalry thoroughly misbehaved. The contingent had been injudiciously reduced, and further, the change of name from "Nizam" to "Hyderabad" did not come to the contingent as it suits the rose:

> "What's in a name? that which we call a rose,
> By any other name would smell as sweet."

Sir Charles Napier was succeeded by Sir William Gomm (born 1780), a fine soldier in his time. He had been present in every battle from Vimiera to Waterloo. This made him aged seventy—an age too squeezable for a Governor-General like Dalhousie. General Mountain, his Adjutant-General, was dead, and General Tucker now succeeded.

After Gomm (1855), a General named Anson was moved up from a Presidency command and became Commander-in-Chief, with Havelock as Adjutant-General. Lord Dalhousie went home in 1856. He died in December, 1860.

The Persian War soon was over (1856-57). There Havelock, Outram, and Jacob had been the Generals. The latter took the cavalry, and unfortunately went ill, and died in 1857.

This brings us to early 1857, when Commander-in-Chief Anson was at Simla, and Outram and Havelock were *en route* from Persia, so soon to learn the news which had sent all hearts quailing. Their first move was to reach Cawnpore. One ray of light was on the ocean in a few troops bound for China, destined now for India; but all things were so scattered.

England's day, indeed, was dark! For every homeward mail brought terrible tales of slaughter. Mothers and their children were hurled to death in a manner too appalling. "Who can stay all this?" now became the cry at home. The wish of all went out to Sir Colin—a real patriot, for he harboured naught of his past ill-treatment, and only thought of the crisis. Call him a noble man! for, with a physique much shaken from the terrible hardships of Crimea, in twenty-four hours he left for India. He commenced his march up country on November 13th, 1857. Rash demons now forbear; and soon to the starving and besieged at Lucknow, it meant comfort—not despair.

CHAPTER XL.

THE GREAT REVOLT, 1857.

"ENVIRONMENT" shapes character for good or evil, quite as much as natural temperament and inherited qualities. It had its influence with Wellington and Napoleon, and perhaps it may have influenced those three men: Dost Mahomed, Jung Bahadur, and Nana Sahib. The former we have already shown to be a man much influenced by his surroundings—he loved the discipline, the drill, and the method of our rule, and remained a trusty friend.

Jung Bahadur was the second of eight sons of a "Kazi," or General commanding in North-West Nepaul. He was cool, courageous, and an expert in all intrigues and methods for seizing sovereignty or power. The methods always meant murder, or ill-treatment amounting to the same; his *coup d'etat* alone sent 150 Sirdars off the scene—all were shot or murdered in one night, and by dawn Jung was in charge of the treasury, arsenal, and palace, and Prime Minister of Nepaul. This happened in 1846, when he was only thirty years of age. Prior to this he had been caught at Benares, planning some war against the British. His personal courage was undoubted. Just two instances to show:—

"It is by no means an uncommon mode of execution in Nepaul to throw the unfortunate victim down a well. Jung had often thought that it was entirely the fault of the aforesaid victim if he did not come up again alive and unhurt. In order to prove the matter satisfactorily, and also be prepared for any case of future emergency, he practised the art of jumping down wells, and finally perfected himself therein.

... (Interior)

To face page 345.

"When, therefore, he heard that it was the intention of the Prince to throw him down a well, he was in no way dismayed, and only made one last request—in a very desponding tone—which was that an exception might be made in his favour as regarded the being cast down, and that he might be permitted to throw himself down.

"This was so reasonable a request that it was at once granted; and, surrounded by a large concourse of people—the Prince himself being present by way of a morning's recreation—Jung repaired to the well, where, divesting himself of all superfluous articles of clothing, and looking very much as if he were bidding adieu for ever to the happy valley of Nepaul, he crossed his legs, and jumping boldly down, was lost to the view of the Prince and nobles, a dull splash alone testifying to his arrival at the bottom.

"Fortunately for Jung, there was plenty of water—a fact of which, most probably, he was well aware. There were, moreover, many chinks and crannies in the porous stone of which the well was built; so, having learnt his lesson, Jung clung dexterously to the side of the well until midnight, when his friends, who had been previously apprized of the part they were to perform, came and rescued him from his uncomfortable position, and secreted him until affairs took such a turn as rendered it safe for Jung Bahadur to resuscitate himself.

"Such was the adventure of the well, which, marvellous as it may appear, was gravely related to me by His Excellency, who would have been very much scandalized if I had doubted it, which, of course, I did not.

"While in a story-telling mood, I may as well relate an account that was given me of the manner in which Jung distinguished himself on one occasion with a *must* elephant. The story is interesting, as it was by such daring feats that he won for himself the reputation of being the most undaunted sportsman in Nepaul.

"The elephant in question had been for some time the terror of the neighbourhood, nor was anyone found hardy enough to attempt the capture of the rabid monster. At last, so notorious became his destruction of life and property that Jung heard of it, and at once determined to encounter him. The animal was in the habit of passing along the narrow street of a village in the course of his nocturnal depredations.

"One night Jung posted himself on the roof of a low outhouse, and as the huge brute walked under the roof, made a vigorous leap, which landed him on the neck of the elephant, and in spite of all the efforts of the infuriated animal, there he maintained his position until he succeeded in blindfolding him with a cloth, and in securing him to a tree, amidst the shouts of the populace."

Jung Bahadur, now in power, accredited himself to England as his own Ambassador—where he soon became spectator as well as spectacle on all great State occasions. He had an interview with the Queen, and received a visit from the Duke of Wellington; and when that great warrior called upon him Jung said: "I felt it to be the proudest moment of my life." When asked, on his return visit, how he would return the visit of a Rajah, he replied: "I shall go just as I went to return the visit of the Duke." He returned to India most wonderfully impressed, and remained right loyal throughout the Mutiny.

With Nana Sahib it was different—for from 1851 he had a grievance with the Government, and the following may account for his fiendish rule:—"In the obituary of the *Times* of January 28th, 1851: 'Died at Bithoor, Bajee Rao, ex-Peishwa of the Mahrattas. His adopted heir, Nana Sahib, made an unsuccessful application to the Company for a continuance of the pension of £80,000.'"

When Bithoor was sacked by Sir Hope Grant in December, 1857, some very singular and incriminating correspondence was found belonging to the Nana's chief counsellor and adviser, Azimoolah Khan. This man Azimoolah was of no birth, but of great cunning. The two worked together, and whilst in England they gained admission to Society, and lived a life of fashion—but (?) was it of the best? It was not the same that had so impressed Runjeet Singh and Jung Bahadur. Then came the Crimean War, when our failures, and not our victories, shone out the most; thus away went our prestige.

The winter of 1856-57 was passed by Azimoolah Khan and a Jesuit, by name de Bauville, at the Hotel Meziri, Constantinople, where any scheme of villainy was easy to mature.

Forgive the soldiers—for surely the Sikhs and the Afghans, so recently our enemies, would have joined the feud and completed the disaster. The Sahibs, whether civil or military, were then as now friendly and most considerate to the home-spun of India. But sedition had been coined from afar—say Constantinople; so when the Mutiny began, like our Mutiny at the Nore, the miscreants in the Bazaar and the riff-raff from off the streets were the sparks that fed the flame.

Certainly some men refused to touch the cartridge, *i.e.*, the bullet with the grease. This occurred with the 3rd Light Cavalry at Meerut in early May. The men (eighty-five in number) were most respectful, but at the same time most decided in refusal. The Court Martial that sat to try them sentenced them all to imprisonment for terms ranging from ten years up to "life." These men all wore the medals of the Sutlej, the Sikh, and other campaigns—all noble men when compared to cringing devils like the Nana and his Khan, and the rabble in Bazaars.

Was not the punishment excessive ? For was not the grease a question of religious sentiment, and one the soldier was entitled to hold opinion on ? It was Lord Clyde, Commander-in-Chief in India, who, when asked to stay the practice of European soldiers marrying native women, replied : "No ! although I disapprove, I never interfere with the soldier in the choice of his religion or the colour of his wife."

Now came the time, in early May, when the Nana gave his solemn oath that if the garrison would but trust him, they should be safely conveyed to the river, and sent down to Allahabad in boats. The agreement was made, and the worn-out garrison, women and children entered the boats—when the perfidious scoundrel withdrew his mask, and all were either drowned or shot ; and, alas ! that awful Well !

Our avengers were the brave and fearless Colonel Neill and General Havelock. The latter wrote to his wife on July 3rd, 1857: "Mutiny and treachery have been gaining ground every day since I last wrote, and you must expect to hear of great calamities. Lawrence still holds Lucknow triumphantly, but has great odds against him. Cawnpore has been entirely destroyed by treachery. I march to-morrow to endeavour to retake Cawnpore and rescue Lucknow."

The morrow came, and Havelock—with his staff of Beatson (A.A.G.), Tytler (Quartermaster-General), and his son as aide-de-camp—hurried off to overtake Major Renaud (Madras Fusiliers) and 820 men, whose position in advance would soon become critical. On July 11th he was enabled to write to the Governor-General: "I have this morning attacked and totally defeated the insurgents, capturing eleven guns, and scattering their forces in utter confusion in the direction of Cawnpore." Amongst the mutineers was the 56th, the very regiment which Havelock led on at Maharajpore (a battle in the Gwalior campaign, where Harry Tombs, Hope Grant, etc., first saw Indian service).

"One of the prayers oft repeated throughout my life has been answered," writes Havelock, "and I have lived to command in a successful action, viz., Futtehpore."

We next hear of Havelock at Bithoor, the stronghold of Nana—which meant his fifth grand victory—and in the words which we see recorded on his statue in Trafalgar Square, he congratulated his troops thus: "Soldiers! your labours, your privations, your sufferings, and your valour, will not be forgotten by a grateful country. You will be acknowledged to have been the stay and prop of British India in the time of her severest trial."

With this action terminated Havelock's first campaign. He had retaken Cawnpore; but to rescue Lucknow meant waiting for Outram and his reinforcements.

Outram on arrival would, in virtue of his superior rank, have at once assumed the command, but knowing how dear to Havelock was the object on which he was bent (and to effect which he had already made one most noble though unsuccessful effort), and with a magnanimity as rare as it was generous, he waived his rank, and left his old companion-in-arms in chief command.

"The important duty of first relieving Lucknow," said he in his Divisional Order of the night of the 16th, "has been entrusted to Major-General Havelock, C.B.; and Major-General Outram feels that it is due to this distinguished officer, and the strenuous and noble exertions which he has already made to effect that object, that to him should accrue the honour of the achievement. Major-General Outram is confident that the great end for which General Havelock and his brave troops have so long and so gloriously fought, will now, under the blessing of Providence, be accomplished."

Havelock thus honoured, he went ahead. His force consisted of two brigades of infantry—the first comprising the 5th Fusiliers, the 84th (Queen's), part of the 64th Foot, and the 1st Madras Fusiliers, with General Neill in command; the second was formed by the 78th, the 90th, and Sikh Regiment of Ferozepore, with Brigadier Hamilton (of the 78th) in command. The third brigade was artillery, and consisted of Captain Maude's, Captain Olphert's, and Brevet-Major Eyre's batteries, under the command of Major Cope. The cavalry were the small body of Volunteers often mentioned, and a few irregular horsemen of a native corps, with Captain Barrow in command. A small body of Engineers under Captain Crommelin was attached.

But to be almost destitute of cavalry was a severe handicap. The shock of the British horse in pursuit would have sent Havelock romping into Cawnpore and Lucknow days earlier. Alas! this "hope deferred" made many a heart sad in the pent-up citadel of Lucknow—for their feelings had been already sorely tried when, on July 2nd, Sir H. Lawrence (from a shell splinter which forced an entrance into his room) fell mortally wounded; he died on July 4th.

On June 30th, Lawrence made a reconnaissance in direction of Chinhut, in hopes of defeating a rebel force on its way to Lucknow; but on first sight of the foe, his Oudh Artillery all turned traitors, and they bolted, first having overturned the guns, etc. This gave an unfortunate ending to the venture, and entailed upon Lawrence a further concentration of his force within the Residency, for Mutchee Bhawm had to be evacuated and its arsenal destroyed. On July 21st, poor Dr. Brydon (the sole survivor from Afghan) was severely wounded in Mr. Gubbins' house. Now came this other chagrin. It was on July 25th that a message arrived from Havelock. The thoughts of all went out to him at once in the hopes of speedy succour. A month later—on August 24th—there arrived another message which told a sadder tale, for no relief could reach them for another month.

On August 27th, the supplies of the late Sir Henry were sold by auction : Beer, £7 a dozen ; a tinned ham, £7 10s. ; a small cake of chocolate, £4 ; brandy, £16 a dozen.

And thus dragged away the weary days until September 25th, when the troops of Havelock were fairly seen. Then, from every post, rose cheer on cheer—for a moment had arrived, the mighty hope, that makes us men.

Havelock and Outram reached their goal at last, but not a day too soon ; for the delay of another few hours might have sealed the fate of Lucknow. Outram wrote on the night of entry :—"We found they had completed six mines in the most artistic manner —one of them ready for loading, and the firing of which must have placed the garrison entirely at their mercy."

By September 27th Havelock had driven off the enemy from all the more advanced positions, and had rendered the garrison independent of native troops ; but until further succour came (November 27th), all was hard, strenuous work. Havelock had been at Jellalabad ; but even there he was not in a tighter grip than his present post at Lucknow.

These were anxieties perhaps only known to the Generals ; but to the garrison, the women and children, the arrival of Havelock and Outram came as an alleviation to that terrible tension of mind and body, which indeed meant relief.

"For this relief much thanks ; 'tis bitter cold, and I am sick at heart," says the sentinel when given rest from his post. So must

Lord Canning, then so sick at heart from weary watching and waiting, have blessed the relief when it came to him in the good news of the succour given to Lucknow; next of the capture of Delhi; and now the arrival from beyond the seas of his first reinforcements.

In No. 1238 of 1857, "The Governor-General *rejoices* to announce that the garrison of Lucknow is saved." The relief to those aching hearts was the good work of Neill, Outram, Havelock; but poor General Neill (1st Madras Fusiliers) had fallen. He had been conspicuous everywhere, ready of resource, and stout of heart. Then from Captain Norman, the young Acting A.G. of the Army before Delhi, came the good news that the city was successfully assaulted on September 14th; but alas! poor Nicholson had fallen. He, supported by Brigadier-General Jones (61st Regiment), had swept the ramparts of the place from the Cashmere to the Cabul Gates, occupying the bastions and defences, capturing the guns, and driving the enemy before him—receiving here his mortal wound, for he died. Harry Newbolt has set to verse the man he was:—

> "It fell in the year of Mutiny,
> At darkest of the night,
> John Nicholson by Jalandhar came
> On his way to Delhi fight."

Many others did mighty things, but how can we tell of all. However, we must mention Major Reid, Sirmoor Battalion; Hope Grant, 9th Lancers; A. S. Jones, V.C., 9th Lancers; Major Harry Tombs, V.C., R.A.; and the gallant corps of Engineers, Alec Taylor and Greathead; then the splendid batch of nine who fired the Delhi magazine, with Willoughby at their head, whose loss the Governor-General specially lamented. Of the nine only four escaped—Willoughby, Raynor, Forrest, and Buckley. The heroic Scully and four British soldiers (alas! whose names we know not) were never heard of again. Willoughby afterwards was killed, fighting in the jungle.

The fights now along the Delhi Ridge are graphically told by Colonel Alfred S. Jones, late of the 9th Lancers, a V.C.:—

"I bought my first charger from Sir Hope on joining the 9th Lancers, and he was never happy on a field day if I did not ride him. I used to like to hear Grant play the violoncello, but I think the horse was the best bond between us; though I used to draw maps and plans for him. He encouraged me to pass an examination in civil engineering, for employment in the Public Works, as I was then living on my pay alone. During the hot weather in 1856, Dr. Clifford, our assistant surgeon, returned from Netley Hospital with a new book, 'Parkes' Military Hygiene,' which I read with the greatest interest, as it is to this day the best and earliest authority on hygiene; and with Dr. Parkes' nephew, now Medical Officer of Health at Chelsea, I have long been associated (as Member of Council of the Royal Sanitary Institute) in controlling the Parkes Museum, built in memory of the founder of Hygiene.

"In January, 1857, I was sent to Meerut with forty Rough Riders, to take charge of and to break in some 300 young horses, collected from the Indian stud for the use of the 6th Carabineers (Dragoon Guards), expected to arrive from England. I do not know who proposed this unusual piece of foresight, but it was a happy thought, because it enabled the new regiment to send two squadrons to join the Delhi Field Force in June, mounted on the most advanced horses I could hand over after some three months' training, and a month or so under their own riding-masters in hot weather.

"On my return from Meerut in April, the regiment was almost every night turned out for incendiary fires at Umballa; but on May 1st, Captain Drysdale and I started for the hills, leave being granted as usual that year during the hot weather to about half the number of officers present in India.

"After dinner I made my way in moonlight down to our tent, and early next morning started by forced march to Umballa, with a light load on one coolie, telling my bearer to follow with my gun and other things; and the faithful* servant turned up with everything safe at the camp before Delhi, about the end of

* Thousands of other instances are known of this devotion to their "Sahib."

June—but Drysdale did not arrive till nearly a month later, having waited for further news.

"We passed Simla and proceeded up the Kulu Valley, intending to cross the Pass into Thibet, but called on the Civil Commissioner (Mr. Hay), to learn if the snow was sufficiently melted to allow of our crossing the Rotung Pass. He advised us to pitch our tent at the foot of the Pass, and take only a small one up to where we could shoot bears, and wait about till we found it safe to send for our heavy baggage to cross the Pass.

"Following that advice, Drysdale and I had a long day's walking and a shot at a bear, and on our return we found a newspaper from Hay with a letter describing the Meerut Sunday outbreak. I at once declared I would rejoin the 9th, and was laughed at by my companion, who did not want to give up his six months' leave until he heard the later news. When I reached Narkunda, three marches short of Simla, I found in the dak bungalow book, accounts of a panic at the latter place; but the refugees who had written them had returned, and when I got to Simla all was quiet there, news having arrived that a force had marched from Umballa to put down the rising.

"At Kussowlie, however, I met some civilians who had fled from the plains armed to the teeth, and one of them presented me with a pair of single-barrelled pistols when he heard I intended to ride on the postcart from Kalka to Umballa, saying he would be quite repaid if I shot one Pandy, and never returned them; but by the end of the siege I was able to return both, when I had an opportunity of buying a Colt revolver at a dead officer's sale.

"Umballa was almost denuded of troops, and the contents of my bungalow had been transferred to the Quartermaster's store, where I donned my uniform, and again took the postcart to overtake the regiment with which were my chargers. Alighting from the postcart, I received a very warm welcome from Grant, as the regiment was very short of officers, and he had little hope that any of those on leave in the hills would get the orders to return, or come (as I had done) without them.

"General Anson, the Commander-in-Chief, died in camp the same night, and we proceeded under General Barnard to join the Meerut contingent at Allipore, about twelve miles outside Delhi. I was sent forward with my troop as an advanced picket, and, reconnoitring, took a few prisoners, whom I sent in to be examined, but otherwise the country seemed deserted; and on my return to Allipore, the cavalry and horse artillery started at midnight (June 7th and 8th), under cover of a canal bordered by trees. We were hindered once or twice by a gun being upset in crossing big ditches. At first dawn we heard guns to our left or left rear, so we crossed the next bridge over the canal, and the main body (except the 4th Squadron 9th Lancers) went forward, while the 4th Squadron, of which I commanded the right troop, halted waiting for orders.

"A staff officer, Major Curzon (afterwards Lord Howe), pointed to a cloud of dust, saying it was raised by the enemy's guns retreating towards Delhi. We were soon galloping in pursuit among scattered infantry fugitives, one of whom fired at me from a hole in the ground as my horse leaped it, and so close was he that his bayonet scratched my horse's quarter. Looking to my left to keep my dressing with the squadron left-troop leader, I caught sight of a gun with six horses galloping towards Delhi, a little to the left of the direction in which our squadron was heading.

"On the impulse of the moment, I pulled my horse up, and the squadron ranks opened to let me through. Thus, when clear of the rear rank, I started away to the left, and as I passed the left of the squadron, managed to point out the gun to our regimental Sergeant-Major Thonger.

"My Arab charger, with only 10st. on his back, soon overhauled the six artillery horses with a 9-pounder gun to draw, but the plucky little drivers of de Tessier's battery, which had mutinied at Delhi a month before, kept on flogging their horses while they looked back at me, until the off-wheel driver felt a blow from my sword and dropped, clutching his reins convulsively as he fell, and thus brought the gun-team to a standstill just as the men came alongside the led horses.

"Finding my small party with the gun and six dead drivers on a wide open plain, with no clear idea in which direction the enemy or our people might be, my first idea was to get a spike ready to disable the gun if we had to abandon it, three such spikes having been served out the previous day. As I took one out of my pouch, we saw our 4th Squadron returning, having failed to catch the cloud of dust; and a little later Colonel Yule, with the other three squadrons of the 9th, turned up from the other direction—so we took powder and shot from the limber, and fired at a village still occupied by the retreating enemy.

"Soon after, the infantry of our force advanced, and on attempting to limber up and move on my gun, I found to my dismay that the four led horses had been wounded in the fore-legs from the 9th Lancers' pistols fired at the drivers, who crouched down to take cover under their horses, and my men had shot instead of spearing them in their excitement. I had only the two wheel-horses intact, and borrowed a spare pair from Alfred Light, R.H.A., to help drag my gun in triumph through the camp.

"I was proud when Hope Grant said to me: 'Hollo, Davy! I hear you have been taking a gun to your own cheek,' but he did not say anything about the Victoria Cross, nor did any such idea enter my head at the time. I was sufficiently elated by the knowledge that one of the six-pounder guns of a field battery, which had mutinied at Delhi and had fired against our people that day, could never be of use again.

"Conscious as I always have been that my powers of eyesight were below average, it was satisfactory to know that I had seen what had escaped the notice of my squadron leader, and to have such a tangible prize to excuse my impulsive breach of discipline, which, had I failed to overtake the galloping prey, would have entailed my trial by court-martial on a charge of having deserted my troop in face of the enemy.

"During the following month I was constantly, with my troop, turning out at all hours to assist in repelling sallies from Delhi, or on picquet at the Mound, whence I could reconnoitre in and beyond the Subzeemundi; and on my report of what I had observed one morning, I had a visit from Norman, the A.G., that afternoon. The enemy kept close behind the city walls at night, and only on one occasion did I encounter a single scout on my early morning rides, and he was silently and quickly disposed of.

"From that first meeting with Henry Norman to this day, I have been perfectly satisfied that his cheery countenance and devotion to duty day by day saved the situation in those depressing times; for except Hope Grant and Neville Chamberlain, our superior officers went about the camp like 'knights of rueful countenance,' in great contrast to the Captains and Subalterns, who had less insight into the situation and less responsibility for the fate of the British Raj, which we all understood turned on the capture of Delhi.

"Perhaps the most critical affairs during the siege took place on the afternoon of the 19th, and during the whole of the 23rd June, from dawn to dusk, on the hundredth anniversary of the battle of Plassey, which the mutineers hoped soon was to terminate the British Raj in India. On that day Colonel Yule, of my regiment, was killed.

"Although we wore only forage caps with white cotton covers, we suffered very little from sunstroke, and a good *bheesti* was always at hand to relieve a headache by pouring water from his mussock; but when turned out without any breakfast, and kept all day without any food, one suffered a good deal. This did not occur in my case, as I made a point of getting a plate of porridge on first wakening, and my syce generally carried a bottle of cold tea.

"When the rainy season began there was much fever and some cholera, and I thought much of the 'Hygiene' book I had read a year earlier, for European lives were precious; so Hope Grant got me appointed D.A.Q.G. to the Cavalry, and I set to work getting drains cut to carry off the water, and seeing to the cleanliness of the camp, for the plague of flies and smells soon became something terrible from the corpses of all kinds in the rear of the camp. Men and horses slain were mostly at some distance away, and disappeared rapidly under the influence of a hot sun, jackals, dogs and vultures; but with camels, who in the wet weather died in camp, the disposal of carcases was a serious matter. I used to borrow an elephant early each morning, to drag dead camels to certain spots behind trees in a hollow out of the direction of the prevailing wind, and as far removed as possible from the camp, in order to mitigate the evil; but when a slight shift of wind came across one of my depots, there was a great outcry. Under the circumstances, I think the health of the camp was improved—at any rate, I did my best; and when Lord Roberts came off the sick list, he adopted my system in the artillery camp, of which he held the same appointment (D.A.Q.G.) so graphically described in his book, 'Forty-one Years in India.'

"I was always anxious for any reconnoitring or other Quartermaster-General's work, but in a standing camp Grant could not find me much beyond scavenging, and going round the batteries to keep him informed of what was going on at the front, which was not much until the siege train arrived, towards the end of August. Then Major Anson, aide-de-camp to Grant (and afterwards the M.P. who fought for the Purchase System so doggedly in the House of Commons in 1874), used to accompany me in most interesting moonlight excursions in the Kudshia Bagha, and later on watch the building and mounting of Major Brind's battery (entirely of fascines aud gabions) in one night—September 7th and 8th—then of Edwin Johnson's, near Ludlow Castle; and Fagan's, near the Custom House.

"That was, indeed, a week of excitement, when we passed most of our time in the natural nullah which formed the first and only parallel connecting the breaching batteries, and we helped to extinguish the fires which broke out in the embrasure from time to time. The orders for the assault assigned the cavalry and horse artillery to the duty of descending from the ridge as soon as the attack on the breach had succeeded, and to connect the infantry attack on the city with General Reid's on the Subzeemundi; and in the event of the enemy quitting the city by the Lahore and westward routes, the cavalry were to pursue.

"Before day-break, therefore, I was in my position as Quartermaster-General, descending just in the rear of Brind's battery, the cavalry following about one hundred yards behind in column of threes, when I saw Brind rush out of his battery to meet and stop me—and at the same moment there came a volley of grape shot, which cut the brass buckle of my chain reins and wounded my charger slightly. Major Brind told me to bring on the cavalry, as our storming party had then reached the Moree Bastion under General Nicholson. True enough! for there I saw our flag; but, alas! too sad! for soon we heard Nicholson had been mortally wounded.

"Grant then sent me with a verbal message to Sir Archdale Wilson, asking for a company or two of infantry to clear the trenches, or check the galling fire under which his men had to sit for some hours or so. While I was away on the quest of infantry, I met Wilson close to the Cashmere Gate, and he shook hands with me in excitement and joy; but he said I must go for infantry from Reid's right attack, as he could not spare any out of those already in the city, into the square of which I had entered, riding by Wilson's side, through the Cashmere Gate.* Accordingly I galloped back through the Gate, and up to the extreme right of our position on the ridge.

"There I first saw a regiment of Ghoorkas, looking very dejected, and they seemed surprised when I told them that Delhi was taken, and that I had been inside the walls. I soon found Neville Chamberlain with his arm in a sling, and when I told him Grant's need, and that Wilson had sent me on to get infantry from

* In blowing up the Cashmere Gate at Delhi, Lieutenant Home, R.E., won the Victoria Cross. Doubly sad reads the death of this young hero: for in this last explosion at Malaghur, poor Home, the best and cheeriest of good fellows, was destroyed. The blowing up of the Cashmere Gate took place on September 14th, 1857, by Lieutenant Home and Lieutenant Salkeld, R.E., Sergeants Smith and Carmichael, Corporal Burgess, and Bugler Robert Hawthorne. Salkeld lived just long enough to learn he had won the V.C.; Home wore it for a few days, and then was killed; Burgess and Carmichael were killed; Smith and Hawthorne alone survived.

the right attack, he wrote a short note to Grant, with which I galloped back; and though Chamberlain was quite calm, and pleased to hear that I had actually been inside the Cashmere Gate, I felt sure that things had not gone well with the right attack. When I rejoined Grant, he told me to read Chamberlain's note to him. While doing so I felt a blow, and expected my horse to fall; but there was a small hole in the stuffing of the plain saddle, and another under the cantle, where a bullet had entered and left the saddle stuffing without touching my horse's back—one of those curious courses bullets *do* take. Another just then lodged in Major Rogers' brain as he sat on his horse in front of the Carabineers' Squadron.

"The position was indeed a trying one to cavalry and horse artillery discipline, and Grant sent me with another urgent message to Chamberlain, who at once sent some companies, who cleared the enemy out of the trenches; and by about noon we were able to get out of range from the walls. But our people who had occupied Moree Bastion were firing the guns, with ammunition they found there, against the part of the city occupied by the enemy, and I was hailed and asked to get them a further supply of ammunition. An artillery officer threw me a note of what he wanted. About 2 or 3 p.m., after some delays, I handed it up to him over the broken parapet. By that time everybody was exhausted; and I remember sleeping soundly all night on the roof of Ludlow Castle, when perfect silence succeeded a week's breaching fire with our heavy guns.

"A week later the whole city was vacated by the enemy, and by the end of the month a column was formed to march down to Lucknow—three squadrons of the 9th, the 8th and 75th Foot, with Colonel Greathead in command, which justly should have fallen to Grant, who was left stranded at Delhi with only one squadron of his regiment and two squadrons of Carabineers remaining as a cavalry brigade.

"I therefore went to Grant to resign my Staff appointment, in order to march with my troop; and he gave me a lecture on the wickedness of volunteering, and said he wished to go with the regiment as I did, but it was his duty to submit to remain, as I ought to do. Before I left, my horse having recovered from a wound he got by a grape shot on September 14th, 1857 (the assault and capture of Delhi)—the very last hostile discharge from the model bastion—I sent him back as a present to Sir Hope, and Grant rode him all through his campaign with Lord Clyde for the relief of Lucknow. Grant had been the only man who could ride him in his hot youth, and I think my care for his old favourite was a great bond of friendship between us.

"The 8th at this time had for Adjutant, Lieutenant G. Walker, as smart and kind an officer as ever lived. He was soon after promoted into the 12th, which regiment he commanded through the second Afghan War.

"In the affairs at Bolundshuhur and Alighur, I had an early opportunity of noting the want of enterprise of our commander, Greathead; for under the former we were held at bay for two or three hours by a weak rearguard, with guns extemporised out of iron sockets for telegraph poles and mounted on buggy wheels, while the enemy were getting away with their baggage intact—and at last, instead of employing his infantry to take the town, it had to be done by the 9th Lancers, who might have been more usefully employed in the open country on either side.

"At Alighur I happened to be on baggage guard, and finding the road blocked by infantry as I approached the end of the day's march, I rode forward to ascertain the cause of the delay, and was invited to take a cup of tea by the General, who was breakfasting by the roadside. Just then Norman rode up, and was accosted by Greathead: 'When shall we advance to attack?' (Alighur)—and this supine General received his Adjutant-General's reply, 'I have ridden through the place long ago. Did you not hear the magazine in the town blown up?'

"A few hours later an 'express' arrived from Agra, begging that our force would come there, instead of going direct to Lucknow, because they were expecting a force of mutineers from Gwalior. A forced march was therefore made, bringing us to Agra Fort about

8 a.m. on October 10th, 1857, where we found all peaceful—the ladies and children riding and driving outside the fort—and General Greathead was told by the civil authorities that the enemy was twelve miles away across the Kulla Nuddee.

"The column was then conducted to the parade ground, about two miles beyond Agra Fort. I heard Greathead himself proclaim that any officers who wished to visit the fort might do so provided they returned by noon, when he intended to advance to the attack. There were high crops of cane bordering the further side of the parade ground, so that nothing could be seen beyond, and no reconnaissance or picquet was thrown out to test the civilians' information. A visit to friends who had been shut up in the Agra Fort all the hot weather, was the opportunity for pleasant gossip with their rescuers coming from Delhi; and it is not surprising that most of the officers in the column took advantage of Greathead's offer, and had to come scrambling back when disturbed at breakfast in the fort by heavy firing in the camp they had left.

"Their return was much impeded by the stream of dhoolies taking our wounded to a hospital, which the forethought of Civil Commissioner C. Raikes' good wife had prepared in the Ali Musjed, a mosque built of pure white marble in the fort at Agra.

"Captain French (the Adjutant), wounded Captain Drysdale, and Clifford (Assistant Surgeon), with myself, preferred to have breakfast in the shade of an old graveyard wall against our camp, where the mess dhoolie was established; but Clifford and I intended to ride to the fort after breakfast, and I had ordered a spare troop horse to be saddled with my plain saddle for that purpose, leaving my charger prepared for the march when we should return to camp at noon.

"We had just finished breakfast, when round shot fired from behind the high crops began to drop about us. The troop horses had been picketed, but not off-saddled, so by the time I could get my sword and pistol and mount the spare troop horse, the 4th Squadron had formed up (many men in shirt sleeves), and some of our guns had opened fire at the smoke of the enemy's guns.

"The 4th Squadron then went off as escort for the first troop of Horse Artillery limbered up. It was commanded by Blunt, and we went straight to our front through the high standing crop, which proved to be but a narrow strip, and then we had a good view of the enemy's guns firing at our camp. Blunt's guns at once came into action, but very soon we saw a body of cavalry moving round our left, and Blunt said they would get into Agra if we did not return; so he limbered up, and we went back through the crops. Blunt immediately stopped two of his guns, and asked for one troop to be left with them to show a front to the enemy there, so French stopped his troop and left it under a Sergeant-Major, as he knew my troop would have a fight on the parade ground, although when the troops were separated he ought to have stayed with his own.

"The guns found an old cart track through the crops, and so got through rather in front of the escort, which had to form up as the men struggled through to the open ground. There were but twenty-three men in my troop, and I wanted to form single rank, when I saw a strong squadron coming down upon the guns; but French said 'No time,' and gave the order to 'Trot—March,' he and I riding close together in front of the troop. I saw his horse turn to the left and pass the flank of the troop, when I gave the word 'Gallop.' The opposing squadron halted, and I saw another squadron *en echelon* behind it, the first being 'Red Irregulars,' and the second in light blue Company Regular uniform.

"As I noticed the squadron leader aiming his carbine at me, I rode at him, and had the hilt of my sword on my hip ready to pass him 'third point,' when he fired; and feeling my bridle arm useless, I took the reins in my sword hand, and pushed my horse into the rank behind him, when my horse reared straight up and slipped me over his tail. I lighted on my feet with the stirrup come away from the body of the plain saddle I was riding on. I distinctly remember kicking the stirrup off my foot on the ground, and then keeping a '7th guard' to my head with my sword.

"The force of many cuts was broken, and those horsemen were looking about them in great anxiety, all the time they were hustling each other to get a cut at me. One man on the left of my troop tried to get at me, but was driven off by the crowd, and when I sank on my side from loss of blood, I pulled out my revolver and shot a man with the first barrel, afterwards cocking it and firing one or two more times. All that time my enemies were dropping off, and at last there were only two left, who were bent upon taking my sword, which hung by the leather sword-knot from my wrist; and when I opened my eyes after feeling cuts on my right elbow, I knew they were trying to cut the knot; then I shook it off my hand, and neither heard nor felt any more enemies.

"I was carried in a dhoolie into the Ali Musjed, very kindly nursed by two ladies, and in two months' time was well enough to travel home. Brigadier-General Sir Hope Grant now joined, and took over command from Greathead, and the Force proceeded down country in pursuit of its original plan to join Sir Colin Campbell for the relief of Lucknow."

* * *

It is my good fortune to have a fellow resident at St. Helens, Isle of Wight, one J. Whittington, a retired Master Gunner of Royal Artillery, aged seventy-seven. He is the happy possessor of seven medals, including the one recently created for Distinguished Conduct on Service. It was for his service in the Mutiny that he earned this last reward. Even at his advanced age, he is the life and soul of the village—the No. 1; his offices extending from instructor on the rifle range to custodian of the village club and church, in fact the handy man.

The column led by Grant arrived at Bunnee Bridge, there to await Sir Colin, who was hastening up country with troops from England—each one worth his weight in gold—so let us join the battery to which Whittington belonged. It was commanded by Captain Middleton, and left England on April 18th, 1857, for service in China. This expedition was abandoned for a time, and all troops then *en route* were diverted by Lord Elgin to the succour of our countrymen, women, and children, in their hour of danger and despair. The battery reached Calcutta on September 17th, the day that Delhi fell, and was greatly welcomed by Lord Canning, the headquarter staff, and others. The gun sheds were handed over for their habitation, whilst the hotels behaved most hospitably, giving to all who could avail themselves of it, a free tariff.

The battery, however, was only too eager to get up country and be at work. At Allahabad, they took over four guns, two howitzers, and 142 horses, all so recently the property of the rebels, and started on November 5th for Lucknow, the outskirts of which were reached on the 15th, enabling them to join in a fight around the Martiniere, where Lord Clyde (Sir Colin) was hard at it for the rescue of the still imprisoned garrison at the Residency. The fight continued up to dusk, when orders were given to discontinue and to seek some rest, preparatory to the advance on Lucknow in the morning.

The next day broke—it was the Sabbath, November 16th. The enemy appeared in great force and in position around the Secundra Bagh, the garden walls of which were all loopholed and entrenched. Blunt's Horse Artillery and Peel's 24-pounders were soon in action, the latter exciting great admiration, for they were the first seen of the new arrivals. Soon the shout of "A breach! a breach!" was heard, and the scramble came as to who should enter first. The 93rd and the Sikhs (4th Punjaub, Major Barnstone) ran like madmen; and what went on inside is hard to tell. Certainly, Cawnpore had been avenged!

Next came the Shah Nujjeef, a large domed mosque with garden walls similarly loopholed as above. A regular work of masonry had been thrown out from the entrance, the top of which was crowned with a parapet; this took three hours to take, and cost to Whittington's battery twenty-one killed and wounded and forty-two horses.

Peel's* battery was caught within the range of some secreted

* Captain Sir W. Peel, K.C.B. (honours received a day or two before death), was the third and favourite son of Sir Robert. He brought the "Shannon" from China in 1857, and accompanied Sir Colin, with his tars and heavy guns, to the relief of Lucknow, where he was wounded. He died of confluent small-pox on April 27th, 1858. His eminent ability, daring and thoughtful courage delighted everyone.

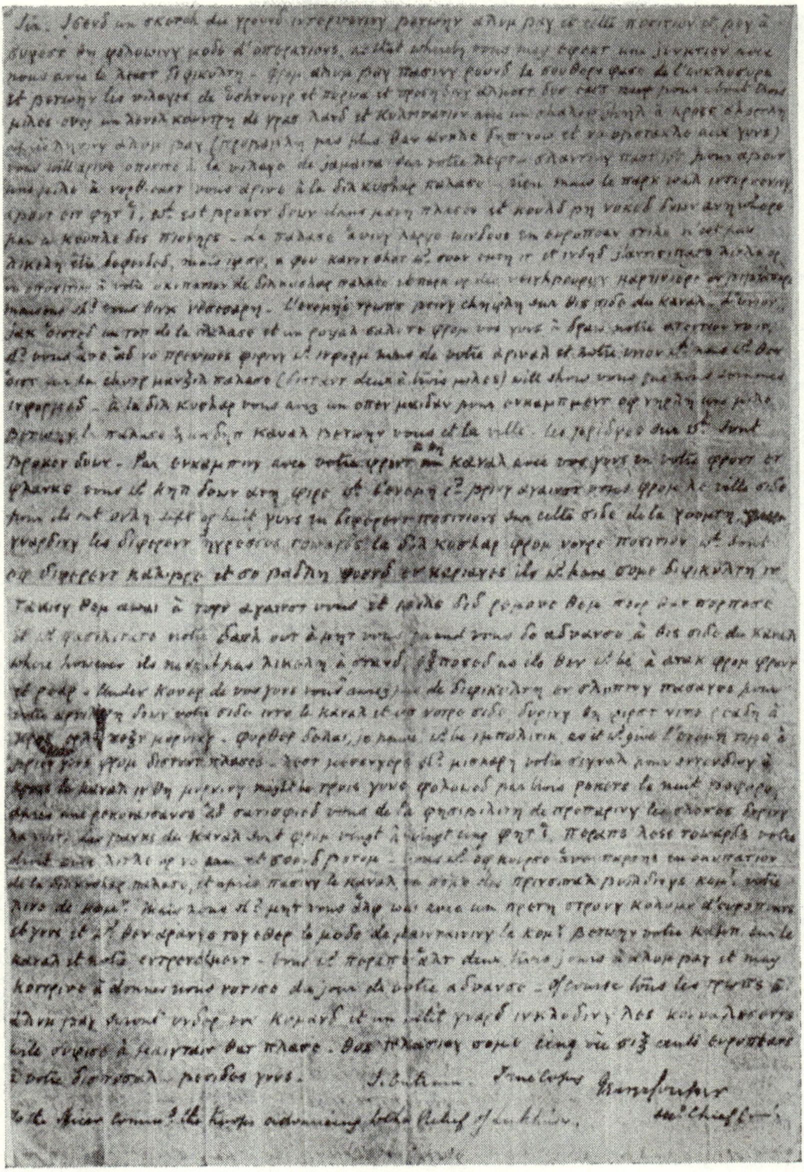
To the Officer Comm^d. the Troops advancing to the Relief of Lucknow.

Secret Despatch of Sir George Couper to Sir Colin Campbell.

expert rebel marksmen. It meant their withdrawal for a time, when up came in their place the guns of Middleton, the right gun of which was left soon after with Whittington and one other, to work and feed with shot and shell; and thus they won their Distinguished Service medal. The howitzers, with plunging fire, soon emptied the enclosure, for a ricochet or two had breached the further or rearmost wall, by which the rebels then escaped.

The Brigade Mess-house now blocked the way, but not for long. Young Wolseley (Lord Wolseley), with his 90th men, was about the first to enter—which meant he did not leave till the coast was clear and eventide arrived.

At nightfall, the battery well deserved some food and rest, for they were all dead beat; so sentries from the 93rd Highlanders protected the guns whilst the men slept. It was no sooner dark than some rebels in hiding commenced sniping from the adjacent trees (killing or wounding certainly a dozen men). It was hard to locate the sepoys by night, but at the first break of day, our guard (the 93rd) quickly espied them, perched like so many monkeys, amidst the branches—and brought them down, one by one, with a good resounding thud.

From this deadly hand-to-hand encounter, the scene now quickly changed; for Outram and Havelock had watched with eagerness the coming of Sir Colin*, and slowly as he carves his way with fire and sword, so surely do their pent-up feelings yield, and these three brave men soon quickly meet hand in hand together.

Now, what was uppermost in all their thoughts was to get the women and children, and the wounded, into other scenes; their battered, beaten, wretched tenements must know them now no longer. The troops, still blood-stained from off their march, formed up, and lined the way, to protect and help their weakened comrades —refreshened, let us hope, at last with thoughts of home. Whispers

---

* Herewith is given the reproduction of a secret despatch sent out by Sir George Couper, Secretary to the Chief Commissioner, to aid Sir Colin on his way.

now went round the camp that Havelock was ill. He was with us, but on the sick list; and all lived in expectation of good news from day to day—it was not to be!

> "In joys of conquest he resigns his breath,
> And, filled with England's glory, smiles in death."
> —(November 25th, 1857).

The troops around the Alumbagh rested on the 26th, for it was Havelock's funeral. "The Funeral" of Shakespeare, as written by Hamley, remains a masterpiece. Would that the same pen could depict the scene around the bier of Havelock. English, Irish, the Highlander, the Sikh, the rescued and the rescuers, from Lord Clyde to the youngest child, all realised their loss.

Time, however, was precious, and on the following morning (the 27th), Clyde and Outram parted. The latter was given a division of 4000 strong to hold the position from Alumbagh to Bunnee Bridge; whilst the former and Sir Hope Grant (now recently arrived from Delhi), with the wounded, the ladies, the children, in a long never-ending tail, moved off for Cawnpore, and not too soon; for the distant sound of guns was heard, and what it meant was quickly learnt—the Gwalior mutineers, an intact and separate force, had marched in from Calpee to Cawnpore, and thus General Windham, hard pressed, had fallen back from outside the city into his entrenchments (see plates).

Lord Clyde moved slowly on: it took three days to reach the Ganges, and then thirty hours to make the passage. When once the convoy was well ensured, Lord Clyde went hard to work. He captured the enemy's camp on December 6th, and followed up this success by the victory at Bithoor, utterly dispersing the Nana's troops. It was here that Whittington laid the gun which brought the Nana down; his elephant was killed, but the miscreant was able to mount a horse and bolt.

Flying columns were now sent in pursuit of the rebels *via* Allahabad and Benares. On February 18th, the rebels made a stand at Sultanpore, but were soon routed with a loss of thirty-five guns. Whittington now returned to Lucknow with his battery, to

Gallant attack of Windham's small force on the Gwalior contingent, Nov. 26th, 1857, at the Pandoo Nudee.

To FACE PAGE 366.

assist at the final capture of the city, when all hoped that the day was at hand when these sickening scenes would cease.

On March 19th, twenty-two Royal Engineers, in carrying out Sir Colin's rule of searching each captured position for fear of secreted powder, came to sad disaster, which, briefly told, is this: In the Jumma Musjid, no less than nine cart-loads of gunpowder were discovered. The powder was packed in tin cases, and it was resolved to destroy it by flinging it down a well. A line of men was formed and the cases quickly passed from hand to hand. The first case flung down struck against the side of the well and exploded. The flame ran from case to case along the whole line till it reached the carts. The cases in the very hands of the men exploded, the nine cart-loads went off in one terrific blast of flame and sound, and with one exception, the whole party (numbering twenty-two men, with two Engineer officers in command) was killed. The only man who escaped was the one who threw the fatal first case down the well! This sad disaster, on what we hoped to be our finishing day, came as a great disappointment and grief to us all—and to no one more than Sir Colin.

March 20th and 21st were desultory days. On the 22nd, Lord Clyde was prepared to thrust an attack home by an assault on the rear of this stronghold—slightly weakened by this explosion. It was in the early stages of this advance that the aide-de-camp to Lord Clyde had his horse killed under him, apparently by one on the fringe of the wood or compound, for a heavy-paunched Sowar emerged from thence and attacked the now dismounted aide-de-camp. On seeing this, Gunner McCarthy rushed to his aid, but he was severely mauled and cut about, and died of lock-jaw a few days after.

Captain Middleton then joined in; his sword broke, and his pistol mis-fired. A 9th Lancer escort to the Chief took two turns at the Sowar, but each point was parried. At last he was shot from afar by an infantry soldier. On his portly form were found at least 400 gold mohurs (the hoarded wealth of India was then most marked—gold mohurs were hidden everywhere).

Shortly after the above quaint incident, Lord Clyde had occasion to pass the battery, when his grey horse was shot under him. Captain Middleton at once offered him his horse. "No, sir! damn it; you will require it yourself, for your battery is badly wanted from where I have just come." Off we went. Lord Clyde then laughingly said "I will get a spare horse from my groom."

The Mutiny, in one hour after this, was over as far as Whittington and his battery were concerned. The final capture of the city and dispersion of the rebels became *un fait accompli* on March 22nd, 1858. It was, however, left to Sir Hugh Rose to give the final *coup de grace* in his brilliant campaign in Central India—when on April 1st, whilst besieging Jhansi, he was attacked in force by Tantia Topee. Without losing grip of Jhansi, he defeated Tantia, capturing eighteen guns, all his elephants and camp equipage, and took Jhansi on the following day. The Ranee fled to fight another day, for on June 19th she fell gallantly fighting at the head of her troops in defence of Gwalior, which was captured (twenty-seven guns) by Sir Hugh Rose, when Maharajah Scindia was restored to his capital. Sir Hugh Rose was in after years created Lord Strathnairn for his services.

It is impossible to think over the atrocities committed by the rebels in this great revolt, and of the heroism displayed by men, women, and children, the victims and survivors, without a shudder at the enormity of the one, and a thrill of pride in the glory of the other. But what was England doing all this time? Her recent terrible display of unpreparedness for war, as witnessed in the Crimea of 1854-55, was again exhibited in 1857.

On January 23rd, 1857, General Hearsay reported from Calcutta that his native troops were out of hand and unreliable. On January 22nd, a whole regiment entirely broke, to be followed by others in February. On March 2nd, Mangal Pandy, a ringleader, was shot. In early May, Meerut and Lucknow broke into open revolt, and massacres were frequent. In early June came the mutiny and massacres at Cawnpore.

See page 366.

Arrival at Cawnpore of the relieved garrison of Lucknow, Nov. 28th, 1857 (from outside the entrenchment).

It was not until June 29th, when Lord Ellenborough and Mr. Disraeli called the attention of the country to the state of India, that Mr. Vernon Smith, President of the Board of Control for India, replied "That by the middle of July 14,000 may reach India, but that he could not concur with some Members who had spoken as to our Indian Empire being imperilled by the present disaster." At the same time Palmerston was hotly opposing the construction of the Suez Canal.

England's pulse is sluggish, and hard to quicken. The trial of a fashionable and pretty young lady, Madeline Smith, for the murder of her betrothed (which commenced on June 30th and lasted ten days)—verdict, non-proven—seemed to overshadow the tale of horrors that came daily pouring in. And even when the war was over, and the time came for Sir Archdale Wilson and other heroes to be honoured at the Guildhall for the part they had taken in the capture and fall of Delhi, the Lord Mayor had not exactly learnt his part—for the toast was given: "To Sir Alexander Wilson, the Defender of Delhi."

Like the late Lord Hill, whose invitations to the Mansion House always included a Lady Hill—a phantom Lord Hill had been looking for all his life, but never found; again, in later years, with Holman Hunt, O.M., the officials found it easier to announce him always as Mr. Alderman Hunt.

CHAPTER XLI.

THE "SIXTIES" AND "SEVENTIES" AT HOME AND ABROAD.

"No game was ever worth a rap
For a rational man to play,
Into which no accident, no mishap,
Could possibly find its way."

IT was in June, 1860, that H.M.S. "Actæon" completed her surveys in the Gulf of Liantung, and on rounding the promontory of Lian-li-Shang, came on an indentation or harbour. This was carefully surveyed. The first ship to anchor there was H.M.S. "Algerine," under Lieutenant-Commander Arthur. The surveyor at once christened the haven Port Arthur.

"And I saw two armies close—
I could almost hear the clarions
And the shouting of the foes,"

when Russia and Japan contested for the same in 1904.

Lord Canning (after his seven years of Indian rule—years which spanned the Mutiny and the loss by death of his noble wife) reached England in April, 1862. Those who knew him, at once saw that the labourer's task was done. He died on June 20th, 1862. Canning was the first, and perhaps the best of Viceroys; for India, after the Mutiny, was handed over to the Queen. She received the country in full Queenly sympathy: "Firmly relying ourselves on the truth of Christianity, and acknowledging with gratitude the solace of religion, we disclaim alike the right and the desire to impose our convictions on any of our subjects."

Canning was the youngest child of the great Earl (1770-1827). It was on the death of Pitt, when England was sore depressed—

for Napoleon had overthrown Austria, had annihilated Prussia, and was keeping Russia under his thumb—that the Earl anticipated the "Wonderful Man" and seized the Danish Fleet, thereby preventing further spoil.

At this same time, however, Lord Castlereagh, as War Minister, launched the country into that most unfortunate Walcheren Expedition, upon which policy the two Ministers, Earl Canning and Lord Castlereagh, agreed so much to differ that a duel ensued. Canning retired into private life for a time, and the country lost what they sadly wanted—his services at Vienna in the Congress of 1814; whilst Castlereagh kept hard at work until 1822, when his death was sad, for it was by his own hand.

Lord Elgin (born 1811, died 1864) now succeeded. He made his mark when Governor of Jamaica and afterwards of Canada; and later on he won distinction in the Chinese complication arising out of the affair of the "Arrow," which required a diplomatic as well as a martial solution—for the Chinese had in their wrath torn down the British flag, and carried off the Chinese crew, refusing to listen either to Captain or to Consul. Whilst on this Mission in 1857, he took the responsibility of deciding that China must wait for her chastisement, by diverting all troops of the Expeditionary Force from Hong Kong into Calcutta, to meet the terrible crisis of the Mutiny. Within twenty months of reaching Calcutta as Viceroy, he was dead.

In the Vale of Cashmere, the dying Viceroy quietly made his will; asked Colonel Strachey, R.E. (Sir Richard) to design a tomb, and approved of the design when submitted to him; dictated the words of a telegram to the Queen, expressing his duty and requesting her to appoint a successor; gave instructions respecting the return of his family to England; took an affectionate leave of all present; and waited calmly for the end.

"Send out John Lawrence!" was now the cry from India. He arrived early in 1864, when India was finishing off a most serious frontier war.

For two months had a British force under Sir Neville Chamberlain been hemmed in, in the Umbeyla Pass—a pass nine miles in length, with Mahabun on its right, and the Gurroo Mountain on its left. When the troops were well within the Pass, the neighbouring tribes, backed by their priests, resisted them at every point. Chamberlain, thus thwarted, entrenched himself on the Gurroo (the Eagle nest), and on the Mahabun (the Crag) he placed two pickets of considerable strength. Round these posts all the fighting took place. The Crag picket was wrested from us on three different occasions, giving us a loss of 349 officers and men. Each tribe went at it in turn.

On November 21st, 1862 (having been driven out on the 20th), the General assembled all, and recovered it in style. The 71st, 5th Ghoorkas, 5th and 6th Punjaub Infantry, and a strong force of mountain guns, with the 101st in reserve, made the enemy vacate. The General and Colonel Hope were wounded, Saunderson and Dr. Pyle (101st) were killed, and many others maimed.

General Garvock now succeeded, and reinforcements were awaited (7th Fusiliers and the 93rd). When all had arrived, Garvock took the offensive, and after two most spirited fights, completely dispersed the enemy, but only just in time; for the Government were on the "Majuba" game, and had decided to withdraw.

Now followed an incident so worthy of remembrance. It was bloodless, but gave that wholesome lesson which all ruthless fanatics deserve. It was to make one tribe destroy the fastness of another. The Bonair Khans were to set fire to the cantonments of Mulkah—the headquarters of the fanatics, and the seat of all the trouble.

Rennell Taylor, a man of wisdom, courage and firmness, was in charge; and with him went six British officers—Colonel Adye, Colonel A. Taylor, R.E., Major Roberts, V.C., Majors Wright and Johnstone, and Lieutenant Carter—but no English troops. The distance was twenty-six miles, and the result was complete; for Reynell Taylor wrote: "The spectacle of a tribe like the Bonairs doing our bidding, and destroying the stronghold of their own allies in the war, at a distant spot, with British witnesses looking on, must have been a thoroughly convincing proof to the surrounding country of the reality of our success, and of the indubitable character of the prostration felt by the tribe which had been the foremost in opposing us."

The reign of Lawrence was in a sense uneventful, if we exclude cyclones, famines and a commercial crisis. Sir Hugh Rose (Lord Straithnairn) was Commander-in-Chief, and Sir Bartle Frere the Governor of Bombay.

The reliefs for India in 1864 (of which our old friends, the 12th Regiment, formed a part) still took the route round the Cape, in the same old three-decker Indiamen, for those days were still Napoleonic-Wellingtonic. The inspections of the troops embarking that year (1864) were taken by Sir George Brown, Commander-in-Chief in Ireland (died 1865, aged seventy-five) and General Sir J. Bloomfield, of Cork (died 1880, aged eighty-seven)—both Peninsula and Waterloo officers.

On the voyage, the first incident of note was the passing of the "Kearsage;" and from her we learned that she had just sunk the "Alabama" in Cherbourg-Southampton waters. An English yacht, the "Deerhound," and the whalers from off the "Kearsage" had rescued the drowning crew. The Stars and Stripes now sheltered both belligerents, and now, practically, the bitter feud between North and South was over.

The "Alabama" had been built at Birkenhead in 1862, obviously for a man-of-war; but between the letters that passed between the Consuls, the Commissioners of Customs, and the Law Officers of the Crown, the orders for her detention arrived just a day too late. The next we heard of her was with the Confederate Flag; so with her gallant Captain Semmes, and with armament of the latest pattern, she became the terror of the seas for a good two years. The last we heard of her was greatly to our cost, for her depredations were summed up against us, at Geneva (June 15th, 1872), for 15,500,000 dollars.

Then the cyclone mentioned as a feature in Lawrence's reign was of a most terrible and destructive nature, and raged its fury chiefly off the Sandheads near Calcutta, on October 26th. The regiment ran the gauntlet, and one ship, the "Alnwick Castle" (with four companies on board), became the show ship at Calcutta on her arrival—for all her superstructure had been fairly swept away, and the bare hull alone was left.

The ship "Aliquis," with the headquarters of the 12th (of which I formed a part), was reported lost, for her voyage had taken 103 days, which meant we were at sea when the great cyclone occurred; but, good luck for us! we were only on the fringe of it, and so reached Calcutta safely. The word "Aliquis" tempted my curiosity, and not wishing my classical education entirely to go for naught, I asked of the captain the reason for the name. "Oh!" replied the captain, "Marshal won her from Greene (the two leading shipowners of the time) in a bet, when her name was changed, and the new owner thought 'Anything' would do."

Another voyage in another ship, the "Megæra," also took the headquarters of the 12th from Portsmouth into Cork. The old ship soon came to great disaster, for when bound to Sydney, with thirty-three officers and 350 men—overcrowded and quite unseaworthy—her captain had to beach her off the island of St. Paul. The stranded troops were at last sighted by a passing ship, and so conveyed to Melbourne. Her end thus seemed pathetic, for, tired out, weary and worn, she seemed to ask her captain (what is written on many a tomb): "Requiescat in pace!" Surely old-age pensions might be extended to the carriers of the deep, thus affording them a peaceful retirement, before going to the bottom with all their living freight.

It is needless to say that all officers and men were fully acquitted by court-martial for the loss of the ship, and it rested with Lord Lawrence to proportion the blame for sending such a coffin-ship to sea; for he sat on a Royal Commission to enquire into same.

The scene now changed to the House of Commons. "I will unmask the villains who have sent brave men to death," said Mr. Plimsoll, when Mr. Disraeli said the Government proposed to withdraw the Shipping Bill. Mr. Speaker then asked the hon. Member to withdraw. The Member said "Yes, I do mean so to apply it, and I don't mean to withdraw it;" and walking up the floor of the House, he threw a paper on the table, remarking "That is my protest against the withdrawal of the Bill," and shook his fist at Disraeli.

Expressions of regret (for this little ebullition) soon put matters right, and Plimsoll gained the day. All honour to Plimsoll! for now his mark is universal; and, thank God! it saves annually many, many a life at sea—he is our "Jenner" of the deep.

Then the wreckage on the banks of the Hooghly baffled description. The hospital ship H.M.S. "Bentinck," a three-decker, with "P. and O." and other ships, were high and dry; tigers lay dead and dying, washed from their lairs by this huge tidal wave; whilst in the city the uprooted trees and battered buildings all bespoke the havoc. The famines of this year and after were especially hard to combat. Sir Richard Strachey, R.E., worked hard, and gave all he knew of irrigation to mitigate this great distress.

Next came the commercial crisis. India was at that time the great cotton market of the world. The American War had made her so. Enormous fortunes were made, and everyone was rich; and the company-monger was hard at work, for cotton stood at £189 a ton. But, alas! when the "Kearsage" fired the shot that sank the "Alabama" and closed the war, it also burst a bubble far greater than the one they call "The South Sea." Cotton at once went down to £56 a ton, whilst banks (including the great one of Bombay) and most commercial houses fell with a terrific crash; and the rupee, which then stood well above par, could no longer stand the strain, and fell into a decline passing all hopes of recovery.

Pardon for bringing in once more the name of Cotton, but a great domestic loss occurred in the same year, in the sudden death of the Bishop of that name. It was in October, 1866, whilst travelling in his Province, that he branched off on purpose to meet the 12th Regiment, then on the line of march. Colonel Arthur Ponsonby rejoiced at hearing this, and a halt in camp was soon arranged.

The Bishop in his younger days had been second to Dr. Arnold, of Rugby fame. We recognise him so often in that delightful book "Tom Brown," and in later years he was head of Marlborough —thus he was well known to many of our officers. The text of his sermon seemed so appropriate: "How shall we sing the Lord's song in a strange land?"

One false step on his homeward journey, between shore and ship, as he was entering his yacht to cross the Hoogly, caused the death of our much-beloved Primate of India, Dr. Cotton. His body was never found. The 12th Regiment on arrival was sent to Seetapore, where it soon took honours at the butts, for in 1865 many individuals won prizes in the National Indian Rifle Association, whilst in 1866 the regiment, being again successful in the above meeting, was victorious in the Grand Inter-Regimental Rifle Match, officers and sergeants winning respectively first and second prizes in contests open to all officers and sergeants of European regiments quartered in Bengal. Captain Warren, Lieutenants Reed, Kyle, Ferris, and Turner were the victors. My post in the regiment was then I. of M.—hence I took some pride in these performances.

It was now time for Lord Lawrence (for a Peerage and other honours were conferred on his arrival home) to put off his armour. He had worn it in many a conflict from early 1830 to the end of 1868; and on the day of Lord Mayo's arrival (January 12th, 1869), he took ship for England, leaving behind a policy of "Not too much interference in the affairs of Afghanistan," to be referred to later. Lord Lawrence died on June 27th, 1879. He lies close to Clyde and Outram in Westminster. On his tomb he leaves a simple but earnest message: "Be Ready."

Colonel Arthur V. Ponsonby died from cholera at Jubbulpore in 1868, whilst in command of the 12th Regiment. He was universally beloved, and deeply regretted. Lieutenant-Colonel Ponsonby served with the 43rd Light Infantry in the Kaffir War of 1851 to 1853 (medal), including the battle of the Berea. He served in the Crimea as aide-de-camp to Sir George Brown, during the expedition to Kertch; attack of June 18th; and as aide-de-camp to Sir William Codrington on September 8th, and fall of Sebastopol (medal and clasp, Sardinian and Turkish medals, and 5th Class of the Medjidie). He joined the 12th Regiment from the Grenadier Guards, in May, 1863, and by his death the regiment had to regret the loss of a man who was genial in all the best meanings of the word; who delighted to don the Zingari ribbon, and lead his regimental eleven; ever ready to join in the amusements of his officers, and promote the comfort of his men; but though he unbent in private life, and was a Subaltern with Subalterns, none knew better where to draw the line between the social and military duties of a commanding officer. If no one feared him, he was respected by all; his reproof was dreaded as the reproof of one who was, indeed, the father of his regiment. None of those who attended him during his brief but terrible illness of forty hours, are likely to forget the lesson taught them by the manner of his death. Beloved by all who knew him, making life pleasant to himself as he wished to make it pleasant for others, if he had no fear of death, he had much to bind him to life, and he tried hard to live; but worn out at last, after thanking those around him, and with a touch of the old humour which had not yet deserted him, excusing himself for giving in, he resigned himself bravely to the fate against which he could no longer struggle—and so passed away Lieutenant-Colonel A. E. V. Ponsonby at 8 a.m., June 16th, 1868.

With Ponsonby went others—one in every rank: Major Espinasse (an Eton boy, a fine soldier, learned and refined), Captain Medhurst, Lieutenant Stoton, Ensign Knox, and then our Sergeant-Major Crane.

Early in 1867-8, the voyage to India was shortened by means of five splendid iron steam troopships, the "Jumna," "Malabar," "Euphrates," "Serapis," and "Crocodile." One was the exact counterpart of the other; so two on either side, *i.e.*, Portsmouth and

Bombay (one kept in reserve), only left the trip across the Desert to be a pleasant change.

M. de Lesseps, in June, 1857, at a Mansion House meeting, explained his project to cut this narrow strip of land, but politicians (Lord Palmerston), capitalists and engineers went dead against him. He persevered, however, and in November, 1869, a pageant of forty vessels, headed by the Empress Eugenie in her yacht, the "Aigle," supported by the Emperor of Austria, Frederick of Prussia, and other potentates or representatives, glided from the Mediterranean into the Red Sea. The festivities next day were further heightened by the marriage of M. de Lesseps, a ceremony attended by the assembled royalties.

How came it that Napoleon III. was an absentee? The whisper went round that His Majesty was ill and greatly worried, for a Hohenzollern had been put forward for the vacant crown of Spain. Sadowa had cast a gloom, but this meant a shock—so much so, that Napoleon, as he sat in his quiet study at the Tuilleries, saw pictures in the fire of "Breakers Ahead!"

It was during this pretty water picnic that other clouds appeared in far-off Afghanistan, so soon to bank up and give a deluge. At first the storm was in dissensions amongst themselves.

Dost Mahomed, the King of Afghanistan—whose name is still familiar to us from the war of 1837—died at an advanced age in May, 1863, leaving Shere Ali as his heir, who as such was recognised by Canning, Elgin, and Lawrence.

Shere Ali submitted his son, Muhammed Ali, to Lawrence as his heir, and affairs ran smoothly until Afzal and Azim (his brothers) and Abdul Rahman (son of Afzal) declared their independence.

For this Afzal was kept a prisoner in Cabul; Azim hid himself in the Waziri hills; and Abdul Rahman fled to Bokhara; whilst Armin and Muhammed Ali (uncle and nephew) were both killed in battle. The death of Muhammed Ali preyed on his father, Shere Ali, for a time; for nine months, they say, he was insane—which accounts for Abdul Rahman marching on Cabul to release his captive father. Azim joined him on the way, and these two chiefs soon made themselves masters of Afghanistan.

At last Shere Ali roused himself and entered the field again, only to meet with worse reverse at the battle of Sheikabad. Thus the star of Shere Ali seemed eclipsed. Shortly after Afzal died, and Azim's troops revolted; so Shere Ali entered into his own again as the undisputed ruler of Afghanistan.

Thus sadly had the family of the great Dost Mahomed been broken up into contending factions; but now that Shere Ali had beaten down all his enemies, the time seemed propitious for England to secure his good-will and recognise his rule. So it was with great heart that Lord Mayo received Shere Ali in Grand Durbar at Umballa in March, 1869—handing him over £60,000 and 3800 stands of arms, on condition that there was to be no coquetting with Russia or invocation with any foreign Power.

We must now stretch hands across the sea, and stay awhile with the 1st/12th Regiment—now in Sydney, although on the eve of starting for the war in New Zealand; also to bid farewell to Colonel Percival, whom we met in the Cape War in 1852. Now he seeks retirement, only to accept a little later the full Colonelcy of his corps.

Sir William Denison (Colonel, Royal Engineers), brother of the Speaker, was then Governor of New South Wales. His civil work was grand, for he left that post to become Governor of Madras, and on the death of Lord Elgin he assumed the grand office of Viceroy, until a successor in Lord Lawrence eventually arrived. Then from all the pomp and majesty of England's Viceroy in the East, he reverted to a Colonel of Engineers in Spike Island, Cork. It was, of course, only to fill in some corps service; but it seemed, as Napoleon would say: "Il n'y a qu'un pas du sublime au ridicule."

In New Zealand, the 12th Regiment came under the greatest Colonial Governor of the day, Sir George Grey. He had been Governor before, in 1845, as Captain Grey. Born at Lisbon, he was only eight days old when his father, Colonel Grey (commanding 30th Regiment), was killed at Badajoz. The son became a soldier, only to leave the Service to take flight to higher things. Thus on his second visit his experience was great, for he knew the language' the manners, the customs, and could think with a Maori brain; and like Gordon in his early Soudan days, he was idolized and was considered by his loving children as "their Father the Governor." But times had moved since 1845; and now in 1862, the Maoris had their arms and ammunition, and other warlike stores, and were no longer children in a sense—but still Grey had his magic touch, which Generals Cameron and Shute tried to over-ride.

The war went on ingloriously, and the British troops had their disasters (for instance the 43rd). At last there came a test case—the Wakoruo Pah had to be taken; Major Hutchens (of the 12th) was the officer selected, and put in command of 500 men. The General said "Impossible;" but Grey said "Try." Major Hutchens was not in any way hampered; the time and all particulars were left entirely to his judgment. He had an able aide-de-camp in one Lieutenant Mair; and when all was ready, the little army commenced their march. The Pah was duly reached, surprised and surrounded, when the Maoris surrendered without firing a shot.

The War Office at home did not approve of these tactics—they must have a butcher's bill; so Grey was recalled, and the war went slowly on. Here is another instance of the skill and tact of the few excelling the violence of the many. Major Hutchens and Lieutenant Mair (both 12th men) no doubt were proud of their share in this pleasing little episode. Sir George Grey lived till 1898, and was buried at St. Paul's. The 1st/12th Regiment returned to England in 1868.

Sir George Bowen replaced Grey as Governor. The finish of the war commenced when one Te Kooti, the real villain of the play, escaped from the Chatham Isles where he was an exile, and seized a schooner (the "Rifleman"), which was filled with money, arms, provisions, etc. Te Kooti further compelled the mate to take him with others to New Zealand, and to land him at North Island, where he seized a mountain fastness, and became a grand freebooter and murderer as well—for he ransacked a little hamlet at Poverty Bay, killing off Major Briggs, Captain Wilson and thirty-three besides.

This meant vengeance with a will. The command of this avenging force was given to a native chief Ropata, who had been all through this war, and right loyal to the British cause. It must be said that although the British force numbered between 15,000 to 20,000 men, it had always been baffled by a few hundreds of the Maori tribe—who for skill, courage, and cunning, could not be surpassed. Ropata had with him Lieutenant Hurst (late of the 12th), a man well versed in all their scouting ways.

Te Kooti's name became a terror, and Government placed £5000 on his head, which kept Ropata and young Hurst on the chase for a good two years or more, when Te Kooti—after terribly rough handling, and deserted by his men—crept into a sort of sanctuary, and there was allowed to rest. Ropata received a sword of honour from Queen Victoria, and Hurst took up his residence in New Zealand (1870) until his death in 1907.

New Zealand soon went ahead by leaps and bounds, and for years it has been a Colony where all Suffragettes from home should go—for there women have the vote.

The Maoris were a merry race. "Hawke pakeha; hawke ngahois" ("Come on, Stranger; come on, Soldier") was a sort of challenge they loved to give. Another trick they had, when hard up for lead, was to show a dummy man in the bush; when of course it was immediately fired at. A man with a string in the background pulled it down. "Oh!" thought our fellows, "we've done for him." Up came the dummy again, cautiously—bang! bang! bang! went the rifles of the British troops—down went dummy; and this went on till the earth bank they had made behind the tree was filled with lead. This method of obtaining lead was not discovered for a long time.

This *ruse de guerre* at once recalls to me a naval *ruse de la paix*:—A certain Admiral was fond of signalling to his fleet when at exercise, "Out bower anchors." An anchor, as we know, is not a feather weight; so a wily Captain thought he would save all tear and fret by concocting one of wood. The Admiral soon noticed how very smart this particular ship was, in going through this drill. One day, with his glass, he thought something seemed suspicious, so he promptly signalled "Drop bower anchors," when all went down with a thud and a mighty splash—except the anchor of this particular ship, which of course would only float, like a huge monster crab with two claws of anchor shape. In the first case, the Maori got his lead; history fails to say what happened in the second.

The late "Sixties" and early "Seventies" were, in India and at home, years of great military reform—for the recent war of 1866, between Austria and Prussia (when the former, after seven weeks, was glad of Peace), had been followed by a far wider reaching disturbance in the Franco-German Campaign of 1870-71.

The fortune of war had set in hard against the French, as the following note will show:—

"Monsieur mon frere,—N'ayant pas du mourir du milieu de mes troupes, il me qu'a remettre mon épee entre les mains de votre Majesté.

"Je suis de votre Majesté,

"le bon frére,

"Sedan, le 1 Sept., 1870." "NAPOLEON.*

Thus ended the Second Empire. The war smouldered on under a Committee of Public Safety, headed by Léon Gambetta—a noble patriot; but when Paris was reached, then the siege was terrible, as the menu for Xmas, 1870, will show:—

"CHRISTMAS.

"Paris, 1870.

"Christmas in the Siege of Paris! It was the climax of the forlorn. Thermometer at zero, snow dribbling, scowling heavens,

---

* An autograph letter from Napoleon, whilst prisoner of war, now hangs in the ante-room of the 1st/12th Regiment.

slippery pavements, ominous silence all round the city; failure of another sortie; troops ordered in to get warm; thousands of people lying in bed to save food and fuel; long line taking their turn for hoofs of horses at the butcheries at a franc per lb., and for dirt-coloured bread at the bakeries; a thousand soldiers lost from cold; growing growl of indignation against Trochu, against the Government, against everybody, the Deity not excepted; coal nearly gone, wood ditto; National Guards protecting the woodyards from the freezing women; freezing women succeed in a raid, and carry off armloads; at other places National Guards do the carrying off; soldiers denouncing their officers in the clubs; complaints of Commissariat Department; complaints of every department; wounded dying hourly of cold, of bad food, or no food, and of infectious atmosphere (as at the Grand Hotel, where it is said 'a man cannot cut his finger and reach the door alive'); small-pox deaths nearly 400 a week, typhoid fever 220—total 2728; horse-meat getting scarce; hares 40 francs apiece, cats 15, chickens 60, turkeys 100, dogs 50, dogs 80, rats 2; wood a penny a pound, and hard to find; no coal; gas gone, lamps instead—dismal boulevards in consequence. Such is Christmas in the Siege of 'Vanity Fair!'

"There was no midnight mass. On Christmas Eve we hurried shivering through the nipping air to the Church of St. Roch. The doors were closed. The lamps threw a sickly illumination on the sacred mass of revolutionary history. Lonely! lonely! we turned away with a sense of bereavement. It would have been a comfort to have heard the music, and joined in the prayers of the midnight mass, on this gloomy Christmas Eve."

At last Peace was signed with the foe beyond the gates; but within the city (where the writer was), the mob, under the name of

> "Liberty! thou Goddess heavenly bright,
> Profuse of bliss and pregnant with delight,"

had consigned Paris, the heart of France, to one consuming flame.

"That day is a day of wrath, a day of trouble and distress, a day of wasteness and desolation, a day of darkness and gloominess, a day of clouds and thick darkness"—for that is what I felt in Paris on the Derby Day of 1871. My being there is accounted for thus:—

When attached to the Royal Artillery at Aldershot, on completion of my Staff College course, I mentioned to some friends at mess that the Derby we might be spared to see again—but Paris, as then, never!

Three at once agreed with me, viz., Gillespie, R.A., Crawford, R.A. (grandson of the Light Infantry General), and Costobadie (my great chum). The next thing was to apply for the usual five days' Derby Leave, which in due course was granted (but wickedly we had concealed all mention of our roundabout route for Epsom. The same night, armed with four commissions, we set off from Farnborough to Havre *via* Southampton, and so on to Paris, where we made tracks to a hotel in the Rue Sebastopol, known to me previously in the year of 1869, when Paris was on the brink, for Pierre Buonaparte, cousin of the Emperor, had shot Victor Noir, a journalist. This set all tongues talking; and I so well remember seeing Ollivier at some railway station awaiting for his royal master, and then afterwards watching the two in deep earnest conversation. The same evening, when dining with the Royal Artillery officers at Vincennes, I learnt sufficient to form my own conclusions. Then at the funeral of Victor Noir, things commenced to totter, for Paris that night was filled with troops, and the reception Napoleon met was not exactly thrilling. Readers will remember that a few weeks later Napoleon called a plebiscite, when the votes against him in the Army reached 41,000—the "ayes" were 249,000.

To return now to our "Derby trip." The landlord was pleased to unbolt his door, especially to a customer he knew, and to provide a menu (which it were best not to inquire into). I crept out to Place Vendôme, where the stately Column was in a thousand pieces, and fires and other horrors at every turn and corner—for that day (May 28th) McMahon had finished off some heavy fighting around Buttes Chaumont, and was now in Paris, which meant that the Commune game was up, but the Cause was dying hard and desperate.

We visited Versailles, where all those grand conservatories were stacked with women named the "Petroleuses," for they had poured the oil and lit the flames that now encircled Paris. All this time mysterious little parties were being marched about, I fear to execution; in fact, the days which Napoleon I. wrote about (and are given in our early pages) seemed now to be repeated —a sad encore.

Thus we passed our "Derby Leave," and Favonius had won! Another Derby Day of different experience came to me in India. On winding up the "Cricket Account" of the past cold season, my company stood about sixteen rupees to the good—a balance of no moment when shared by so many—so I proposed the purchase of a ticket in the great Calcutta Derby "Sweep" (£8000). This was received with acclamation, so off went our Gold Mohur—we drew a horse (thank goodness, a rank outsider, but still he was to start). As the time approached the excitement became great, and I noticed small *coteries* of my men about—evidently each an architect of some castle in the air. I confided in my colonel, and he thought with me, that if the unexpected happened, my company would disappear, and I should be left alone and stranded—result: the horse was "scratched" before he started.

A great amount of mental stock-taking was now necessary, if only to render our military education up to date, and to renovate our "tactics." The drill and manœuvre to meet the new fire tactics, the rapid breechloader, had now to be shaped and learnt; each one (from General to Captain) was allowed his say. The simplicity of "Fours" was good ground to work upon. "Fours deep" would place the regiment or company in four ranks—the first to commence attack, the second to re-inforce, the third and fourth in support or reserve, until the point of impact came, when rank fed rank again; and the regiment (or what was left of it) went on in fire and battle shock to acclaim the victory!

Like as in "Onward Christian Soldiers," the Church has one Foundation into which we enlist on Baptism ; but the "Onward" process is attended with much nonconformity. Some sects or sections get too high, some keep too low, others lag behind and get out of hand ; but one and all are, in their way, pressing on for the hill top, to re-assemble once again as the real and only Church.

A Royal Warrant had abolished "Purchase" (October 18th, 1871), but it still remained an open question as to what would be the substitute, and how and when the officer would be repaid his purchase money—for each officer held a separate stake, and (like shareholders in the winding-up of a great company) was keen to know his monetary worth ; hence many Committees sat.

"Purchase" was evidently the survival of the days when the wars of Europe were fought in great measure by mercenary bands —in which bands the purchase of a certain number of shares (for the basis of these mercenary organizations was wholly commercial) entitled the holder of the shares to a certain rank in the field. The over-regulation (or what one may call a premium on these shares) crept in, I imagine, when sickness pervaded our Army, as it did in the Marlborough campaigns, Walcheren, the Netherlands, etc., when extra sums were offered to purchase out the sick.

"Purchase" had its vagaries, and may be likened to our English climate—very variable in its workings. For instance, the command of Aldershot was at that time in the hands of that fine old soldier, General Sir James Yorke Scarlett—the type of Lord Gough, kind, courteous, and of similar lion courage, as the handling of the Heavy Brigade in the Crimea fully showed. Scarlett was born in 1799. He was M.P. for Guildford from 1835 to 1840 ; after which date he commanded the 5th Dragoon Guards for fourteen years, so his advancement by purchase was rapid. He died in 1871.

Colonel V. Baker, 10th Hussars, obtained his Colonelcy in eleven-and-a-half years, and kept the command for fourteen years ; whereas Henry Havelock (a non-"Purchase" officer) required twenty-three years before he could even get his Captaincy. General

Dacres, Royal Artillery (a non-"Purchase" corps) was over twenty years a Subaltern, and died a Field Marshal.

Another instance was a Colonel McKay, of the 12th Regiment, for he obtained command of the Battalion (the value of which was £7000) in seventeen years' commissioned service, without having purchased a single step, and had never served out of England for a day until he went to India in 1871. He had previously been in the ranks, until appointed Ensign and Quartermaster at the School of Musketry at Hythe in 1854.

A few weeks since (October, 1910), there died in Brighton Mr. Rassan, the last of the eight prisoners kept as hostages by King Theodore of Abyssinia, on account of an unanswered letter to the Queen, supposed to have been lost in the Foreign Office. This led to a war for their release, the command in which was given to Sir Robert Napier in India, from which country the expedition started on April 1st, 1868, the 33rd and 45th being the European troops. Napier telegraphed thirty miles from Magdala: "Prisoners all well; troops all well." (April 10th): "Release of the prisoners; King Theodore defeated." Theodore sent back Mr. Rassan and the other prisoners with 1000 cows and 500 sheep as an Easter Egg, but would not tender his submission; so Magdala was assaulted, and entered on April 13th, when Theodore destroyed himself.

This expedition, from start to finish, was in every way a success. Torrential rains and precipitous rocks had to be encountered *en route*. The Queen sent her congratulations thus: "I congratulate your Excellency with all my heart; you have taught us once more what is meant by an army that can go anywhere and do anything. From first to last all has been done well. I must ask leave for a motto. My own suggestion would be 'Qualis ab incepto.'"

Another equally well conceived and carried out Expedition was that to Ashanti by Sir Garnet Wolseley, in October, 1873, to punish the King of Ashanti for his attack on the Fantees—a friendly tribe of our own along the coast.

Governor Pim, in 1864, had been ordered to proceed against him; but with a force inadequate, the Expedition returned unsuccessful and dispirited, and all down with disease. Discussion in the House on this ill-fated affair brought the Government majority down to seven, in a house of 459.

In 1873, however, when the same cause arose again, the Expedition was well equipped, and most ably commanded. They went, they saw, they conquered. Coomassie had been entered, and the Royal Umbrella sent home intact, by February 5th, 1874.

The seamen and marines here rendered most excellent service. Commander Glover, R.N., was given independent command of the Friendly Natives in the Eastern District. Captain Sartorius, one of his staff, here gained the V.C., and Giffard (the other recipient) was aide-de-camp to Wolseley.

The three British regiments engaged were the 23rd Welsh Fusiliers, a Battalion of the Rifle Brigade, and the 42nd Highlanders. The officers then fighting were in this private plight: "Purchase" had been abolished, no pensions had been provided, Assurance Companies refused their lives by reason of the deadly climate—so their monetary worth, professionally, was only the clothes they fought in! Thank goodness, the loss of life was small!

Then there were frequent War Games, and lighter pastimes in the shape of Polo, Cricket, Theatricals, and Balls. As an instance of the latter, we will change the scene from Aldershot unto our Manœuvre Camp at Delhi, in November, 1872: for there—with a Viceroy like Lord Mayo, a king amongst men; with Lord Napier as his Commander-in-Chief; and a Staff mustering Sir Harry Tombs, Sir C. Watson, Charles and Hugh Gough, Fraser Tytler, Roberts ("Bobs") (all of whom were V.C.'s), Sir Peter Lumsden, Chapman, etc., etc., the military fabric stood forth in great strength. It was on the break up of these manœuvres, with a view of wishing Godspeed to Lord and Lady Mayo (so shortly starting South *en route* to Andaman), that the Army Corps gave a Ball; and by request of the Viceroy we were to appear in *petite tenue*, as her ladyship and he must be in travelling attire, for they left that very night. The Ball

The Calcutta Gazette
EXTRAORDINARY.

TUESDAY, FEBRUARY 13, 1872.

Government of Bengal.

NOTIFICATION.

FORT WILLIAM, THE 13TH FEBRUARY 1872.

THE LIEUTENANT-GOVERNOR announces with inexpressible grief and pain to the people of these Provinces that the VICEROY and GOVERNOR GENERAL OF INDIA died on the 8th instant, at the Andamans, from wounds inflicted by a convict.

This sad event was announced by a Gazette Extraordinary of the Government of India, a copy of which is annexed.

The LIEUTENANT-GOVERNOR feels sure that not only all Officers of Government, but all private subjects, European and Native, will unite with him in deploring the untimely end of one who was not only a distinguished Statesman and a most able and successful representative of HER MAJESTY, but was endeared as an individual to all who knew him from his rare personal qualities.

The Hon'ble J. STRACHEY, under the provisions of the law, acts as Governor-General till the arrival of HIS EXCELLENCY THE RIGHT HON'BLE FRANCIS BARON NAPIER OF MERCHISTOUN. Copy of the Notification on this subject is annexed.

The Acting Governor General in Council has directed a general mourning for the late VICEROY in a Notification of which a copy is also annexed.

The LIEUTENANT-GOVERNOR is sure that he need do no more than make known this mode in which respect may be shown for the memory of the deceased VICEROY.

RIVERS THOMPSON,
Secretary to the Government of Bengal.

The Gazette of India Extraordinary, February 12, 1872.

HOME DEPARTMENT.

NOTIFICATION.—PUBLIC.

Fort William, the 12th February 1872.

THE Government of India announces with inexpressible grief that the VICEROY AND GOVERNOR GENERAL OF INDIA was assassinated at Port Blair at 7 P.M. on the 8th instant.

The assassin was a convict under sentence of transportation for life. He broke through the guard and stabbed the VICEROY as HIS EXCELLENCY was on the point of embarking after inspecting the station.

"The country has lost a Statesman who discharged the highest duties which HER MAJESTY can intrust to any of her subjects with entire self-devotion, and with abilities equal to the task.

Those who were honoured by the EARL OF MAYO's friendship, and especially those whose pride it was to be associated with him in public affairs, have sustained a loss of which they cannot trust themselves to speak.

The Government of India therefore abstains at present from saying anything on this great calamity.

Under the provisions of the Indian Councils' Act, section 50, the office of VICEROY AND GOVERNOR GENERAL devolves upon HIS EXCELLENCY THE RIGHT HON'BLE FRANCIS BARON NAPIER OF MERCHISTOUN. Orders will be given in a separate Notification as to the marks of respect to be shown to the memory of the EARL OF MAYO.

By Order,

E. C. BAYLEY,

Secretary to the Government of India.

The Gazette of India Extraordinary, February, 13, 1872.

HOME DEPARTMENT.

NOTIFICATION.—PUBLIC.

Fort William, the 13th February 1872.

No. 751.

WHEREAS HIS EXCELLENCY THE RIGHT HON'BLE LORD NAPIER OF MERCHISTOUN, upon whom the office of Governor General of India has devolved under section 50 of the Indian Councils' Act, has declared that it is not his intention to assume the said office until he shall have taken his seat in Council, it is hereby notified, in accordance with section 51 of the said Act, and section 62 of the Act of 3rd and 4th years of King William the Fourth, cap. 85, that the office of Acting Governor General has devolved upon the HON'BLE JOHN STRACHEY, Senior Ordinary Member of Council, until HIS EXCELLENCY's arrival.

No. 752.

WITH reference to the great public calamity announced in yesterday's *Gazette Extraordinary* the Acting Governor General in Council directs that the Flag of Fort William be hoisted half-mast high until further orders.

Forty-nine Minute Guns will be fired from the Ramparts of Fort William this afternoon, the last gun to be fired at sunset.

Similar marks of respect will be paid on receipt of this Notification at the respective Seats of Government, and at all the principal Military Stations in India.

The Acting Governor General in Council directs that all the Officers of Her Majesty's Civil, Military, and Marine services do put themselves into mourning for a period of one month, and invites all classes of Her Majesty's subjects in India to join in this tribute of respect to the memory of the late EARL MAYO.

Further orders will be issued on the arrival of the remains of the late VICEROY, now on their way to Calcutta.

By Order,

E. C. BAYLEY,

Secretary to the Government of India.

had a touch of that historic dance at Brussels, but with us our military work was over; and little did we think, as we saw our Viceregal guests depart, that there was to be no return of the Earl of Mayo. The *Calcutta Gazette* of February 13th, 1872, tells the rest.

This great loss was soon followed by another, in our much-loved soldier Harry Tombs, the seventh son in a family of ten. Tombs* (like Lord Roberts) was educated both at Addiscombe and Sandhurst, and at the age of eighteen joined the Bombay Gunners. Within the year he was on service in the Gwalior Campaign, winning mention in despatches, and the Bronze Star for Punniar. In quick succession came Moodkee, Ferozeshah, Budiwal, and Aliwal—where he fought either with his guns or as aide-de-camp to Sir Harry Smith (other medals and mention). When the great Sikh army again showed restlessness, he was in the thick of Chillianwallah and Goojerat, then at Delhi, Oudh, and Lucknow. Grant and Tombs were men Lord Clyde could not do without.

It was from Delhi (the historic Ridge) that Tombs penned his last despatch (Manœuvres, 1872), giving the names of troops captured in his last fight:—"Headquarters, Beraree, 11th, 8 a.m. Troops captured: two guns 40th Native Infantry, one squadron 12th (?), 1st Native Infantry, band and colours 92nd, one squadron 11th Hussars, one 12th B.C., one 6th B.C., Brigadier-General Fraser and his whole Staff. None of these troops should be allowed on the field to-day.—(Sd.) H. Tombs." Sir Harry Tombs resigned the Oudh Division only when cancer, the malady which no human skill can cure, had laid him prostrate. He was spared to reach England, and

---

* It was Sir Charles Napier who got the Ghoorkas to volunteer for general service, which meant they would get higher pay than if only localised in the hills. Napier sent young Tombs to read and explain the terms; and when asked on his return if they fully understood the proposals: "Why, yes! they volunteered not merely with alacrity, but a joy, evinced," said Tombs, "by extraordinary screams of delight, unlike anything he had ever heard." Poor fellows, they were starving, and vehemently hailed the means of sustaining life.—October 7th, 1849.

to die peacefully at Ryde, Isle of Wight, in 1873. It was my great privilege to have been—a good year and more—upon his Staff.

After Scarlett, the command at Aldershot was given to Sir Hope Grant. No one could gainsay his qualifications—his soldier-like instincts and ability shone everywhere; he was happy in his amusements, with hunting, his violoncello, and golf. It took a fortnight then to arrange a "foursome," as Aldershot could only muster four who had ever seen a club—Sir Hope Grant, Colin Moncrieff (Scots Guards), Anstruther (R.H.A.), and myself. The North Camp Racecourse formed our Links, improvised for the day only. Rumour soon took forcible shape, by asserting that if the General and his *confreres* ever sought another billet, it would surely be Broadmoor (a home for lunatics).

Whist (with low points) was then our after-dinner recreation. Society—(how it leads us to live only for what others will say of us)—has now decreed that it must be "Golf everywhere, Whist nowhere." Give us both! Whist, in hopes of living long enough to hold the thirteen trumps!

"Has any whist-player ever held the thirteen trumps in one hand? The phenomenon was seen at the United Service Club, Calcutta, on the 9th inst. The players—we trust they will forgive us 'naming' them, but whist history must be above suspicion—were Mr. Justice Norris, Dr. Harvey, Dr. Sanders, and Dr. Reeves. Two new packs were opened, and were 'trayed' and shuffled in the usual way. Dr. Sanders had one of the packs cut to him, and proceeded to deal. He turned up the Knave of Clubs, and on sorting his hand found that he had the other twelve trumps. The other three suits were unevenly divided in the other hands, but in the excitement of the moment no record was taken of them. The fact was duly recorded in writing, the six gentlemen signing their names to the document. The odds against this combination are, we believe, according to Dr. Pole, 158,750,000,000 to one; the probability of a given player holding thirteen cards of a particular suit, named before the deal is concluded, is put by the same authority as once in 635,000,000,000 deals."—*Englishman*, 1876.

" Hour after hour, the cards were fairly shuffled
And fairly dealt, but still I got no hand;
The morning came, but I, with mind unruffled,
Did simply say—' I do not understand.'

" Life is a game of Whist, from unseen sources
The cards were shuffled, and the hands were dealt;
Blind are our efforts to control the forces
That, though unseen, are no less strongly felt.

" I do not like the way the cards are shuffled,
But still I like the game, and want to play;
And through the long, long night will I, unruffled,
Play what I get until the break of day."

Cricket at Aldershot always gave great pleasure. The ground had just been re-made, and the matches were frequent and interesting. One of special interest (in which I played) occurred in May, 1874, for on one side was a Gentleman Cadet from Sandhurst, and on the other a Subaltern of Royal Artillery. It was only a few weeks after this that the Gentleman Cadet applied for leave of absence, in order to accept an invitation from his country to become their King Alfonso I. of Spain.

The Subaltern of Artillery was no other than the popular Prince Imperial, Louis Napoleon, the last inheritor of this mighty name, who five years later was killed in the Zulu War (June 1st, 1879).

" It was on June 1st, I started with General Newdigate's Column, about six a.m., from Kopje Alleen, and when crossing Blood River I met the Prince with Captain Carey (98th Regiment), four men of Bettington's Horse, and one native. The Prince was in the uniform of the Royal Artillery, and seemed to be in very high spirits, laughing and talking with his comrades as he went along. I was on horseback and about to take a wrong road, to join the first division of the Column, when I was hailed by Captain Carey, and informed of the road I had better take; and after a few kind and cheerful words from the Prince, I bade him 'Bon jour. Au revoir.' The Prince was riding a grey horse named 'Tommy.'

" I got to camp about five p.m. One of the party (a Dutchman) whom I had seen in the morning with the Prince, came into laager and said, 'I have brought the Prince's horse, but not the Prince.'

"The sequel is known. The remains of this much lamented Prince passed through Durban, and were embarked for England on H.M.S. 'Boadicea.' 'You could not have done more had it been the Queen herself,' was the remark made by a Frenchman at the time; and it was quite true. The demonstrative resources of a town like Durban are not great, but they were put forth to the utmost possible extent in honour of the Prince Imperial. There were Englishmen and Frenchmen, Germans and Dutchmen, nor were Americans of the Great Republic wanting. There were the olive-hued sons of Asia, represented by the turbanned Hindoo, the long-robed Arab, and the keen-faced Chinaman. There were the dusky children of this dark continent of Africa in large abundance, typified by the negroesque Krooman of the West Coast and his equivalent at Zanzibar, the liberated slaves from Mozambique, and the free labourers from Delagoa Bay, etc. No doubt, amidst this swarthy crowd were many Zulus, wondering what manner of man this could be whom their compatriots had so unsuspectingly stricken down, and whose scarred body was being borne away over the great water."

—*Natal Mercury.*

Ballooning and voyaging in the air was at this time much talked of and speculated upon. If Britannia can rule the waves, why not try for the mastery of the air? In August of this year, M. Jules Duruof, a courageous young Frenchman, ascended with his wife from Calais, intending to cross to England. The balloon was, however, carried over the German Ocean, and the aeronauts were rescued the next morning by a Grimsby smack that happened to be fishing on the Dogger bank, 170 miles off the mouth of the Humber.

Captain Burnaby of the "Blues," with Lord A. Manners, made an ascent from the Crystal Palace in November, to try instruments and conduct experiments in high latitudes, thus recalling to one the experiments made by M. Gay-Lussac (in the early days of 1800), who on one occasion, wishing to rise very high, threw over from the car a strong deal chair. The balloon was out of sight, so those who watched the chair descending (for it fell into a field at harvest-time) naturally thought it to be the property of some Archangel. This miracle remained unsolved for some time, when at last the evicted chair was claimed by the balloonist.

The high regions, as described by Mr. Glaisher, must indeed be fascinating :—" Above the clouds, the balloon occupies the centre of a vast hollow sphere, of which the lower portion is generally cut off by a horizontal plane. This section is in appearance a vast continent, often without intervals or breaks, and separating us completely from the earth. No isolated clouds hover above this plane. We seem to be citizens of the sky, separated from the earth by a barrier which seems impassable. We are free from all apprehension, such as may exist when nothing separates us from the earth. We can suppose the laws of gravitation are for a time suspended, and in the upper world to which we seem now to belong, the silence and quiet are so intense that peace and calm seem to reign alone. Above our heads arises a noble roof, a vast dome of the deepest blue; in the East may be seen the tints of a rainbow on the point of vanishing; in the West the sun silvering the edges of broken clouds. Below these light vapours may rise a chain of mountains, the Alps of the sky, rearing themselves one above the other, mountain above mountain, till the highest peaks are coloured by the setting sun. Some of these compact masses look as if they have been ravaged by avalanches, or rent by the irresistible movements of glaciers. Some clouds seem built up of quartz, or even diamonds; some, like immense cones, boldly rise upwards; others resemble Pyramids, whose sides are in rough outline. These scenes are so varied and so beautiful, that we could remain for ever to wander above these boundless planes. . . . But we must quit these regions to approach the earth; our revolt against gravity has lasted long enough; we must now obey its laws again. As we descend, the summits of the silvery mountains approach us fast, and appear to ascend towards us; we are already entering deep valleys, and glaciers all flee upward. We enter the clouds and soon see the earth: we must make the descent, and in a few minutes the balloon lies helpless, and half empty, on the ground."

Mr. Green, who died in 1870, had a grand record. He made 1400 ascents; he crossed the sea three times, and twice fell into it; he took up 700 persons, among whom were 120 ladies and many persons of high rank. On one occasion he ascended sitting on a favourite pony suspended from the hoop in the place of the car. The animal, which had been trained at Astley's, did not manifest the least uneasiness, but ate freely during the excursion some beans given him by his rider.

M. Henri Giffard, as early as 1855, made most successful progress through the air at the rate of seven miles an hour, in his steam airship of 3 horse-power, 150 feet in length and 40 feet in diameter, and weighing three cwt. in all.

The airship of the present day seems a "Rip van Winkle" of the past—lying by for sixty years, whilst the scientists took some sleep—when enters "Rivalry" and wakes them with a start, urging them to fly again before they are quite wide awake. For assuredly the nation mourns the loss of too many brave and clever men who so hurriedly have given their lives to Science in this year of 1910. Time must be given for one to reflect, to wonder, and to turn to best account, this their grand sacrifice of life.

On March 7th, 1873, our Chief, General Sir Hope Grant, G.C.B., died. His soldier-like instincts and saving common-sense won the confidence and affection of all; and behind all this was his bravery and ability in war—Maharajpur, Sobraon, battles of the Punjaub, and the great Mutiny. "I always feel safe," exclaimed Lord Clyde, "when Hope Grant is at work."

Lord Derby (then Stanley), in moving the Vote of Thanks within the House of Commons on April 14th, 1859, said:—"Sir Hope Grant has perhaps been more constantly and actively engaged than any other officer. He has been mentioned more frequently than any other officer in despatches; always in the front, always in the post of difficulty; a complete narrative of the engagements in which he has taken part, would in itself furnish a history, not very imperfect, of the whole of the Mutiny operations" (see "Mutiny" Chapter).

Lord Granville, on the same day in the House of Lords, said:—"The achievements performed by Sir Hope Grant and Sir Hugh Rose seem more like the prodigies of valour recounted in the pages of an ancient romance, than actual historical events occurring in our own times."

In Lady Grant, the men, their wives and children, knew

"To woman's gentle kind we owe
What comfort and delights us here."

Her care was for the sick, and those not on the married roll; for to them, so often, help in work, clothes, and food came as a gift twice given and equally so blessed.

There was another lady then in Camp—the wife of the Staff Paymaster, Major Ewing, one who has given so much delight to young and old in her books, "Jackanapes," "The Story of a Short Life," "Daddy Darwin's Dovecote," etc. It was a treat to see, at Aldershot, the play of "Helping Hands," when staged by Mrs. Ewing (afterwards Mrs. Scott Gatty).

Sir Thomas Steele (a Guardsman of Crimean repute, Military Secretary to Lord Raglan, etc.) now succeeded Hope Grant, and around him were other worthies—Sir Daniel Lysons, Sir Arthur Herbert, Sir Archibald Alison, George Byng, Harman, Pomeroy Colley, Valentine Baker, Sir Thomas McMahon, etc.

In the Military Colleges and neighbourhood were other men of mark—Sir Edward Hamley*; Colonel C. C. Chesney; Major Adams, late Austrian Army and Professor Staff College; Professor Dowson, a renowned Oriental scholar and Professor of Hindustani at the Staff College (the examinations in which, at that time, were taken by the late Archbishop of York—the Rev. McLagan, late Lieutenant Indian Army); Lieutenant Evelyn Baring, now Lord Cromer; Lieutenant, now Sir Frederick Maurice*; Sir F. Chapman, etc., were

* Sir Edward Hamley and Sir Frederick Maurice will always be known as the Military Historians of the Century. There was one to follow on in Colonel Henderson. His life, however, has not been spared; he has left the world richer in his one volume, "The Science of War."

students. Colonel A. S. Jones, V.C. (whose reminiscences of the Mutiny, as Staff Officer to Sir Hope Grant, have been so fully given) was the Adjutant, and another V.C. hero (Lendrum) was Quartermaster.

"He that knows not, and knows not that he knows not, is a fool—shun him.
"He that knows not, and knows that he knows not, is simple—teach him.
"He that knows, and knows not that he knows, is asleep—rouse him.
"And he that knows, and knows that he knows, is a wise man—follow him."

—*Arabic.*

A lecture was given at the United Service Institution, in early 1872, by Mr. Vernon Harcourt, Q.C., M.P., to support the views he held relative to the impossibility of the invasion of England. I apologise to my readers for asking them to follow some remarks made by Colonel Home, C.B., R.E., in discussion. In 1872, the German Navy was only 5 ironclads, 9 frigates, 8 first class gunboats, 1 Admiral, 1 Vice-Admiral, 1 Rear-Admiral, 27 Captains, 217 Lieutenants, and their flag was hardly known; Japan, *nil*. If invasion was worth consideration then, surely it need not be forgotten now.

Colonel Home, R.E.*: "In his lecture, Mr. Harcourt told us he was a lawyer, and accustomed to elicit truth by the examination of friendly, and the cross-examination of hostile witnesses. Facts are, of course, the great things to be sought in dealing with such a question as that before us; but facts coming from a legal mind do not, I think, bear the same face as when they come from a judicial mind. Captain Hoseason has termed Vernon Harcourt's lecture a 'brief,' and a great Judge has drawn attention to the fact that a legal mind seeks the triumph of the cause it espouses—a judicial mind seeks the triumph of truth.

* Colonel R. Home, C.B., R.E., died February, 1879.

"Mr. Harcourt has put certain questions, to which he wishes answers to be given, and amongst the rest, he asks, 'What are the character and extent of the preparations necessary in order to carry out such an enterprise as the invasion by sea of a country possessed of a powerful fleet?'

"Well, sir! I am perfectly free to acknowledge that if a fleet is present and ready to dispute the passage, I do not think a passage could be forced except by a very much more powerful fleet than that which attempts to defend it. But the question is, will the fleet be there? Now, Mr. Harcourt went through the ironclads: he told them off, one by one to each other, and this one was to take that one, and he said 'Here are ten left.'

"But I argue that that is not war. The Emperor Alexander of Russia appears, before the battle of Austerlitz, to have reasoned precisely as Mr. Harcourt has done. He was a young man entering life, and amongst other things he is represented as being very ignorant of war. He appears to have gone into the question, and to have said: 'Here is a Russian soldier; he carries so many rounds of ammunition and a firelock; he is clothed the same as a French soldier; we put that Russian against that Frenchman, and there is another, and so on,' and as he told off the Austro-Russian Army, he said: 'Here are 20,000 men left.'

"After Austerlitz, what was the result? When the French General Savary went to see the Emperor Alexander, the latter said: 'How was it, whenever we met you, you were three to one, although you were over 20,000 inferior to us?'

"The reason of this was that the Emperor Napoleon, who commanded the French Army, had not the same idea of war that Mr. Harcourt has. Mr. Harcourt's idea of war is, that it is the art of being strongest; but Napoleon says war is the art of being strongest at a given point, and that point the decisive one—or, in other words, 'the right men in the right place at the right time.' And unless Mr. Harcourt can make sure that he always will have his spare ironclads upon the decisive point whenever they are wanted, I think his argument falls to the ground.

"There was a remark made by a gentleman on 'the unforeseen.' Mr. Harcourt referred to that, and said he really could not provide for what was 'unforeseen.' Well, now! I do not dispute with Mr. Harcourt for the meaning of the word 'unforeseen'—I think that is open to discussion; but I should like to remind the meeting of the words of a statesman in England, who for many years has filled a considerable place in the minds of his countrymen. Mr. Disraeli, the master of many happy phrases, says that there is nothing so possible, nothing so probable, as the 'unforeseen;' and I think that is rather *against* Mr. Harcourt's idea. Take the case of one of our great races—who wins? the 'foreseen' horse of the papers, or the 'unforeseen' horse? Generally it is the outsider—the 'unforeseen' horse. Mr. Harcourt has really put himself in the position of laying tremendous odds on the favourite against the field."

Mr. Harcourt, in reply, had naught to say to Colonel Home, but continuing: "There is one point upon which Colonel Baker has, I must say, convinced me, and has brought the matter very strongly before my mind—and that is the very great difficulty of preventing a landing by troops; that is to say, that the power that troops embarked have of moving along the shore, and so baffling the pursuit of land troops, is very great. I think Colonel Baker has brought out that fact in a very strong manner."

This point of Colonel Baker was remarkably emphasised by the fleet attending Wolfe at Quebec. To quote from the Hon. John Fortescue's letter on the battlefield of Quebec (see *Times*, May 16th, 1908): "Montcalm was obliged to withdraw battalion after battalion from the entrenched camp, to watch the fifteen or twenty miles of shore which were daily threatened at one point or another by the fleet. Most of it, indeed, was rendered inaccessible by cliffs, but there were occasional paths by which they could be scaled, and in the first week of September, Wolfe noted that one of these, a mile and a half above the city, was weakly guarded. There he resolved to ascend, if he could, to the Plains above.

"For six days the fleet went up and down the river above Quebec with the tide, threatening all points, and keeping the unlucky French officers in torturing suspense, as they marched their troops to and fro.

"On the night of the sixth day, the line-of-battle ships made feigned preparations for an attack on the entrenched camp below the city, while Wolfe, with 4500 men in boats, dropped down the river from above to his chosen landing place. All went well; and when Montcalm, after a troubled night, rode up towards Quebec at 5 o'clock in the morning of September 13th, he saw an ominous line of scarlet to the west of the city. And then, it must sorrowfully be confessed, that the poor man, worn out with long anxiety and strain, completely lost his head."

In January, 1847, the Duke of Wellington wrote to Sir John Burgoyne the following *re* "Invasion":—"You are aware that I have for years been sensible of the alteration produced in maritime warfare and operations, by the application of steam to the propelling of ships at sea. . . . I have in vain endeavoured to awaken the attention of different Administrations as to this state of things, as well known to our neighbours as it is to ourselves. . . . I have examined and reconnoitred, over and over again, the whole coast from the North Foreland, by Dover, Folkestone, Beachy Head, Brighton, Arundel, to Selsea Bill, near Portsmouth; and I may say that, excepting immediately under the fire of Dover Castle, there is not a spot on the coast at which infantry might not be thrown on shore at any time of the tide, with any wind, and in any weather."

Several Peace meetings shortly followed for the purpose of neutralising, if possible, the effect of the Duke of Wellington's warlike words.

Referring once again to the "unforeseen"—take 1908. Was not the Grand National won by an 18-guinea horse from out of a cab, or some light vehicle of draught? Then the mare Signorinetta—did not she win the Derby, starting at 100 to 1 against her, and two days after distance all her compeers for the Oaks?

Now, later, have I not found a hero and a heroine ? for to such-like is the dedication of this book—for where could there be a finer example of doing in need some decisive thing, than Mr. R. L. Fowler in his up-hill game at cricket (Eton v. Harrow, 1910), or in Miss Leitch's splendid win at golf (1910), for in both cases there was "invincible pluck"—which accounted for the "unforeseen."

CHAPTER XLII.

THE career of a good soldier is always interesting, so we will accompany Valentine Baker from the United Service Institution to Northern Persia for a while. He was now on half-pay after twenty-five years' regimental service, half of which had been passed in command of the 10th Hussars; and failing to see

> "What pleasure, Sir, find we in life
> To lock it from action and adventure,"

he left with Dick Clayton, 9th Lancers, and Lieutenant Gill, R.E., on a tour to Merv, Herat, Cabul, and home through India. But when Baker, on arrival at Teheran, wrote to Lord Northbrook, Viceroy of India, for permission to do the latter, Northbrook requested it of Shere Ali, the Ameer of Cabul and son of our old friend Dost Mahomed. The Ameer refused it, giving for his reason that his people were such rude mountaineers, and to travel amongst them was to court danger.

Baker was disappointed. He, however, reached Merv, the source of the Oxus, and then thought fit to return. No lives were claimed on this Expedition, but the tragic death of his two comrades, Clayton and Gill, in later years, is here given.

Poor Dick Clayton was the first to go. Lord Lytton, Viceroy of India, writing to Queen Victoria from Delhi at Christmas, 1877, says:

"Sunday and Christmas Day were days of rest. Divine Service was performed in the Viceroy's Camp by the Bishop of Madras and Archdeacon Baly, and special prayers were offered up for your Majesty in reference to the event we were about to celebrate.

"Our Christmas Day was saddened by a sudden and deeply felt bereavement—Captain Clayton, of your Majesty's 9th Lancers, who was attached to my Staff as an extra aide-de-camp at Delhi, broke his neck by a fall from his pony whilst playing at polo, and expired in the course of the night. This excellent and most efficient officer was warmly beloved by all who knew him. His untimely death is a great loss to your Majesty's service, and a lasting sorrow to his fellow officers and many friends.

"To poor Lord William Beresford,* who from boyhood had known and loved him as a brother, the shock and grief of it have been quite heart-rending to witness. I have written to express my deep sympathy to the officers and men of his regiment. He has been buried in the Camp at Delhi."

Lieutenant Gill's tragic fate is recorded below, as inscribed on a Tablet in St. Paul's :—

In Memory of Three Brave Men—

PROFESSOR EDWARD HENRY PALMER, Fellow of St. John's College, Cambridge, Lord Almoner's Reader in Arabic, and a scholar and linguist of rare genius;

CAPTAIN WILLIAM JOHN GILL, R.E., an ardent and accomplished soldier, and a distinguished explorer;

LIEUTENANT HAROLD CHARRINGTON, R.N., of H.M.S. "Euryalus," a young officer of high promise;

Who, when travelling on public duty into the Sinai Desert, were treacherously and cruelly slain in the Wady Sadr, August 11th, 1882.

Their remains, after a lapse of many weeks, having been partially recovered and brought to England, were deposited here with Christian rites, April 6th, 1883.

That tragic fate was shared by two faithful attendants, the Syrian Khalil Atik and the Hebrew Bakhor Hassum, whose remains lie with theirs.

"Our bones lie scattered before the pit: but mine eyes look unto thee, O Lord God."—Psalm cxli.

"Manners maketh man" is the motto of our oldest public school, but it holds good for all, as Gill became the happy recipient of a fortune, through the simple fact of his courtesy and kindness in assisting an old gentleman to a seat in a Brighton church, which enabled him to undertake this journey, and afterwards to seek a seat

* Their ponies collided.

in Parliament. Gill was a Brighton College boy; likewise was Raymond Margery, of the Consular Service, who was selected for special service to assist a British Expedition sent up the Irrawaddy to explore some trade routes, and was cruelly murdered, close upon the finish of his useful work, on February 21st, 1875.

Baker (*i.e.*, the trio) reached home almost on the eve of the Russo-Turkish War, and nothing would prevent Baker from taking part in the same; and as his soldiering there was so interesting and instructive, it is given here somewhat fully in hopes a lesson may be learnt.

At that time, the greatest indignation existed in England and elsewhere against tbe Turkish misrule. Turkey was then at war with Servia and Montenegro. A European Conference was held in December, 1876, to put matters straight—it was an utter failure; then Midhat Pasha appeared as a Reformer, and proclaimed a Constitution. In the meantime, Russian troops were concentrating on the Pruth, preparatory to declaring war in April, 1877. It was when the British fleet was off Constantinople in 1878, that Abdul Hamid, the Sultan, annulled this Constitution, which the Young Turks of to-day (1908) have determined to revive.

One of the earliest appointments on the declaration of war between Russia and Turkey, was that of Colonel Valentine Baker to the Staff of Marshal Mehemet Ali, Commander-in-Chief of the Army of the Lom at Shimla; and he accordingly left Constantinople—with Major Sartorius,* Captain Molloy, and other English officers—for Varna on August 14th, 1877.

Baker had brought with him from Montenegro, two aides-de-camp, Salim and Mehemet. Both were gigantic specimens of men, but Mehemet was quite a character—very young, with a frank, good-tempered face; he was as active as a deer, and had the most extraordinary intuitive military instinct ever met with. He seemed to know everything that was passing, and could detect the enemy's

* Major Sartorius and Captain Molloy were both officers in the Indian Army. They preferred spending their furlough in fighting with Valentine Baker in Bulgaria.

movements almost as far off as we could with our field glasses. He was as bold as a lion.

On one occasion during the Servian War, Sali Pasha was very anxious that Servian prisoners should be taken, in order that some information might be gained relative to the troops opposed to the Turks at that point.

Without saying anything, Mehemet mounted his horse. Presently he was seen quietly cantering towards the enemy's lines. The Servian sentry, seeing him approach, levelled at him and fired. Dashing at the man Mehemet caught him, with one hand lifted him from the ground, threw him across his saddle, and, under a shower of balls from the astonished Servians, cantered quietly back, bringing his struggling prisoner with him.

The Marshal, a man of kind heart and good nature, welcomed Baker and his staff most heartily at dinner, after which many operations were discussed. At first operations went satisfactorily, but soon everything went contrary, so Major S—— was sent to Constantinople to interview the Sultan and to plead for common sense; also to call on the English Ambassador, the celebrated Sir Henry Layard.

But before going on, there must be no mistake as to whom Major S—— refers to when he mentions the "Military Attaché" of the English Embassy; for there were other military attachés, namely, those following the armies during the war. The one Major S—— had to do with was a Major-General and a well-known British officer, who had won the V.C. during the Crimean War. He spoke Turkish very well, and no doubt for that reason had great influence with the Sultan.

Returning to what Major S—— had to bring before the Sultan, the following were the most important:—

1.—The exaggerations and downright falsehoods of Suleiman Pasha.

2.—The way the Turkish Generals objected to help one another.

3.—The refusals and objections of any officers (who had attained the rank of Colonel or higher) to obey orders, without first telegraphing or writing to Constantinople and receiving the answer.

4.—The division of the Army into three parts, each commanded by an independent General. The promise to each of these Generals, that whoever of the three became most successful should get the much-wished-for "Ministry of War" appointment.

5.—Baker Pasha very strongly objected to this latter, and gave Major S—— instances proving how ruinously it was acting in the war, all of which he could lay before the Sultan.

6.—The splendid way all officers (of lower ranks) and the whole of the soldiers behaved.

7.—Having placed as much as possible of the above before the Sultan, and the rest before Sir Henry Layard, Major S—— was then to let both the Sultan and Sir Henry Layard know the full ideas and complete intentions of Mehemet Ali Pasha and Baker Pasha as regarded the future of the war. These were to re-unite the armies into one command, and at once proceed to attack the Russians before the latter's reinforcements could come up.

The Turkish soldiers were fighting splendidly, and no one of those who had been, and were still, watching the war in Bulgaria, had any doubt whatever as to what would happen if Baker's and Mehemet Ali Pasha's plans were agreed to.

Lastly, Major S—— was to find out, and let Baker Pasha know, what was really going on at Constantinople. The principal person he quite rightly suspected was Suleiman Pasha, although there were many others. Major S—— accordingly agreed with his friends, Mehemet Ali Pasha and Baker Pasha, to do all possible.

As soon as he arrived at Constantinople, he called upon Sir Henry Layard, who was, of course, very anxious to know how matters stood in Bulgaria. He was very much disappointed at Major S——'s news, but he read the letters carefully, and then said, that he as an Ambassador could not, and not being a military man he felt himself incompetent to, make suggestions; but nevertheless he advised Major S—— to call on the Military Attaché. Possibly Major S—— might get some advice from him, as he, Sir Henry Layard, knew that his Attaché was most anxious to know how the war was really going on.

Major S—— accordingly went to the Attaché, naturally expecting that an English Major-General who had been through the Crimean war, and had won the V.C., would be a most competent officer to help him, and also through his knowledge of the Turkish language, he would make the interview with the Sultan comparatively easy. Still more was Major S—— delighted, because Sir Henry had also told him that the Sultan had become a very great friend of his Attaché, and trusted him in any advice he might give.

Major S——, with great pleasure, gave the Attaché all the news he could. To Major S——'s great astonishment, the General's only answer was that "It was he, himself, who had persuaded the Sultan to divide his Army into three separate parts, with an independent General to each of them, and to give them the promise that the most successful of the three should have the appointment of Minister of War."

Major S——, although very much disappointed, put before the Attaché all those incidents Baker Pasha had carefully made known to him, how the three Generals were not helping one another, and how Suleiman Pasha was at the present moment acting in that way at the Shipka Pass.

Major S—— had intended to give further examples, but seeing how angry the Attaché was, he (Major S——) stopped in order to get some explanation, and to find out why he was so angry. The General did not hesitate a moment, but declared that he in every point concurred with Suleiman Pasha, particularly as regarded his refusal to move from the Shipka Pass. That he considered, politically, it was most dangerous for the Russians to be allowed to cross the mountains, and that Major S—— knew nothing about it, and could not judge.

Major S—— again quietly put before him the opinions of Baker Pasha, whom he must have known was one of the best officers the British Army had had, so far as military knowledge was concerned.

But the General only got more and more angry, and although Major S—— tried once more to go over the reasons given by Baker Pasha and Mehemet Ali Pasha, it was no use. He only very angrily asked Major S—— how he dared to criticise his plans, for was he not older, and had he not far more experience than a young officer like him? Major S—— saw it was no use talking any longer, but he could not help saying that "He was greatly astonished, although he had not the experience of the Crimea, yet in all his studies on military subjects, he had never come across the name of any English officer who had such curious military views."

There could now be no doubt as to who was responsible for the misfortunes likely to fall on the unfortunate Turks; for although they had fallen under the hands of a very brave soldier, yet that soldier, in spite of his great experience in the Crimea, was evidently quite useless in such things as giving military advice to the Sultan, or indeed to any military officer of any kind. There was no doubt that the Turks were his special friends, and so *vice versa*, but that was all. If he had only made up with Baker Pasha and taken the latter's advice, he would have got all that he most wished for, and would very soon have seen the Russian Army (as it was) promptly driven across the Danube before the Russian reinforcements could come in time to help their countrymen.

Besides all this, the Turks, who were fighting splendidly, would have not only seen the Russians on the other side of the Danube, but also seen them fixed there for the remainder of the year—indeed, till the end of the next Spring; and it was almost certain that the finance of the Russians would hold out as little as their fighting men could.

Major S—— paid a last visit to Sir Henry Layard, but the latter gave him no hope, as he said that his Military Attaché was far too powerfully in the good graces of the Sultan, who always remembered how the Attaché had served in the Crimea on the Turkish side and had won his V.C. there.

Altogether it was no use trying any more. Major S—— could only regret and only shrug (in the way the Turks do) and ejaculate "Kismet!" (Fate), "Inshallah!" (God's will).

The hope is now realised to Baker of pressing one day some well-fought and hard-won field of battle:—

[Lord Wolseley—Gentlemen, Colonel Maurice is about to lecture on the Battle of Tashkessen, one of the most important events in the war between the Turks and the Russians in 1877, where a man who, I am glad to say, was an English officer, and one for whom his friends, when he lived in this country, had the highest regard, Colonel Valentine Baker, was the General Officer commanding the Turkish forces. I daresay Colonel Maurice will tell you that it was a matter of very great regret to those who took an interest in the war, and who sided with the Turks, who were the weaker side, that his advice had not been followed earlier in the war, because, I think, those who study the campaign with attention will come to the conclusion that if his advice had been followed the result would have been a very different one, indeed.]

Colonel Maurice: My Lord, and Gentlemen,—Probably to all of you it will be familiar that the beginning of the 1877 and '78 campaign of Russia against Turkey consisted in the successful passage of the Danube by the Russians, after which the rapid progress of a body of troops, pushed forward under Gourko across the Balkans, was brought to a check because the Russian armies, to their great surprise, discovered that a considerable Turkish force had established itself at Plevna, on the flank of their line of advance. You will also remember that the Russian armies made a series of violent attacks upon the works of Plevna, in all of which they were disastrously defeated. From that time onwards for many months, towards the end of the year 1877, the main force of the Russian armies was concentrated upon an effort to capture Plevna by closely investing it. Therefore, it became equally important for the Turks to relieve Plevna. Various bodies of troops were gathered, and vast supplies were collected at different points in order to gain that end. The treachery or imbecility of the different Turkish commanders prevented them from taking advantage of the many opportunities which the success of Osman Pasha in Plevna afforded them.

Apropos of what Lord Wolseley has said, I may mention that Valentine Baker told me himself that day after day he had implored Mehemet Ali, with whom he during the earlier time was acting, to advance. Had he done so at the right moment the Russians must have been driven back over the Danube. Mehemet Ali, on the Lom, and Suleiman Pasha, in the Shipka Pass, had failed to make any successful attack upon the Russians, or push their troops forward to co-operate with Osman. One convoy had certainly reached Plevna, but that was before the Russian investment was complete. Towards the end of November, 1877, the situation was as follows :—Mehemet Ali had been moved round to Sofia, and charged with the duty of carrying relief to Plevna. A Turkish force under his orders lay in a certain position in the Balkans, where a pass, known as the Orchanie Pass, gives access from Sofia to Plevna. Stores had been collected there as in other places for the relief of Plevna, but Mehemet Ali, rashly for his own interests, committed himself to a frank statement of the difficulties that stood in his way. He was immediately relieved by the authorities in Constantinople, and a general named Shakir was appointed to command the forces in what are known as the "Etropol Balkans," through which the Orchanie Pass leads. Suleiman was put in general command of the armies outside Plevna, and when, late in the year, Plevna fell, Suleiman, not realising the hopelessness of still maintaining the extended line of the Balkans, continued to cling to them. His own headquarters were in Sofia. Shakir, in the Etropol Balkans, acted under his orders.

Shakir had his camp in the neighbourhood of Kamarli. Opposed to him was a large Russian force under Gourko, gathered along his immediate front. The Turkish forces at Kamarli had been reduced to about 16,000 men. Baker Pasha, who was with Shakir at that point, had endeavoured to persuade the Turks, while there was yet time, to retire from the Balkans. In this he had failed. He began, therefore, to be excessively nervous lest the Russians, now that Plevna had fallen, should push up large bodies of troops to the Balkans, and turning Shakir's position, cut him completely off from Adrianople. Independently of the direct pass which runs by way of Kamarli towards Aka Kamarli, and so down south, there is another road more to the west which would lead them to the villages of Tchouriak and Potop. If they made their way over the Balkans towards Tchouriak they could easily reach the main Sofia road, and so cut off both Shakir from Suleiman, and intercept the retreat of either army.

On the 24th of December, Baker succeeded in persuading Shakir Pasha to allow him to make a reconnaissance with a party of men to see what the Russians were doing in the neighbourhood of Tchouriak. Getting to Potop, a little village in the neighbourhood of Tchouriak, he succeeded in playing with his cavalry, in his usual clever way, so as to draw out the forces of the Russians. First of all he found out that Tchouriak, instead of being, as it was represented in the maps on which they had relied, at the top of the Balkans, was, in fact, at the foot of the southern slopes. The Russians, therefore, once in position at Tchouriak had virtually crossed the main chain of the Balkans, and had only to pass the low lying hills on the further side. He found out that they had already a detachment of all arms in Tchouriak. They were manifestly engaged in improving the road over the Balkans towards Tchouriak from Orchanie, where their forces were already established. It was evident that the Russians were only awaiting the arrival of the forces relieved by the fall of Plevna. They had made all preparations for their movement by way of Tchouriak, and, as soon as the new troops closed up, the storm would burst from that side.

It was merely a question of time, now that Plevna had fallen, how soon, in addition to those Russian forces which were already engaged against Shakir, a further army from Plevna, probably estimated too high by Baker at 120,000 men, would be pushed across the more westerly passes of the Balkans, cutting off the retreat, first of the Kamarli force, and then of that under Suleiman, on the Sofia side. In the course of his reconnaissance on that day, Baker observed that there was one position which would give him an opportunity of at least checking the Russians if they made good their passage across the Pass. He had failed entirely in persuading either Shakir or Suleiman to occupy the Tchouriak Pass. Both of them said they could not spare troops for it. But although the Tchouriak Pass had not been occupied except by a post of observation, there was yet a possibility of saving the Kamarli force from being taken in rear. A road comes down across the Balkans towards Aka Kamarli, which lies in rear of Kamarli, and gives an opportunity of retreat towards the south—towards Adrianople. The main Sofia road, before it joins this line of retreat, crosses near Tashkessen, a small under-feature of the Balkans sufficiently considerable to give an opportunity for occupying it as a military post. This position struck Baker as being so good that, if he held it with a small force, he might, at all events, check the Russians sufficiently long to give Shakir Pasha time to retreat. In vain both Shakir and Baker had endeavoured to press on Suleiman the necessity for the force at Kamarli to make their retreat whilst there was yet time. Even after they had obtained the information about the occupation of Tchouriak by the Russians, Suleiman refused to allow Shakir to retreat. But on the 27th December, three days after the reconnaissance, the telegraph communication between Sofia and Kamarli was cut. From Kamarli they could see right across to the line between Tchouriak and Tashkessen, a distance of about six miles. They could see the movements of the Russian troops down the passes of the Balkans. They could see, further, that a small Turkish cavalry force, which had been sent out to watch the Russians on that side, was being driven back, and that the Russians were steadily advancing towards the Sofia road. In fact, Baker and Shakir were in the greatest anxiety lest Tashkessen might have already fallen. Under these circumstances Shakir, for the first time, was ready to allow Baker to take off with him a body of troops in order to occupy Tashkessen, to protect his rear from being turned. Now, at last, Shakir anxiously implored Baker to do so. It so happened that Baker had under his command three very fine battalions of Bosnian and Albanian troops, in whom he had the greatest confidence. I will give you the strengths of the several battalions at once, because what makes the whole story so specially interesting, is the extraordinary disproportion that there was at each stage of this little affair between the Russian force which was against him, and the force which he had under his command. The three little battalions that he had with him when he moved from Kamarli were of these strengths: the Prizrend were 580 strong, the Touzla 450 strong, and the Uskubs 350 strong. These were the three little battalions which formed the nucleus of his force throughout, and the whole infantry force with which he was able to move in the first instance. On this day, December 27th, he took with him, in addition to these three little battalions, two field guns, two mountain guns, and three squadrons of cavalry. When he came on the evening of the 27th, towards Tashkessen, he was met by fugitives of all kinds who told him that the Russians were already in possession of the village. He pushed on, and was fortunate enough to find that these reports were false, and that as yet the Russians had not entered the village. Nevertheless the danger was imminent. He found that the Russians had spread over all the neighbouring hills. He made up his mind at midnight to ride back over the ice and snow in the dark of the night—a most troublesome ride—to Kamarli, in order to implore Shakir to retreat at once, and to reinforce him at Tashkessen with as many men as possible. On his arrival, after his dangerous ride, he pointed out to Shakir that from the moment this fresh Russian army had taken up the ground it now held, the defence of the position at Kamarli became altogether a secondary matter—Tashkessen was the point of all importance.

For if Tashkessen fell the whole Kamarli force must be captured if it remained where it was. Suleiman's troops at Sofia must be also compromised. As far as Kamarli was concerned, there was always, as long as the Russians had not passed Tashkessen, an opportunity for making a tolerably safe retreat, if proper precautions were taken, even from the immediate presence of the Russians. Shakir pleaded that Suleiman had always objected to his retreat. To which Baker answered that Shakir was now absolutely responsible for his own conduct, since his telegraphic communication with Suleiman had been cut off. Ultimately Shakir gave him his promise that he would, if no orders came from Suleiman on the following day, begin his retreat that night without fail.

Another dangerous night-ride brought Baker back to Tashkessen. On the following morning, the 28th December, Baker prepared at grey dawn for the occupation of the position. It was a very peculiar one. In the first place there was a mountain in the rear, exceedingly strong in itself; but the whole of the view from the front of it was concealed by a hill marked by three knolls of lower but commanding ground. There was a spur from the mountain which gave pretty easy connection with this lower hill. He found that these knolls themselves gave him an excellent view over the country in front. At foot of them was the little village of Tashkessen. About a mile further west there was a line of small low-lying hills, favourable for the working of cavalry. Baker arranged his troops in this way: He disposed his cavalry with their videttes on the crest of these low lying hills. Thence they could see most perfectly every movement of the Russians. He placed the main body of the little force of cavalry just behind the same rising ground. He was joined early in the morning by a bad and weak battalion, the Edirné, 300 strong. He placed his infantry thus: his Touzla battalion on the right knoll, his Uskubs on the centre knoll, and the Prizrend on the left. The Edirné were in reserve in rear of the centre. He placed his two mountain guns with the Touzla on the right, bearing down the road which led to the little village of Daout-Koi, which was too far off for him to occupy. He had placed his two field guns on a little hill on the left, near the chaussée, so that they could retire by it to the main position if necessary, could bring their fire to bear down the main road, and, this part of the ground being somewhat thrown forward, could fire at extreme range across the front towards the right beyond Daout-Koi. He determined to retire to the main position in rear as soon as he should be seriously attacked, but in the first instance he hoped to gain time by making the Russians believe that the advanced position was strongly held. He hoped that they would be induced at least to delay their movements until they had had enough time to prepare attack. He had for this purpose one advantage in dealing with the Russians. At the beginning of the war of 1877-78 they had utterly despised entrenchments. They had gone at the Turkish entrenchments at Plevna with the wildest fury, and had suffered frightfully in doing so. A tremendous reaction had followed. They had gradually come themselves to adopt entrenchments on almost every possible occasion.

If, therefore, he could induce them to believe that the Turks were in a well-prepared position, it was likely that they would not attack without long preparation. About an hour after daybreak the Russians were under arms, and began to move down towards Daout-Koi. Though the range from the field guns on the extreme left cannot have been less, if the map be correct, than 5,500 yards, Baker, in order to make an impression on the Russians, ordered them to open at what he himself calls "extreme range." He showed the troops as much as possible in their position. He also ordered the little mountain guns on the right to fire on the Cossack parties that had preceded the advance. The ruse succeeded. They had been steadily advancing against his right in a solid column. The first shot fell near them. The whole column halted. For some minutes they hesitated, whilst the fire continued. Then, to his infinite joy, he saw that, instead of seriously attacking, they turned off to their right towards some high ground about 6,000 yards from the front of his position. On that high ground they began to entrench. He felt he had gained a day, and had already given opportunity for the forces at Kamarli to be withdrawn. That evening he received a reinforcement of two battalions and three field guns. Of the two battalions one, the El Bassan, 320 strong, were good Albanians; the other, the Eski Cheir, 220 strong, were miserable recruits, who could not be trusted. He placed the El Bassan in the village, and made them loophole all the outer walls. The Eski Cheir were placed on a hill south of the Sofia road, so that they commanded it. During the course of the day after the Russian attack, he had, with great difficulty, succeeded in dragging one of the field guns, by the help of two companies of infantry, on to the steep peak B on the right, where the mountain guns and Touzla were. The mountain guns had had no effective range beyond Daout-Koi. There was great danger lest the Russians should occupy it, and thence direct a night attack. It was with the object of preventing this that the field guns had been dragged over the deep snow on to the rocky hilltop. The other field gun had remained on the hill marked D; the three others which now arrived were placed with it there close to the chaussée. The following day, the 29th, heavy snow fell. It produced such a darkening effect, and fell so steadily all day, making all movement so difficult, that he felt morally certain that the Russians for that day would not attack. They did not. Therefore yet another day had been gained. In the course of the day he received the further reinforcement of an indifferent battalion, the Tchengueri, 320 strong. He placed them close to the main position on the right overlooking the Daout-Koi road. At this time, *i.e.*, by the evening of the 29th, he had received in all about 2,000 men.

On the morning of the 30th, he saw that the Russians were moving with part of one of their divisions round his right flank, apparently intending to occupy a lofty and precipitous hill on that side, which he had all along observed, as threatening to outflank the advanced position he held on the three knolls. It was too distant for him to occupy. It was, he thought, too precipitous to be very useful to them. After apparently threatening to attack from that side they merely completed their movement, and began to entrench. The troops on this flank were clearly not going to make an attack on that day. Shortly afterwards a very curious movement took place. The Russian cavalry, which had been showing in considerable force on the high ground, where the infantry were entrenched 6,000 yards off in his front, now moved down on to the main Sofia road, and along it in a solid column, as if they intended to march straight into the village of Tashkessen.

On the previous evening, when Baker had received his reinforcement of three field guns, it will be remembered that he had placed four of his five guns on the left, so as to be able to bear down the main road. He now gave the guns orders not to open fire until the cavalry column had approached within 1800 yards of them, and then to fire down the road into the middle of the column. They did so, with the effect that after a moment's delay the whole column broke up and tumbled back in confusion right away to the rear. His little force of three squadrons of cavalry, which had been lying behind the low ground in front, as soon as they saw the Russian cavalry turn tail, galloped out after them. Naturally, it greatly raised the spirits of the whole of Baker's infantry to see this enormous mass of cavalry chased by these little squadrons. The incident had a very valuable moral effect in giving the whole force a confidence which, under the circumstances, was badly needed if they were to hold their own against such odds. As a result of these operations yet another day—that of December 30th—was gained. But the day had not been altogether satisfactory to him. Ever since he had reached Tashkessen and had taken up his position there, he had feared that the Russian cavalry must sooner or later make a discovery very dangerous to him. There was a road quite independent of that which his position closed. It led to the little village of Kokantia, thence to Makatch, and so by an independent pass to Aka-Kamarli. He feared greatly lest the Russians, instead of making their whole attempt upon his position, which directly blocked the main road to Aka-Kamarli, should hold him in front with one division, while they threw another completely around his left flank. If they did this he could not prevent them from intercepting

Shakir's retreat from Kamarli upon Aka-Kamarli. Further, his own force would be cut off, and whilst assailed in front by superior numbers would be in danger of complete destruction. Therefore, it was with no small anxiety that, after this imbecile movement of the great cavalry column had broken down, he saw a small detachment of them move into the village of Kokantia. He saw its inhabitants come out and bow down before the Russian officers. He saw them point over towards the village of Makatch, manifestly giving the very information which he was so anxious to keep from the enemy. He at once sent one of his companies of Prizrend towards Kokantia to turn the Russians out of it. As soon as the cavalry saw the Turkish infantry approach, they cleared out of the village and went away. But the mischief had been done. He felt certain that full information had been given to the Russians.

There was yet one other important incident on that day—the day preceding that of the most wonderful rear-guard action of our times, if not of all times. Personally I do not know anything like it to have ever taken place before. The incident I speak of was that Baker had a communication from Shakir, in which the Turk told him that he was now quite ready to retreat, and was most anxious to do so. He asked Baker whether it would be possible for him to retreat on that evening. Baker enquired whether any preparation had been made, and received for answer that none *had* been made. Baker then said that it would be impossible to withdraw from such close neighbourhood to the Russians, the sentries at Kamarli being only twenty yards apart, and the shelter trenches only 400 yards apart, unless the great mass of the stores and guns, and everything else which might impede the movement of the fighting force over the snow-clad mountains, were first removed. Therefore, he strongly recommended Shakir to move during the night all his superfluous guns and stores down the road upon Aka-Kamarli, and get them clear out of the way before morning. He then could move his main army on the following evening, that of the 31st. He implored him that both movements should be made under the cover of darkness, in order not to draw the attention of the Russian force in Baker's front.

The movement from Kamarli to Aka-Kamarli, though not visible during its progress to the Russian army in Shakir's front, would, if it took place by daylight, be seen by those with whom Baker was engaged. It was certain that if they became aware that Shakir's force was escaping from them, that knowledge would induce them to hasten their attack upon the position at Tashkessen. The dilatoriness they had hitherto shown would vanish at once, and they would attack at all cost and any sacrifice. Shakir duly promised to carry out all Baker's requests, which included an urgent demand that Shakir would send out a reconnaissance towards Kokantia, and would send two battalions to occupy Makatch, in order to check the Russians should they make an attempt on that side. The total force which, after his various reinforcements, Baker had in hand, including a good little battalion, the Chasseurs (250 strong), who arrived on the evening of December 30th, consisted in all of 2790 infantry, 180 cavalry in three squadrons, five field guns and two mountain guns. He now disposed them in the following way:—He had placed his little battalion of Tchengueri on the high ground of the main position in rear of the three knolls, in order to establish connection between the advanced position on the three knolls and the high ground. He had the El Bassans, in the village of Tashkessen, in front; the Touzla and the Uskubs, with the two mountain guns and one field gun, on the right knoll. He had the Edirné on the central knoll. Four companies (half of the Prizrend) were entrenched on the left knoll. Four field guns were on their left, close to the Chaussée. The Eski Cheir, entrenched, held a small hill on the south of the Chaussée. He detailed the other four companies of Prizrend and his good little battalion, the Chasseurs, as a movable force to work on the left of the position. The left was well entrenched. That part of the position admitted of good entrenchments, and the movable force on the extreme left had favourable ground for acting on. On the other hand, on the right the ground was so rocky that it was quite impossible for the Turks to entrench it. Here they had piled stones, one on top of another, to give a certain amount of protection from infantry fire, though the rocks gave increased effects to the shells from the enemy's guns. In that position they awaited the attack of the Russians. There was a "Han," a very solid building, which effectively commanded with one of its faces the spur that gave access from the line of the three lower knolls on to the main position in rear. It happened to be so placed, that one of its faces gave also a very effective fire down the Daout-Koi road. As his intention was only to occupy the three knolls in front until the Russians were fairly committed to attack upon this his false position, and then to fall back and receive the attack on the main position in rear, this "Han" was an important point for him, as covering the movement from the false advanced position to the real ground he had preposed to defend in rear on the main mountain. In taking up the advanced position, he would have at least a chance of inducing the Russians to make their earlier movements consist of an attempt to turn this false front on one, perhaps on both flanks. He knew that their terrible losses from frontal attacks had led to the regular practice among them of wide turning movements for attack. These turning movements were apt to be made in a very regulated mechanical sort of way. If he could thus induce them to attempt to turn this front position from one or both flanks, then, by falling back on to the main position, which was in itself very strong, he might hope to find that the outflanking forces, already broken up in the movement of attack, were huddled together, out of order and well exposed, in the very front of his main position. If that were so, their trained dread of a frontal attack might much assist him in keeping them under the fire of his main position, and they would be in no condition to attempt an effective turning movement against it.

As soon as day dawned, the Russians were to be seen in movement on all sides. A division of the Guards, which had been all along before them, began to occupy his right front and right, and to detach a small force for occupation of the high hill on his extreme right already described. They were evidently engaged in an attempt to get a gun up to his right rear. They also, with the right of their left division, occupied the ground which had hitherto been held by the Turkish cavalry. They pressed the cavalry back into the village of Tashkessen. Then an entire fresh division of the Guards moved down to the Sofia road, and thence in the general direction of Makatch. Naturally, he began to be afraid that, after all, what he most dreaded was about to take place, viz., that whilst they held him in front with one division, they would, with another, turn him by way of Aka-Kamarli, avoiding with this second division an approach towards his position itself. So wide became the separation between this Russian division on his extreme left and the one which had already begun to turn him on his right, that for a time it seemed to him by no means impossible for him to make good his escape by passing through the gap between them. But to have done so would have meant deserting Shakir. The purpose for which he had moved on Tashkessen would have been abandoned by the act. Nevertheless, with such odds against him, the whole thing looked so hopeless that he could not help thus turning over in his mind any opportunity that seemed to offer a chance of saving the troops he had with him. Whatever speculations he had indulged in on this subject were very soon brought to an end. A third division of the Russian Guards soon afterwards appeared, filling up the wide gap between the other two. The Russians themselves confirm, in all substantial particulars, Baker's own account of their forces. They admit that they had on this occasion forty-eight guns and forty-five battalions of the Guards engaged in the attack. But the Russians say, no doubt with truth, that these battalions had been very much reduced in numbers. No one probably in Russia, or out of it, knows what their actual strength was, but that they were not nearly up to their proper strength is certain. The Russians had suffered appallingly in the course of the war, and although large reinforcements had repeatedly reached them, the battalions were certainly even now very weak. They were, from exposure and disease, daily growing weaker. Baker's estimate may therefore be exaggerated, when he says that the proportion of troops against him was as twelve to one, but it must have been nearly that.

To resume the story. As the division of the Russians on the Sofia-Makatch road began to approach Makatch, and in doing so, came within rifle range of the high ground, the Prizrend troops, who were on this side following their march, opened fire upon them. In a moment the whole direction of the Russian movement was changed. Instead of continuing their advance upon Makatch, they pushed up to attack the Prizrend, and, just as Baker had hoped, became involved in an attempt to turn the false position. In the movement upon Makatch, and in turning up to make this new attack, it almost inevitably happened that their right was completely thrown forward. They were met by the Prizrend and the Chasseurs very effectively on each of a series of little hills which continued to the left the general line of the three knolls. The centre was not even threatened. On the other flank, the left division of the Guards had pushed up three battalions to turn that flank. They managed with much difficulty to get up a gun on to the ground to his right rear, towards which they had been struggling for some time. But, as the position was an impossible one for effective action, their shot missed the ground on which Baker's men were, and positively assisted him by sending shells which burst among their friends on the further side. He very soon prepared for retiring on to the main position, the attack being now fairly committed to a movement against his false front.

Before the next attack was made, he had proceeded with this retirement so far as to withdraw one battalion, the Touzla, from the right flank, and to add it to the Tchengueri, whom he had already placed in occupation of the main position behind. The El Bassans, who had been previously in the village, replaced the Touzla. A squadron of cavalry now arrived as a reinforcement, and he sent forward the whole of his cavalry, now one complete regiment, to cover the retreat to the main position. On the left centre his miserable Eski Cheir battalion bolted, and the four guns near them having lost their commander, retired also, but were promptly made to occupy their old position. On the right, a brigade of the Guards made a direct frontal attack, fortunately for him, on the right knoll held by the Uskubs and El Bassans. As he had anticipated, the Russians were driven back with heavy loss.

The Russian guns could not be brought in any way to bear on this part of the position; consequently the numbers of the Russians only added to their losses in the attack. As soon as this attack on him had failed, Baker withdrew his guns from this side on to the main position in rear, and continued the movement for occupying that prepared ground. He first withdrew the Edirné from the central knoll, whence troops could be best spared. Then in succession he drew back the remainder. So far everything was proceeding well. He had, especially on the left of the "Han," solidly occupied the position with two of his best battalions. He now had his men so disposed as to be ready to receive the troops engaged against the Russians on the lower ground to the left, as they fell back on the main position. But unfortunately, in the course of the movement, the Russians had succeeded in establishing themselves about 700 yards from the "Han," and they began to bring an effective fire upon it. The Edirné, who were to have occupied the "Han," had not yet entered it. Under the Russian fire they cowered outside of it. Nothing that could be done prevented them from breaking and bolting back down the side of the hill into the middle of the Uskubs, who, after their gallant fight below, had just arrived. Fortunately the guns withdrawn from the lower ground were able to bring a pretty effective fire to bear on the Russian column, but not without serious loss from close infantry fire. The gunners, alarmed by the flight of the Edirné and the near approach of the Russians, began also to retire. With difficulty they were made to return to their places, and four companies of the Uskubs were pushed into the "Han." The effect of this fire down both roads of approach made the Russian columns recoil. Baker now formed up his more useless battalions as far as possible out of the way, merely to show a certain amount of force, so as to leave the impression that he was occupying parts of the ground that he could not with efficiency hold. Meantime the fight was going on between the Russian right column and the four companies of Prizrend, with the other little battalion of Chasseurs, as these fell slowly back towards the higher ground of the main position.

Baker had begun to be very uneasy as to what had been done to secure the retreat of Shakir's force from the Kamarli position. He sent back Colonel Allex to communicate with Shakir Pasha, to ask him to send all the reinforcements he could, and to ascertain what had been done up to this time, especially whether the steps which were to have been taken the previous evening had been taken or not. Colonel Allex had not been gone very long before he sent back a message to Baker to implore him to give up the fight, and retire at once. They had been, he said, betrayed by Shakir. Shakir was withdrawing his troops. Baker would have the whole force of the Russians hitherto opposed to Shakir on his back. Baker felt the difficulty of his position. He thought of his duty to his own troops, but, on the other hand, it was not certain that Shakir was deserting him ; it was possible that there was some misunderstanding as to what was going on. He waited for half-an-hour. Then he received another message from Colonel Allex repeating the same thing, and looking towards the hills behind, Baker could see the retreating column moving on Aka-Kamarli. It seemed to him that it was composed mainly of baggage horses. He sent a detachment of cavalry with an officer to Araba Konak to inquire what brigade was retreating. After another half-hour's anxious delay he received news that what had been done was bad enough, but was not quite as bad as Colonel Allex had announced. What had occurred was this—Shakir, instead of getting off with his stores and guns on the previous evening, as had been arranged, had been so much delayed by Turkish procrastination that he was now moving them back in broad daylight. There was no intentional treachery. This misfortune was that the retreat now taking place in full view of the Russians would show them the necessity for making a decisive attack to prevent the escape of Shakir's troops. Accordingly a fresh and furious assault was made immediately.

The position was, however, very strong, and the troops that now held it were very good. Like the others, this attack was driven back. In the very crisis of the Russian attack, however, another serious danger had occurred. The Russians were armed with the Berdan rifle. A certain number of the newly-raised Turkish troops were armed with the Peabody-Martini. Unfortunately, with the exception of the Chasseurs, only the militia were armed with it. All of the better Turkish regiments, the Prizrend, Touzla, and Uskubs, were armed with the Schneider. The Berdan was distinctly superior to the Schneider, but inferior to the Peabody-Martini. At this moment the Chasseur battalion, which was fighting well on Baker's left, sent back a message to say they could hold out well if they were only supplied with ammunition, but their ammunition had failed them. The Chasseurs were armed with the Peabody. They could not, therefore, get any supply from their neighbours, the Prizrend, armed with the Schneider. They could not get their own ammunition up from the ground behind to the exceptionally steep part of the ground they were on. For a moment it seemed as if the whole defence must collapse. Fortunately, as it happened, a certain detachment of militia had been left at the village of Tashkessen some time earlier. Having been employed on various detached duties, it had been broken up. Its pack-horses, with ammunition, had remained behind. They were sent up in hot haste to the Chasseurs, the approach from the village being apparently of easier access than that from behind. So the fight continued. The day had been wearing on; the fight having begun at 7 o'clock a.m., it was now between two and three. No serious impression had been made on the main position. Nevertheless, Baker began to be very anxious as to the gradual turning of the Russians round his extreme left flank. He had also to meet renewed attacks on his right. He first placed there his two mountain guns behind a low ruined building, whence they were able to keep up their fire at close range, and further secured this flank with four companies of the Prizrend, hitherto held in reserve since their retirement to the main position. Fearing to produce a bad effect on the men if he himself left the central "Han," he despatched another officer to watch the left. As yet the Russians had made no impression there. As, however, the cavalry was no longer of any use in the front, and might at least watch the extreme outflanking movement of the

Russians beyond the left, he had already placed them in the plain on the reverse side of his position. As the field guns had already suffered heavily, and had little ammunition left, he now also placed them in the same plain to fire upon the mountain tops, should the Russian infantry attempt to turn the position by passing them beyond the left. At length, about half-past three, it appeared as if the Russian left division had given up all hope of forcing the part to the right of the "Han," and were hoping that the other division might be more successful against the left. Both were hampered by the effect of the turning movement against the false front, and were almost forced into costly frontal attacks, little supported by artillery. Feeling his right safe, Baker now rode to the left to see what was going on on that side. He found that there was a small portion of the Russians that was fairly outflanking him on this side. He was, therefore, obliged to send two of his four remaining reserve companies to occupy a hill thrown back, guarding this flank. Before long these companies became sharply engaged, and it was necessary to reinforce this vital point with the last reserve, the remaining two companies. The night, however, was at last approaching. Before the day closed, the Russians made a last desperate effort against the hill. They suffered terribly from the Turkish fire in the advance, but nearly reached the crest. As they approached it, seizing his moment, Baker gave to the whole of the Turkish troops the signal to dash forward.

The cry of "Allah!" sounded along the mountains. The Turks sprang to their feet, and in a moment the Russian Guards were driven headlong down the hill. So complete was the repulse that not even a picquet was left to watch the movements of the Turks. Darkness, therefore, now completely covered the retreat. First the wounded and sick were removed; then the guns. Next the troops were brought in from the outlying hills, and replaced by dummy sentries. Lastly, the whole force was marched back to Aka-Kamarli. At the first halt Baker met Shakir Pasha, who had detailed a brigade under Mehemet Pasha to relieve him of the duties of rear guard. Shakir had made good the retreat of his own force without any loss, except that of some guns whose damaged carriages would not allow them to be withdrawn.

The incidents of the following morning were very curious. The Russians thought that Baker Pasha was in command of the whole force at Kamarli and Tashkessen; and first of all they sent in a *Parliamentaire* to inform the General that he and the force at Kamarli were completely surrounded. The Russians, therefore, demanded their surrender. The party accompanying the *Parliamentaire* came upon the dummy sentries, and found that the Turks had vacated this part of the little camp; next equally in vain they sought the Kamarli Army. When the Russians from both sides finally advanced, the Russian from the north met Russian from the south, and found to their amazement that the Turk was gone. That was the end of Baker's rear-guard action. You will find an account of it in his own book on the 1877-78 campaign. I have taken this description, with very little alteration, from it. I ought, perhaps, to apologise for merely giving you an account of what you can read for yourselves. But if I induce you to look at it, I am sure no one who does so will regret that I have drawn his attention to it. I think it is a shame to us that this, the finest story of a rear-guard action in history—an action successfully carried out by an Englishman in command of Turkish troops in the most recent Continental war—should be forgotten. It is a brilliant example and lesson as to what may be accomplished under such circumstances by a man in the direst difficulty. A careful study of the exact steps Baker took at each phase of the battle is the very best instruction that anyone can have. In my judgment, a knowledge of the facts of this little incident is in itself much more valuable to an officer than any theories we can lay down about it.

A respite of forty hours after this brilliant fight, gave to Baker the rest and sleep so much needed. He had placed in safety the retreating force of Shakir Pasha, before commencing his leisurely march to Petric, where the Headquarters stayed in a house, fit and comfortable—the gift of Lady Strangford to the Bulgarians. Here Baker had to mature further plans for keeping the pursuing force at bay. It was on one of these occasions, when scouting for his foe, that a message reached him saying "The war is over; Shipka position has been surrounded, and Vesoul's army captured; armistice proclaimed."

The end of the war was not yet, and the reader is well advised to follow for himself the continuance of same in Baker's own work, "War in Bulgaria."

After the fall of Plevna, the battle of Senova decided the Russo-Turkish war; for the Turks, without any further struggle, abandoned Adrianople and surrendered the enormously strong positions of the Bujuk Tchekmedgi lines before Constantinople. It was to prevent this last disaster that Baker again proved his mettle, for he knew the Russian move would be from Adrianople to Gallipoli; and to make a rush for the latter, he led Suleiman's force quickly across the Rhodope Mountains to the Ægean Sea, and embarked at Lagos for Gallipoli on January 23rd, 1878. Once at Gallipoli, Baker had secured for England a landing-place, with a friendly force of Turks to greet her on arrival.

It was also on January 23rd that instructions were given regarding movements of the fleet:—"Admiralty, 23rd, 1878, 7 p.m. To Admiral Hornby, Vourla. Most secret. Sail at once for the Dardanelles, and proceed with the fleet now with you to Constantinople. Abstain from taking part in the contest between Russia and Turkey, but the waterway of the Straits is to be kept open; and in the event of tumult at Constantinople, you are to protect life and property of British subjects. Use your judgment in detaching such vessels as you may think necessary to preserve the waterway of the Dardanelles, but do not go above Constantinople. Report your departure, and communicate with Besika Bay for possible further orders, but do not wait if none are there. Keep your destination absolutely secret. Acknowledge."

Thus, side by side with England, Baker was working hard. After landing his men at Gallipoli, he was summoned to

Constantinople; only to learn that the nerve of the Ottoman Empire had gone, and that the lines of Bujuk Tchekmedgi, which made Constantinople so safe, were to be abandoned. Baker had worked hard at these lines—they were of his own creation—first at the close of 1876, and early in 1877, when he urged on the Ottoman Empire their great value, and inspired from England the assistance of two officers of the Royal Engineers.*

This surrender came as the last straw. He applied for leave, and in company with his friend Colonel Burnaby, he left for England on January 29th, 1878. The Press, on Baker's arrival, fully expressed their pride that an Englishman should have done all this.

* Colonel R. Home, C.B., R.E., was the principal Royal Engineer officer assisting in this defence of Constantinople, and sent home most valuable reports, becoming at once the trusted adviser to Lord Beaconsfield on a great deal relating to the Eastern Question.

Colonel Home contracted typhoid fever while Chief Commissioner on the Roumanian Boundary, and never regained his health, passing away in February, 1879. Of him it may be truly said that, though his services and his name were barely known to the public at all, yet during the last few years those services have been of almost the first rank in political or military life. General Hamley, whilst on leave in Italy, received a despatch from the War Office asking him to succeed Home (March 3rd, 1879).

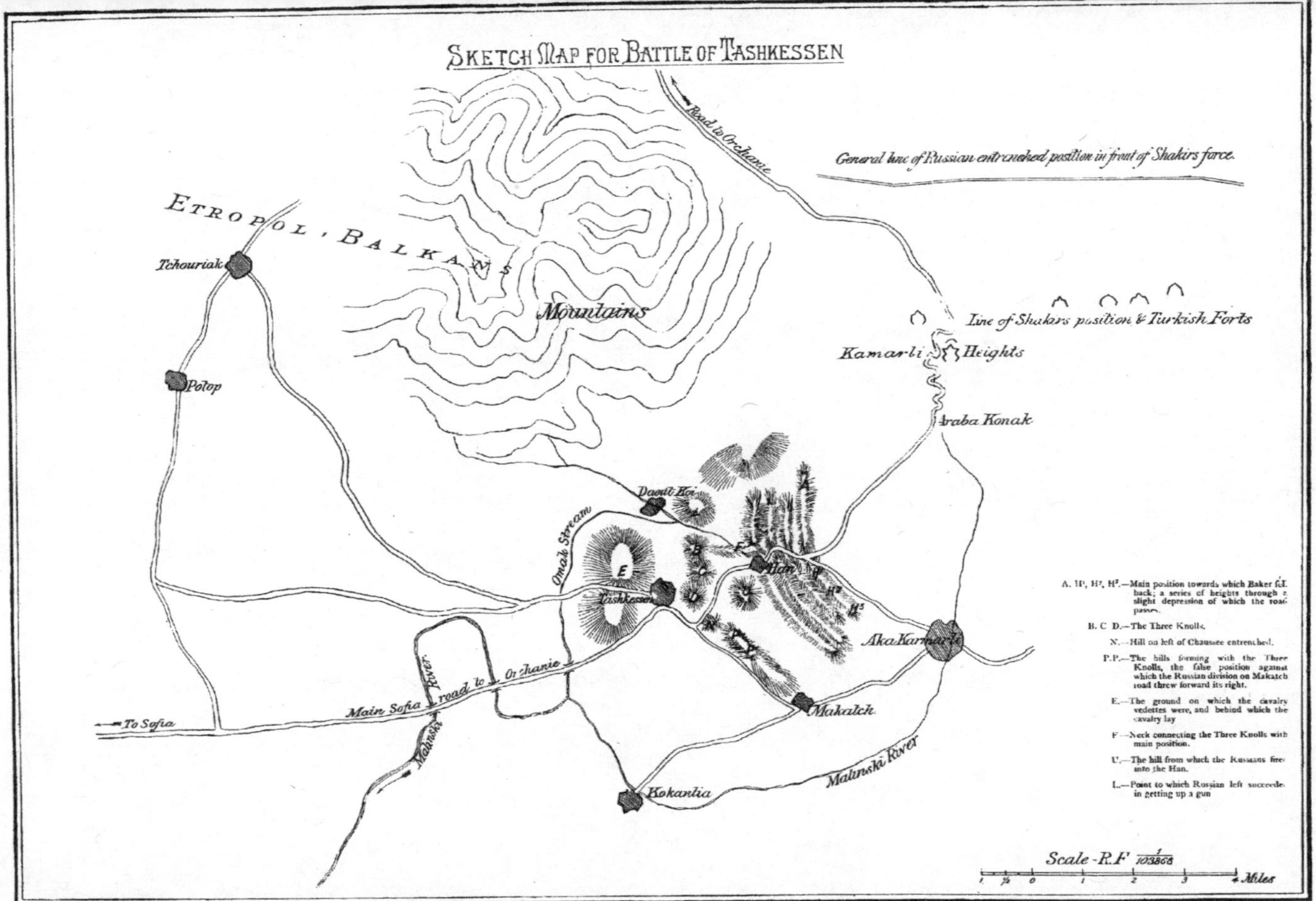

To face page 428.

CHAPTER XLIII.

THE SECOND AFGHAN WAR.

"February 26th, 1806.—At dinner received Europe news. The Austrian Army at Ulm destroyed. The French and Spanish Fleets annihilated, and Nelson no more. I am stupefied with the news, but feel more horror at the prospect of Continental affairs than joy at our naval success."

—*Elphinstone's Journal.*

THE apprehensions of French interference with India greatly increased with Napoleon's successes in Europe. The French were to the Anglo-Indian politician of 1806 what the Russians have been to him for the last fifty or sixty years, or since the Crimea. It seems "the ruling thought" to be too negligent of risks from attacks by enemies close at hand, *i.e.*, just across the water—whereas any project of invasion of India from the wilds of Asia creates endless activity and an Afghan War.

It was on January 28th, 1878, that a Supplementary Estimate of £6,000,000 was voted—because the terms of peace between Russia and Turkey, as submitted by Count Schouvaloff to the English Cabinet, were of a somewhat sweeping character, and needed looking into. This looking into was done by Lord Beaconsfield and Lord Salisbury at the Berlin Congress. In July, 1878, the two Statesmen were received at Charing Cross on their return with great welcome, and from the platform, Lord Beaconsfield addressed the crowd: "Lord Salisbury and myself have brought you back peace, but a peace I hope with honour, which may satisfy our Sovereign and tend to the welfare of the country."

Whilst this Congress sat, a strong Brigade of Indian troops as under, and in full service order, awaited the issue of events at Malta :—9th B.C., 1st Bombay Cavalry, 2nd Ghoorkas, 13th and 31st Bengal Infantry, two regiments from Madras and Bombay, with Artillery and Engineers. These would have been the first send-offs, in case of war with Russia, and whom Baker was in readiness to receive at Gallipoli.

Lord Northbrook had, early in 1876, resigned the rule of India, and Lord Lytton succeeded, with evidently a "waiting brief" for England in case of this expected war; but the lesser of two evils came—a war with Afghanistan.

It is said of Louis XVI. that, on application being made to him by Tippo Saib, not long before he suffered, to assist him in taking possession of some provinces in India from the English, and annexing them to the Crown of France, Louis nobly refused his assent and said: "In the American War, my Ministers took advantage of my youth and inexperience. Every calamity that we have suffered in France took its rise from that event." Surely so, for in joining America against England, France increased her immense debt, and propagated that spirit of revolt which cost Louis his head.

Surely the experience here quoted, might show to Russia the folly of assisting a friendly race to war against her neighbour, when that neighbour is her parent and protector. Is it beyond all hope that this war we are about to narrate may be the last of our Afghan troubles? Afghanistan and Turkey should both be independent countries, resembling Switzerland, a country we all know so well; and like a pitch for cricket or the links for golf, their soil should be placarded, "No war or violence allowed."

On January 1st, 1877, the Queen—who, as a child, in 1837 ascended the English throne, fair, happy, scarce nineteen, "In whose white hands her little sceptre lay"—was proclaimed Empress of India. Lord Lytton, the Viceroy, was a poet, a genius, and ceremonial of all kinds was dear to him—so it was amidst unprecedented splendour that this proclamation took place.

The following year (1878) found Shere Ali across our path. Russia had a Mission at Cabul, and England desired the same; but Neville Chamberlain was stopped at the Khyber Pass, and thus came our *casus belli*. So far Shere Ali was consistent, for he had told Baker the same. The Ultimatum expired on November 20th, 1878, when a force under Sir Sam Browne, V.C., K.C.B., left Peshawur; and after a day's check at Ali Musjid, the troops pushed on rapidly to Jellalabad and Jugdulluk, without any serious opposition, and remained at Gundamuk until May, 1879, when Yakub Khan arrived and entered into peace negotiations with Major Cavagnari, the Political Agent.

This long halt at Gundamuk gave ample time and opportunity to those of thoughtful mind, to ponder over the sad mishaps of 1842. For there in the near horizon, and of easy approach, stood out a conical hill, fairly bleached with bones—for that is where the last stand was made by the remnant of the Cabul army. After this last stand, one—and only one—maddened and bewildered, crawled along, until a stray grass cutter's *tat* tempted him to ride. Thus the one survivor entered Jellalabad, as Lady Butler (*neé* Miss Thompson) has painted for us, in her magic still-life way.

The then onlooker could also picture the infernal fire within the Khoord Cabul Pass, as told by Mrs. Waller:—"The ladies were mounted on horses, and placed in a body together, and made to gallop as fast as possible into and through this path of death. One officer was shot through his head while actually riding beside me, but none of the ladies were hit excepting Lady Sale, the reason of this being that the Afghans had a deep-rooted prejudice against hurting, much more killing, any woman—but Lady Sale was wearing a man's cap and cloak at the time, and was not recognized in the confusion. This led to the eventual salvation of all the ladies, for when Akbar heard of this, he became so concerned that he insisted on taking us all under his protection as hostages. This proposal was accepted on one condition, viz., that our husbands and the wounded officers should be with us, and so saved from the awful privation and almost certain death that otherwise awaited them. Thus my wounded husband and Lieutenant Mein were now allowed to join. My little girl was in the care of another officer, who held her in front of his saddle, and they were lost sight of for some days. Ahead of us was the fast diminishing army. Their path was stopped at intervals by barricades, at each of which the slaughter was terrible, until their last stand on this conical hill. My father, Major Griffiths, Captain Souter (44th), and Mr. Bluett, were on this hill, but had separated themselves for a while to arrange matters with some Chief before the fight began. Thus they were not killed." Captain Souter all this while had the Colours of his Regiment round his waist; the establishment of H.M. 44th that day stood: One Captain, two Colours, remainder dead. The nine ladies, the many children, and the eighteen wounded officers, were now tossed about the country until the day came, when Pollock enabled Sale to receive all the ladies and children back again—but many we know were fatherless and widows.

The sequel to this tragedy is interesting, for in this war of 1879 it fell to the duty of Lieutenant Waller, the eldest son of the lady we have quoted, to assist in the collection of the whitened bones from off the hill, and to give them Christian burial; and when a village was showing signs of animosity in more ways than one, and a fighting unit was despatched to quiet matters down, Lieutenant Waller become one of the party; it was then a Sirdar, or Chieftain, appeared expressing great regret, saying he could produce a document to prove his loyalty and sincerity, although perhaps he could not answer for all the rest. Lieutenant Waller then was shown a "chit," on which his father's and his mother's names were written, testifying to this man's kindness. The following is a copy (the original is at Windsor):—

"Buddeabad, Lughman,

"March 11th, 1842.

"Bahoow Khan, whom Sirdar Mehemed Akbar Khan placed in charge of the ladies and others of the Kabool Force detained at this place, having requested a certificate of his conduct, we the undersigned have much pleasure in stating that he has conducted himself with great kindness and attention, showing every desire to make our situation as little irksome as possible.

" W. Elphinstone, Major-General,	H. Robinson, Captain,
J. Shelton, Brigadier-General,	A. M. Anderson,
C. Mackenzie, Captain (Political Agent),	Fanny Macnaghten,
for self and C. Griffiths, Major,	John Magrath, Surgeon,
J. A. Souter, Captain H.M. 44th,	Emily Eyre,
Vin. Eyre, Lieutenant Artillery	B. Melville,
and D.C.O.	Florentia Sale,*
R. Waller, Lieutenant Horse Artillery,	A. Sturt,†
G. S. P. Lawrence, Captain Cavalry,	A. Waller,
W. Anderson, Captain 2nd Cavalry,	M. Trevor,
E. Pottinger,	G. Mein,
Colin Troup,	J. Trevor,
B. Bygrave,	L. B. Ryley,
W. Wallons,	J. C. Boyd,
T. Bluett,	G. T. Mainwaring."

Permit me to draw a moral. If in 1842, the Afghans and Afridis held the women of our race in quite respectful awe, and considered it an act of great ill-omen either to hurt, wound or kill them, what must they now think, when our daily picture papers portray them brandishing revolvers, and wrestling with the police, like mad jack-a-napes of some foreign type? Our women have *now* more to do with our present Eastern troubles than we are apt to think. Cause and effect move slowly; but there always will remain the fact that women are *not* men, and they must be content to remain as God has made and planned them. If not, they will become, as Queen Victoria said, the most hateful, heartless, and disgusting of human beings, if allowed to unsex themselves.

The Russian Mission left Cabul early in 1879, and with it disappeared Shere Ali, leaving Yakub Khan to succeed him as

* Died at the Cape in 1853. A cutter named after her is still on Brighton beach.

† Colonel Holmes and his wife (late Mrs. Sturt) were shot in their carriage at Meerut in May, 1857.

Amir. Lovers of Marion Crawford's book will read how Mr. Isaacs set Shere Ali free, and put him on the sure road to Tibet and into the Tartar country :—"The great rough warrior, the brave patriot who had shut the gates of Cabul in the face of Sir Neville Chamberlain, and who had faced every danger and defeat, rather than tamely suffer the advance of the all-devouring English into his dominions, was proud and unbending still, through all his captivity, poverty and trouble, and weariness of soul and suffering of body. He could bear his calamities like a man, the unrelenting chief of an unrelenting race. But when Isaacs stretched forth his hand and freed him, and bestowed upon him, moreover, a goodly stock of cash, and bid him go in peace, his gratitude got the better of him, and he fairly broke down. The big tears coursed down over his rough cheeks, and his face sank between his hands, which trembled violently for a moment. Then his habitual calm of outward manner returned."

At Gundamuk Yakub Khan announced to Major Cavagnari, the Political Agent with Headquarters, his accession to the throne. At the same time he expressed his willingness to treat, placing into the hands of the British Government the entire control of his foreign relations, and permission for a British Envoy to reside at Cabul.

It was this last concession which proved fatal to the whole arrangement, for Cavagnari, his staff (Dr. Kelly, Lieutenant Hamilton, V.C., and Mr. Jenkyns) and escort were, shortly after reaching Cabul, all murdered ; and the war, now no longer of prestige, but for revenge, re-opened. On receipt of the news of this massacre, the Government of India at once organized two invading forces. One, under Sir Frederick Roberts, proceeded *viâ* the Kurram direct to Cabul ; the other, under Major-General Bright, C.B. (afterwards Sir Robert), proceeded *viâ* the Khyber, Jellalabad, and the Sataband Pass, to keep open a line of communication for supplies, etc., to Cabul. On the Staff of the latter were Colonel W. Lockhart, C.B. (afterwards Sir William Lockhart, G.C.B., C.-in-C. in India), Major W. J. Boyes, A.A.G. (late Adjutant of 12th), and Captain the Honourable C. Dutton, A.Q.M.G.

All went well for some time. Roberts crossed the Shutargardan Pass and entered Sherpur, when, in the middle of December, 1879, by a sudden rising, he found himself besieged. Bright at once sent up all the troops he could—the Guides and others, under Sir Charles Gough, V.C., etc.—when Roberts was relieved after a most spirited defence of ten days.

Sir Donald Stewart came up a little later, *via* Ghusin, and after a severe fight at Ahmed Khel, entered Cabul and assumed the chief command. Yakub Khan surrendered at once to Roberts on the massacre of Cavagnari, and was sent to India as State prisoner*; and Abdur Rahman, who had been ten years in exile beyond the Oxus, now established himself as Amir, and Ayoub Khan, the son of Shere Ali, marched from Herat, and defeated the English force that went out to meet him. This reverse (the battle of Maiwand) meant the siege of Candahar, until Roberts, with 10,000 men, made a rapid and brilliant march from Cabul (313 miles), drove off Ayoub Khan, and entered Candahar in triumph.†

Sir Donald Stewart, in his quiet, stern, soldier-like way, withdrew from Cabul into India, and not a tribe molested him *en route*. Thus whilst Roberts, with all the able-bodied and only the mule as transport, went off pleasantly, Sir Donald said good naturedly: "I'll pack up and pay." Sir Donald lived to be C.-in-C. in India, and when his death came on April 6th, 1900, the Indian Press thus wrote of this simple, straightforward man :—

"A more modest and kinder heart than Sir Donald's never beat; and as he was himself unchanged by success, so he was incapable of grudging success to others. Of the personal rivalries, jealousies, and intrigues that come to cluster round the names of so many successful soldiers, he was absolutely free. Fair and just to all

---

* Yakub came out to Roberts of his free will, and said, "I am not guilty. Surely I could have escaped to Herat instead of placing myself in your hands."

† The one who upheld our flag at Ladysmith, all through that trying siege (Sir George White, V.C., G.C.B.), was at this time a Major in the 92nd. In the interval he has been C.-in-C. in India, and Q.M.G. in the Army; also a hero of the Mutiny, V.C. Colonel Hill-Johnes, R.A., was present in these fights.

men, we do not suppose that in the course of his long career he can ever have made an enemy; and though many of his closest friends must have passed away from the scene before him, he still leaves hosts behind, in whose minds the tall spare soldier-like form, the handsome features, the clear voice that carried in its tones something of decision and authority, will always remain in vivid and affectionate memory."

From assisting Colonel Valentine Baker in the war in Bulgaria, Major Sartorius rejoined his regiment in Afghanistan, where his younger brother Euston was serving with the 59th, and who now was the hero in the following little fight:—A night attack was planned by Colonel Kennedy on a certain steep cone or hillock, which was occupied by a mad mullah, one Sahib Jan. A force was sent out, composed of 2 mountain guns (Major Harris); 150 sabres 2nd Punjabi (Major Lance); 100 rifles 59th Regiment (Captain Sartorius); 100 rifles Beloochis (native officer). Orders were given for Sartorius to take the hill, and it was no child's play advancing up that path. Captain Sartorius was wounded in both hands, and a man beside him killed; but the hill was taken, and all agreed that by the death of Sahib Jan they were rid of a real firebrand.

Captain Sartorius was given the V.C. His eldest brother had received the same honour in the first Ashanti War, and Major Sartorius—whom we are so soon to meet again in the Sudan—had earned it many a time, if only in Bulgaria, passing down his line teaching his men to fix their sights in the thickest of the fight; whilst the old Admiral, the father, we have read of in former pages.

The deaths from cholera amongst Europeans alone amounted to forty-nine. Colonel Nicholetts, of the 2nd Beloochis, was about the first to go. Captain Stoddart was *in extremis*; but as he had been heard to say, when on a previous visitation of cholera at Morar (where he lost his wife and children, and the doctor also died), that Sauterne had saved him, the same beverage again was given, and it maintained its charm, for he recovered.

The 1st/12th Regiment were on service throughout the second stage of this war, commanded by Colonel Walker—late Adjutant of the 8th (King's) at Delhi—guarding communications between Cabul and Candahar. These duties do not give the same dash and glory as being at the front, but still, our depots provide the provender that makes the war horse "go." There is a story—it bears date from the Crimea—that to each Colonel of a regiment a V.C. was sent, to be given to some worthy whose deeds had escaped official recognition. One Colonel asked his men to be the judges, saying, "You select some comrade who has shown, throughout this campaign, the greatest pluck." Their choice fell on the Corporal of Pioneers. "Well," said the Colonel, "he is, we know, a right good sort, but his duties have been looking after your food and camp, more than in the front." "Ah, sir," said the men, "but *he* serves out the grog." So the 12th were in this post of honour, and took charge of the rum. They tell of the 12th, when under Picton, that on one occasion they made their own grog by emptying three hogsheads of sugar and seven puncheons of rum into a well of water. A bucket and rope were then when wanted.

By Christmas, 1879, the troops returned to India, when the wags in camp would have it—

M stood for Millions so uselessly spent;
Z stood for Zero, the result of the war.

Concurrently with this Afghan War, ran the Zulu fights at the Cape. Lord Chelmsford went out reconnoitring, and the troops he left behind—consisting of the 24th Regiment (thirty officers and 500 men)—to guard his camp and stores at Isandhlwana were on his return dead; for, failing ammunition, these brave men had to succumb one by one to an endless storm of assegais and spears (January 22nd, 1879). The Zulus, thus inflamed with victory, pressed on to Rorke's Drift; but there a weak garrison of Lieutenants Chard, R.E., and Bromhead, and eighty men of the 24th, improvised a grand defence, in which Dr. Reynolds and the Rev. Adams also took a leading part. At dawn of day the assailants withdrew, for they had seen Lord Chelmsford's jaded men approaching. It is supposed that the Zulus (all so superstitious), on seeing a fresh host advancing, thought they were the risen bodies—the ghosts—of those whom they had slain in the morning, and silently one column passed the other, each perhaps afraid, for they both looked the other way. The four officers named above, all received the V.C.

Then, on June 1st, occurred the sad death of the Prince Imperial (see page 391). On July 4th, the Zulu army (20,000 strong) was attacked at Ulundi by Lord Chelmsford and completely routed. Their brothers in arms in far-off Afghanistan, when they heard that the victory of Ulundi had been celebrated with champagne, wondered how they got it, and what transport was allowed—for such a thing had never crossed the border-line of India.

CHAPTER XLIV.

EGYPT.

> "When I consider life, 'tis all a cheat—
> Yet fooled with hope, men favour the deceit;
> Trust on in hopes to-morrow will repay—
> To-morrow's falser than the former day."

IT was in 1879, that Ismail Pacha, the then Khedive of Egypt, got thoroughly beyond control—his general indebtedness to all nations, his frequent changes of Ministers, and his feebleness all around, caused England and France to consult with the Sultan for his dismissal. It takes time for the Sultan to move. However, he was quickened on this occasion by Prince Bismarck, who stated that if England and France would not remove Ismail, Germany would—so Ismail was deposed, and Tewfik reigned in his stead (June, 1879).

Tewfik (a headstrong young man of 24) wished to be his own Prime Minister, but eventually yielded. Riaz Pacha then became Premier, with Evelyn Baring and M. de Bliquieres as Controllers for their respective Governments—and two most valuable servants they proved, for Baring had been hard at work on Egyptian finance since May, 1876, and knew the ropes well; whilst M. de Bliquieres was a French official of the very best type, and had also been a Commissioner on the Egyptian Debt Inquiry.

The trio worked well together, and Egypt seemed at last to have entered on better days, which Father Time would probably have solidified, if the War Minister had only shown a just and firmer hand. The two previous revolts of the Army had been bad enough, but the third resulted in open mutiny, for Arabi and his soldiers had imprisoned the Khedive and all his Ministers, and quite thought himself a second Mehemet Ali.

Such was the situation about May, 1882, when the question of the hour became "Who was to subdue Arabi, and reclaim the Army?" The answer was "Either England, France, or Turkey." The latter Power, tortuous and tricky as usual, would decide upon nothing; France withdrew; it remained for England to act alone. This was promptly done by the bombardment and capture of Alexandria under Admiral Seymour (Lord), and the victory of Tel-el-Kebir by Sir Garnet Wolseley (Lord), September 13th, 1882.

The question of sending Turkish troops to Egypt for the suppression of the revolt had been almost of daily and hourly consideration between Lord Dufferin and Baker in Constantinople—for the latter, although still head of the Turkish Gendarmerie, was to go as second in command, with as many European Officers as he wished to take; and even the day of starting or embarking had been fixed. But sundry quibbles and irritations now arose, which greatly delayed all real business, so it was not until September 13th, that Lord Dufferin was empowered to sign the Convention permitting Turkish troops to go.

The Turks in their delay, however, befooled themselves, for on the very same day of signing this Convention, the guns of Tel-el-Kebir took the place of the pens at Constantinople—and England, unaided, had saved the situation. The insurrection was crushed, and Wolseley slept that night in Cairo, having given to England that foothold on the Nile, which already in its tread has carried civilization to the source.

Baker arrived in Cairo early in October, 1882, and took service with the Khedive. His immediate Chiefs were Sir Auckland Colvin (Baring having left for India, as Finance Minister to the Viceroy) and Sir Edward Malet, who had in the previous year been appointed Consul-General. Both were men of marked ability; and under them General Baker set to work to regain authority and restore order, for a good deal of chaos had arisen out of the bombardment of Alexandria, and the dispersion of Arabi and his army at Tel-el-Kebir.

Lord Dufferin arrived at Cairo on November 7th, 1882, specially commissioned to unravel, if possible, the tangled skein of Egyptian politics, to suggest the best substitute for an English Protectorate, and to lay down the basis upon which self-government might be re-established in the Nile Valley. There he found his three Lieutenants all hard at work, and Baker ready with his scheme for the re-organization of the Egyptian Army and for the establishment of a Gendarmerie. A short summary is here given :—

"The present constitution of the army and gendarmerie, as proposed by General Baker, is as follows :—

"Infantry.—Twelve battalions of 500 men each. Three of these battalions (alien) and three battalions (Egyptian) under English superior officers. Part of this whole force when organized might afterwards be turned into gendarmerie.

"Mounted Infantry.—Two battalions of 500 men each (alien) under English superior officers.

"Cavalry.—One regiment of 500 men (alien) under English superior officers.

"Artillery, Coast.—Two battalions of 500 men each; one under English superior officers. If it is decided that the coast fortresses should not be maintained, this force would not be required.

"Artillery, Field.—Six batteries of 100 men each; half English superior officers.

"Engineers.—One company of 100 men; English superior officers.

"Supply Service, Hospital, etc.—300 men.

"Gendarmerie.—1400 men. To be added to hereafter. English Inspectors."

General Baker was asked by the Khedive to be the first Sirdar or Commander-in-Chief, but to this the Home Government refused consent—a disappointment most keenly felt by Baker.

The ship of State seemed at one time fairly started. Soon, however, appeared rocks ahead, which grew in size as the days crept on. How to blast or escape these rocks became the question of the hour. It was on December 20th, 1882, that Baker was engaged all day with Lord Dufferin on the look-out for a Commander of the Expedition now urgent for the Sudan. And why was this Expedition requisite? To answer this, it is first necessary to set back the clock to the year 1874, when "the Corporation of Brighton entertained Sir Samuel Baker at a banquet in the hall of the Royal Pavilion. Addressing the assembly after dinner, Sir Samuel said there was now a fair prospect for the development of the country of the White Nile, but the future would of course depend upon the energy and intelligence of the Governor, who should be perfectly unfettered by Egypt, and should be furnished with supreme power. He much feared that the Viceroy was rather too impatient for quick returns from his new territories; but as he had now entrusted his (Sir Samuel) late command to an English officer of high reputation—Colonel Gordon—he felt sure all would be done that was possible with the means furnished by the Viceroy. The good work had been begun, and must be continued by that good Englishman."

Thus Colonel Gordon, unfettered, in his bold heroic dramatic way, became ruler of the Sudan from 1874 to 1879. One visit he paid to Cairo, at the request of Khedive Ismail, to assist him out of his financial and other straits. This task was an impossibility, for Ismail was beyond redemption—his reign was practically over.

Gordon returned quickly to the Sudan, for Suleiman, son of Zubehr, was again up in arms, and had seized upon the Province of Bahr Gazelle. Gordon gave to one Romolo Gessi, a valued lieutenant, the task of bringing Suleiman to submission. Suleiman was caught and tried, and with ten others shot. Perhaps the conviction was not, on evidence, what you like to call the very best, for Zubehr later on publicly confronted Gordon with this fact, and enmity arose upon the point, but it soon subsided. So when Gordon left the Sudan in 1879, he was able to report that "not a man could lift his hand without my leave, throughout the whole extent of the Sudan," and he had torn the heart out of the slave trade.

Gordon had only left three years, when the Sudan was in the state Sir Samuel Baker had predicted—the cancer had reappeared. The mantle from Sir Samuel Baker and Gordon had fallen on the Mahdi and Osman Digna, which meant that slavery and villainy were thus again triumphant.

It was to smash the Mahdi, that Valentine Baker submitted the name of Colonel Hicks (Indian Army) to Lord Dufferin as a good leader for the expedition. This selection was made from amongst other eligibles one morning at Shepherd's Hotel. The first name drawn from a hat was to decide. The first business for Hicks to do was to regain Kordofan, where the Mahdi flame was fiercest. His starting point was Khartoum.

Colonel Hicks reached Khartoum on March 4th, 1883, with the following officers :—Lieutenant-Colonel Hon. J. Colborne (11th Foot); Lieutenant-Colonel de Cöetlogen (70th Foot) ; Major Farquhar (15th Foot)—his grandfather we met at Mauritius in 1810 ; Captain Warner (12th Foot)—a fine young officer, a victim to our excess of examinations, who consequently left the Service ; Captain Massy (57th Foot) ; Captain Forestier Walker (late Buffs) ; and Surgeon-Major Rosenberg. On March 18th Valentine Baker received a cypher message from Hicks, begging of him to explain to Lord Dufferin how terribly short he was in men and money. With a squeeze, the Egyptian Treasury found £147,000, barely sufficient to discharge arrears, much less to provide sinews for the future. As regards the men, the Minister of War paraded some in wooden stocks and chained, some with fore-fingers off, some with lime in eyes—all with certificates of thirty-five years' service, which meant their service had expired. And when he was remonstrated with, he calmly said: "They are so old, and cannot run away."

The end came rather quickly, for on November 7th, 1883, the army of Hicks Pasha, 11,000 strong, despatched by the Khedive of Egypt to disperse the insurgent force of the Mahdi and to subdue the Sudan, was utterly destroyed near El Obeid, above the Fifth Cataract. Led by a treacherous guide into a narrow defile, where the guns were useless, the little army, after three days' hard fighting, was massacred to a man; and on November 19th, Captain Moncrieff, the British Consul at Suakim, was killed and his force destroyed by a band of slave-traders at Tokar.

If we from Cairo lay down the history book, and speculate upon what might have happened but for the fatal occurrence of what actually *did* take place, we have no doubt often thought to ourselves what a specially good time it was for the Mahdi to play his wicked pranks, when he knew poor Egypt was without men, money, clothing, food, or camels. But if men like Gordon, Baker and Kitchener had been taken at their full professional worth, they would have been to us our Wellington, before whom the Mahdi would have fled. Then the flags of England and Egypt would have floated side by side, but without the stain upon them of El Obeid, Tokar and Khartoum.

No more of this conjecture! History has to be taken up again, and the Mahdi must be crushed—but not with the same brilliancy as attended the smashing-up of a certain other Mahdi on the Belgium Frontier in 1815; for now England stood aloof, and her Ministers sat with folded arms to watch this wrack and ruin, although Indian troops, Turkish levies, and British soldiers and sailors, were ready and available. Instead of which, Baker was asked to scrape together a few black troops, and to save the Sudan if he could.

The loss of Hicks and his rabble in the Sudan (November, 1883) was the one topic in Cairo at that date, and "on Sunday Valentine Baker consented to retrieve, if possible, the recent disaster. The Khedive, of course, said he was deeply touched by his devotion." [They all say it when they want something.] His office then was literally besieged with query after query, officer after officer. At 7 o'clock the same evening, a Turkish officer rushed in, saying the Turks all refused to march. Alas! and they were to be the backbone of the Egyptian Force!

Baker went to the troops at once, and no more was seen of him until midnight. The troops disclosed their grievances, for their engagement said their services were only for Egypt proper, and not the Sudan; and Baring, Clifford Lloyd,* and the other English anthorities, insisted on their not being military but civilian. Baker employed as much persuasive oratory from others as he could, and in Arabic spoke as follows :—"You are well within your rights, but I have been in your Service for some time, and *I* have to go. Now, I want the brave men to go to the right, and the cowards to the left" —and, taking out his watch, said: "Take ten minutes to decide." Suddenly a tall fair Turk went over, then two or three more, then a good many more; then some of those who had gone first rushed back, seized on the hands of their particular friends, and pulled them over, until nearly all were on the "brave" side. These men were worth recovering, for the others were the refuse of all Egypt, and when chained together looked quite "Les Miserables." It was with this ill-assorted flock Baker left Cairo on December 10th, 1883. The Khedive had allowed for him a special steamer, for the man was overwrought with work and trouble; and this consideration gave to Baker a few extra days by the bedside of a dying daughter.

The greatest enthusiasm attended the departure of General Baker on the morning of the 14th, for the station was crowded with his friends, and all wished him "Luck!" This word *must* be used, for no other element of success, either in men or war *materiel*, went with him—it was cruelly withheld. Zubehr and his Turks were not even allowed; the promised stores and telegraph did not arrive; but Baker's great friend, Burnaby, was with him, whilst Sartorius, Hay (Earl Kinnoull), Giles, etc., and the young Turk (his A.D.C., who was killed), formed a trusty Staff.

Gordon and Colonel Stewart were also in Cairo that day, and left the same evening on their "never-to-return" journey to

---

* Clifford Lloyd was a first-rate Irish Magistrate, and often have I been with him in a Belfast Kidney Riot; but to send him to Egypt on a technical Sudan question was like sending Theodore Hook to Mauritius as First Finance Minister.

Khartoum. Gordon, that night at dinner, requested that his host would dispose of his European clothing, then in a bundle outside, and give the proceeds to the poor.

Christmas Day was one of great quietness, in spite of the hospitality of Sir Evelyn and Lady Baring, and the kind and compassionate bearing of Sir Frederick Stephenson, the Chief of the English Army quartered in Cairo. The dinner at Sir Evelyn's was one after true English fashion, down to the mince pies and brandy; but still, there lurked the feeling that another crisis was at hand.

Colonel Sartorius had managed (being again on leave from India) to rejoin his old Chief, and had with his wife and daughter, left a week before Christmas for Suakim, the rendezvous of this so-called Army of the Sudan. Baker arrived in the "Mansurah" on the 23rd. By permission of Mrs. Sartorius, we extract bits from her diary:—

"Christmas Day! What a difference from home! Here the weather is most lovely, and a beautiful sun shining, while the atmosphere is so delicious that one seems to draw in fresh life at every breath. At our Christmas dinner we had a large party, and we rejoiced in turkeys from Jedda, a sirloin of beef cut from a cow that had belonged to the rebels, a tinned plum pudding from England; the eggs used for the custard were from Suez; and lastly, the cake was one Captain Darwell brought from Aden. Was it not a wonderful assortment of things from different countries to get in such a place? Even the champagne had its story; for my husband had ordered it for poor General Hicks, and owing to some confusion it never reached him.

"General Baker, with Admiral Hewett, started early on the 30th for Massowah—it was so important to see what could be done in opening up a route to Khartoum. Then he stayed at Trinkitat, and as far as the telescope could show, the country was very clear and good for cavalry. The next day Baker took the cavalry out to reconnoitre in the direction of Osman Digna's camp. The same evening we were in a state of great excitement and alarm, for a report had been spread (happily untrue) that Baker had met with hopeless disaster.

"January 31st.—Three Khedivial steamers came back from Trinkitat, with urgent letters from General Baker to my husband, that troops should be embarked and sent off sharp, as the steamers would have to return another time before all was completed.

"February 1st.—My husband left this morning for Trinkitat.

"By 8 p.m. on February 3rd, fires were put out, and silence prevailed over the camp. A few of the European officers were still together, discussing the probable events next day would bring forth, but at last even they retired to rest and so make ready for the fatigue to come. A few shots were fired into camp about 9 p.m., evidently directed at a bright light burning close to General Baker's camp cot; but on its being put out, no more were heard.

"The 'Réveille' was sounded at 4.30 a.m. on the morning of February 4th, and by 6.30 all were ready to move."

In a short hour, this small army (if you will), which General Baker and Sartorius had with Herculean labours got together, was annihilated. Brave Tewfik and his force at Sinkat soon learnt of this disaster, which meant massacre for themselves one week later. It was at 4 p.m. the same afternoon at Westminster, when the Mover and Seconder of the Address to the Crown in either House were purring over things imaginary in an optimistic sort of way, that this news arrived. Soon a gloom overtook both Houses, and Votes of Censure were moved.

In the Lords, the motion of Lord Cairns and Lord Salisbury was affirmed by 181 to 81. In the Commons, Sir Stafford Northcote made a long mild and moderate speech, to which Mr. Gladstone replied vigorously. The moment he sat down, Lord Randolph sprang up to attack him in rhetoric, which can only be sustained by passion in few men on rare occasions. "Too late!" he cried. "'Too late' is an awful cry. From time immemorial it has heralded and proclaimed the slaughter of routed armies, the flight of dethroned monarchs, the crash of falling empires. Wherever human blood has been poured out in torrents, wherever human misery has been accumulated in mountains, wherever disasters have occurred which have shaken the world to its very centre, there—straight and swift, up to heaven or down to hell—has always gone the appalling cry: 'Too late! Too late!' The Opposition cannot but move a Vote of Censure upon a Government whose motto is 'Too late;' and the people of this country will undoubtedly repudiate a Government whose motto is invariably 'Too late!'" The Conservative party, profoundly stirred by tales of blood and shame, continued shouting at this fierce conclusion long after the orator had ceased (311 for Government—292 for Opposition).

"I cannot admit," so Granville wrote to Gladstone, "that either Generals or Statesmen, who have accepted the offer of a man to lead a forlorn hope, are in the least bound to risk the lives of thousands for the uncertain chance of saving the forlorn hope." But what General dare send out a forlorn hope, without the arms and implements the leader asked for. Poor Gordon was denied everything; and had not Baker been treated the same?

On January 17th, the Cabinet at last decided to send out General Gordon. They wired for him from Brussels, for he was in the service of the King of Belgians, asking him to help to *enable* the garrisons to get away. He, with his friend Colonel Stewart, was off that very night; and as we have read, passed through Cairo on the very day Baker left for Tokar. They reached Khartoum on February 18th, 1884, and they were most enthusiastically received. But to get down to Egypt all the Egyptian employés and their belongings, and to replace the same by Soudanese, was no easy work, and required *extempore* arrangements, some of which our Government would not understand, and refused straight off the things he needed most. One request in particular was that Zubehr Pasha might be allowed to help. How could a Ministry, a Cabinet calling themselves English gentlemen, sit at home and hinder, fetter, and at last destroy so grand a man, whose whole life had been spent in doing wondrous things? And if he was at times *un peu* eccentric, it took quite a pleasant form—so different to this Cabinet at home, who were wrong-headed, blunt, beef-witted.

In other affairs of State, the Cabinet seemed equally erratic. Germany, on the question of her Colonies, was censuring Granville for his vagueness; France was pin-pricking Gladstone for remaining on in Egypt, when he promised he would not stay (see Mansion House Speech, November 9th, 1883; when Hicks, with 11,000 men, had been slaughtered only on the 7th); and Russia was perplexing us in India—for she had crossed the borders of Afghan (Pendjeh). Thus England was everywhere estranged, as well as hated.

A story is told of Gordon that, on receipt of an invitation or command from our late King Edward to dine at Marlborough House in May, 1879, he pondered and asked the reason why? A reply was sent that His Royal Highness wished to entertain the Viceroy-elect of India (Lord Ripon) and his staff—for Gordon had accepted, in some impetuous moment, the appointment of Military Secretary to Lord Ripon. Gordon declined the invitation, simply stating that he had been five years Governor-General of the Soudan, and had never before been so honoured, and he felt that he did not deserve that honour now. The dinner passed—no Gordon there. Next day the infinite tact and peace-loving nature of our late King brought a friendly letter, simply asking Gordon to luncheon *en famille* with the Prince and Princess of Wales. Gordon went, and never felt more happy. If our Ministers had but handled Gordon so, there would have been no Day of General Mourning for our slaughtered hero in 1885.

At last the people raised their voice and said "Let loose our soldiers; at least Gordon shall be saved." The 10th Hussars, who had had their send-off from Bombay for home (Sir James Ferguson was the Governor, since killed in the earthquake of Jamaica), were stopped at Aden on February 14th, with orders to take on board camp equipment and disembark at Suakim. This order was received with cheering, and great delight was expressed at the thought of again being on service with their dear old Colonel. Suakim was duly reached on the morning of the 18th.

In the afternoon of their arrival at Suakim, General Valentine Baker went on board the "Jumna" to visit his old regiment, and was received by a guard of honour, the whole of the 10th turning out and cheering him enthusiastically, the band playing "Auld Lang Syne." He then announced that he proposed to offer the horses of his three regiments of Egyptian Gendarmerie to mount the 10th. At the conclusion of this visit General Baker, accompanied by Colonel R. S. Liddell and the field officers of the regiment, landed; the horses were carefully inspected, and a sufficient number of them selected to mount officers, non-commissioned officers and men, 300 in all.

The next morning (the 19th), at daylight, the 10th disembarked, and, marching to the camping ground on the west of the town, pitched the Indian tents, which had been landed at the same time. As soon as the camp was ready and the horse lines prepared, the horses were led down by the Egyptians and picketed. The saddles were of a rough description, mostly an old French cavalry pattern, with no means of attaching the carbines; the bits were of a very severe kind—the Mameluke bit. No heel-ropes and a very few head-ropes came with the horses, no nose-bags, and altogether very little kit. Admiral Sir William Hewett, V.C., and the officers and men of the fleet, rendered the greatest assistance in every matter, and helped to remedy all the deficiencies. Sailmakers were landed, who set to work to make nose-bags, head and heel-ropes.

On February 25th, the remainder of the 10th was rapidly disembarked, the rafts from H.M.S. "Jumna" being used, and some temporary piers were erected on the shore. The remainder of the troops composing the force under General Graham had been brought from Cairo, and consisted of the 19th Hussars, the 60th Rifles, the 42nd and 79th Highlanders. The 89th, which had accompanied the 10th from India, and remained on board the "Jumna" at Suakim, was now disembarked; Major Holley's Battery of Artillery, which had also come from India, being left at Suakim as part of the garrison of that place. The 65th Regiment, stationed at Aden, and ordered to England in H.M.S. "Serapis," was also detained for service in the Soudan, and joined the force at Trinkitat on the 27th. No Artillery accompanied the troops, but a Naval Brigade, with machine guns, was formed, and took a conspicuous part in the campaign.

On the 27th a Cavalry Brigade Field Day took place on the sands outside the earthworks of Trinkitat, under the command of Colonel E. A. Wood. On this day, Lieutenant C. Grenfell, 10th Hussars, who had been seconded for service in the Egyptian army, was attached to the regiment. Lieutenant F. H. Probyn, 9th Bengal Lancers, was also attached. In the evening Brigadier-General Herbert Stewart, C.B., arrived from England, and took command of the Cavalry Brigade. At the same time Generals Redvers Buller and Davies arrived and assumed command of the two Infantry Brigades.

Two days later (the 29th), the force paraded immediately after the men's breakfast, and advanced at 8.30 a.m. The Infantry Division moved off in a large square, covered about a mile in advance by the first squadron of the 10th Hussars, under Major H. S. Gough and Lieutenant Durham. The remainder of the Cavalry Brigade and mounted infantry were *echeloned* from their right in rear of the left corner of the infantry square—the 10th Hussars in the front line, 19th Hussars in the second.

The force passed over the line of Baker Pasha's late retreat only three weeks previously, and the bodies of the Egyptians lay thick about the plain, being 2500 in number; and from the dry climate and hot sun they had been preserved in an extraordinary manner. The infantry square was led, under the guidance of General Baker himself, to the right of this line, in order to avoid this demoralising sight as far as possible. The ships from the harbour opened fire to cover the advance.

The infantry square now made frequent halts, and moved along leisurely, inclining towards the left of the enemy's position. The cavalry moved forward and drew the enemy's musketry fire from the right of his position. The Krupp guns of the enemy opened fire upon the infantry square; and one shell exploding in the centre of it, besides causing other casualties, severely wounded Baker Pasha in the face. The guns also directed their fire on the cavalry, but the range was not accurate, and the shells passed over their heads. The Arabs now descended in considerable numbers from their entrenched position, and rushed with splendid courage on the left corner of the infantry square, howling, firing and throwing boomerangs. The fire of the rifles and machine guns was too deadly to allow of success, and although here and there one or two of these gallant savages may have reached the square, all were shot down as soon as they arrived at a certain point, from 150 to 200 yards distant.

The Cavalry Brigade now received orders to advance, and as it passed on the right or outward flank of the square, the infantry gave a round of cheers. The brigade, under Sir Herbert Stewart, moved forward at a rapid pace—the 10th Hussars in the first line, the 19th Hussars in support—and, having cleared the position to the front, took a few prisoners, and was in the act of driving in some cattle, when intelligence was brought that the left squadron of the 19th, *echeloned* in rear, had been attacked in flank by a body of the Arabs, who had suddenly sprung up out of the grass and nullahs. The brigade was at once brought back nearer to the infantry square, which was hotly engaged while wheeling round and outflanking the left of the position of El Teb, and thus enfilading the Arabs in their entrenchments. The Cavalry Brigade now found large numbers of the enemy concealed among the bush, and the two regiments charged again and again, eventually succeeding in dispersing them, but not before poor Major Slade (10th Hussars), Lieutenant Probyn (attached to the 10th), Sergeant J. Cox and three privates had been speared and killed.

The enemy being thus dispersed, the squadrons were then halted; some troops were dismounted, and opened fire on the enemy retiring through the bush. The infantry continued to advance through the enemy's position round the wells, keeping up a tremendous fire, and, passing over the rifle pits and entrenchments, completed the victory. The British force upon this was massed round the wells; the horses were taken to water; a bivouac was formed, with the cavalry in the centre; and in this manner the night was passed.

Parties went out to search for the killed in the evening, and brought in the bodies of Major M. M. Slade, Lieutenant F. Probyn of the 9th Bengal Cavalry (attached), Sergeant J. Cox and Privates J. Brinsley, J. Douglas and F. Stride. The gallant Major Percy Barrow was most severely wounded (a wound which caused his death shortly after), and many non-commissioned officers and men. The total loss of the British force was twenty-eight killed, two missing, and 142 wounded; that of the enemy was calculated at over 2500, the dead lying in heaps in their entrenchments and rifle pits. Four Krupp guns, two brass howitzers and a Gatling, besides a quantity of arms and ammunition, were captured.*

Before the bivouac was formed, Baker Pasha, severely wounded as he was, with a bullet still in his face below the eye, rode into the 10th lines to inquire after his old regiment and congratulate it on the day. Thus ended a gallant fight.

Many individual acts of gallantry might be recorded of this day, but one of which special mention was made was the conduct of Private Hayes, a bandsman of the 10th Hussars. This man, who was skilled as a pugilist, being annoyed during the action with the difficulty of approaching his active and lithesome adversaries, dismounted from his horse, and, attacking a group of Arabs, knocked them down with his fists, and then again mounted. This soldier was afterwards thanked by General Graham for his courage, and in the following year had the honour of receiving from Her late Majesty's own hands, at Windsor Castle, the Distinguished Service Medal. His name, too, was also brought prominently forward before the public in some lines published in *Punch*, which attracted

---

* All these had been taken by the Arabs from the force under General Baker.

a good deal of notice at the time, as the country felt that the references to General Baker Pasha in the "Tale of the 10th Hussars" were most opportune:—

A Tale of the Tenth Hussars.

When the sand of the lonely desert has covered the plains of strife
Where the English fought for the rescue, and the Arab stood for his life;
When the crash of the battle is over, and healed are our wounds and scars:
There will live in our island story a Tale of the Tenth Hussars!

They had charged in the grand old fashion, with furious shout and swoop,
With a "Follow me, lads!" from the Colonel, and an answering roar from the troop;
On the Staff, as the troopers passed it, in glory of pride and pluck,
They heard, and they never forgot it, one following shout—"Good luck!"

Wounded and worn he sat there, in silence of pride and pain,
The man who'd led them often, but was never to lead again.
Think of the secret anguish! Think of the dull remorse!
To see the Hussars sweep past him, unled by the old white horse!

No alien, not a stranger: with heart of a comrade still,
He had borne his sorrow bravely, as a soldier must and will;
And when the battle was over, in deepening gloom and shade,
He followed the Staff in silence, and rode to the grand parade.

For the Tenth had another hero, all ripe for the General's praise,
Who was called to the front that evening, by the name of Trooper Hayes;
He had slashed his way to fortune, when scattered, unhorsed, alone,
And, in saving the life of a comrade, had managed to guard his own.

The General spoke out bravely as ever a soldier can—
"The Army's proud of your valour, the Regiment's proud of their man."
Then across that lonely desert, at the close of the General's praise,
Came a cheer, then a quick, short tremble on the lips of Trooper Hayes.

"Speak out," said the kindly Colonel, "if you've anything, lad, to say;
Your Queen and your dear old country shall hear what you've done to-day."
But the trooper gnawed his chin-strap, then sheepishly hung his head.
"Speak out, old chap!" said his comrades. With an effort, at last he said—

"I came to the front with my pals here, the boys and the brave old tars;
I've fought for my Queen and country, and rode with the Tenth Hussars;
I'm proud of the fine old regiment!"—then the Colonel shook his hand—
"So I ask one single favour from my Queen and my native land.

There sits by your side on the Staff, sir, a man we are proud to own;
He was struck down first in the battle, but never was heard to groan;
If I've done aught to deserve it;"—then the General smiled, "Of course!"—
"Give back to the Tenth their Colonel, the man on the old white horse.

If ever a man bore up, sir, as a soldier should, with pluck,
And fought with a savage sorrow the demon of cursed ill-luck,
That man, he sits beside you! Give us back, with his wounds and scars,
The man who has sorely suffered, and is loved by the Tenth Hussars."

Then a cheer went up from his comrades, and echoed across the sand,
And was borne on the wings of Mercy to the heart of his native land,
Where the Queen on her throne will hear it, and the Colonel Prince will praise
The words of a simple soldier, just uttered by Trooper Hayes.

Let the moralist stoop to mercy, that balm of all souls that live;
For better than all forgetting, is the wonderful word "Forgive."

On March 1st, the force fell in and continued its march for the relief of Tokar, which was found deserted; but at Dubbah, three miles beyond, were found the whole of the guns, rifles, ammunition, stores and baggage captured by the Arabs from Baker Pasha's force. The 10th destroyed over 1300 Remington rifles and bayonets, buried the ammunition, and carried away on the ten camels provided as transport the property of the English officers of Baker's force. The soldiers throughout had been greatly helped by the Naval Brigade.

The force returned to Trinkitat, which ended for a time this brilliant but useless campaign—for the loss of these valuable lives, and the slaughter of a few Arabs, had in no way helped Gordon at Khartoum.

On November 17th, 1887, a little vessel, gently riding to her anchor in the Sweet Water Canal, was seen to dip her flag half mast —Valentine Baker had suddenly passed away.* When this news

* His soul had taken flight before Dr. Sandwith could reach him; his only daughter was in time to be beside him—for the others, his dear wife and daughter Hermione, had passed into the vast dominion of the dead but two years before. In Cairo Church is a tablet to the memory of Mrs. Baker and Hermione, from one who greatly valued their friendship and faithfully cherishes their memory. It was erected by Major Edward Stuart Wortley.

reached England, a wire replied, "His funeral must be that of a Lieutenant-General," for to that rank in Her Majesty's Army he had been admitted; and Lord Cromer wrote: "Greatly as I feel the personal loss, I consider the public loss, both to myself and indirectly to the English Government, as a still greater one. The benefits General Baker has conferred upon Egypt, irrespective of his distinguished services elsewhere, are of themselves sufficient to show that by his death a man has been lost, whose courage, ability and judgment, not only England, but the country which at the time of his death he was serving, could confidently rely on in the hour of need." A patriot unto the last, for amongst his papers was found a carefully-sealed envelope, which contained, in case his country ever needed them, his surveys and secrets of the Lines of Bulair. They were handed over complete to the then Foreign Minister.

Our thoughts and hearts must now return to Gordon. A few notes from Colonel Sir Charles Watson, late a Staff Officer to Gordon when Governor-General of Soudan, and in touch with him to the last, are here permitted:—

"All through the spring of 1884, the English nation had been watching the action of General Gordon at Khartoum, and loud were the murmurs against the Government for doing nothing to help him in the dangerous position in which they had placed him. Why the Government showed this indisposition to assist Gordon is very hard to understand. At last, in the beginning of August, it was decided to make preparations for sending a force to Dongola. In a despatch dated August 8th, 1884, from the War Office to General Sir F. Stephenson, who was then in command at Cairo, Lord Hartington informed him of the decision of the Government, and also told him that it was proposed to make use of small boats to transport the troops, similar to those which had been employed in the Red River expedition. General Stephenson replied pointing out the disadvantages of the small boat scheme, and on being asked what he proposed, replied on August 11th:—

"'Can move to Wady Halfa, four battalions, 2200 bayonets; two squadrons, 200 sabres; one battery—horse or field artillery;

two batteries, mountain; and mounted infantry. Small boats proposed not suitable. Can procure large amount of water transport, locally.'

"It is interesting to note that General Stephenson's plan seems to have been exactly what General Gordon wanted, and if it had been carried out, the English troops would probably have joined hands with the crews of Gordon's steamers towards the end of October. But his recommendations were not followed, and 400 boats were ordered on August 12th, the day after the receipt of General Stephenson's telegram;* and shortly afterwards, 400 more. Now, if the boats had been built and used to assist the expedition in returning they might have been of great use, but *because* they were being built, the expedition was kept back to wait for them. It will be remembered that it was on August 11th that General Stephenson said he could send a force at once to Dongola, but it was not until November 1st, that the first of the small boats, having passed the Cataract at Wady Halfa, started for Dongola.

"The two months thus lost could never be regained in time to reach Khartoum before it had succumbed to famine. What would Gordon have thought on October 6th, when his steamers at Shendy were all ready for the English troops, had he known that the boats which were to bring those troops up the Nile, had not yet begun to arrive at Assuan!

"And not only was the great distance against the boats, but the Nile was falling, and every day the rapids became worse. As a comparison of General Stephenson's proposal with the War Office plan, it might be mentioned that half of the Royal Sussex Regiment, which was sent up to Dongola in September in native boats, took *thirteen days* to travel from Sarras to Dongola, a distance of about 210 miles, whereas two months later on, the troops in the small boats took about *five weeks* to cover the same distance. Looking

* Admiral Sir John Hay, Captains Molyneux and Hammill, three naval officers who had been deputed to examine the Cataracts of the Nile, all reported adversely to the Hartington-Wolseley scheme of small boats.

back upon the whole transaction, it is sad to think of the responsibility which the English authorities in London incurred by neglecting the advice of their advisers in Egypt. The final result was that the advance guard of the English expedition was not assembled at Korti until the latter part of December, 1884. Lord Wolseley reached Korti on December 16th, three days *after* General Gordon had written :—'*If some effort is not made before ten days' time the town will fall.* It is inexplicable, this delay. If the expeditionary forces have reached the river and met my steamers, one hundred men are all that we require, just to show themselves.' But *thirty-nine* days elapsed after he wrote this before the troops met the steamers, and Gordon's worst predictions were verified. The gallant effort of Sir Charles Wilson to reach Khartoum in time, failed because the expedition reached Matammeh too late."

Gordon, in full expectancy of earlier relief, had steamers all in readiness—between Matammeh, Shendy, and Khartoum.

The following morning, December 17th, the steamers again started up the river, and shortly afterwards saw the "Bordeen," which it will be remembered had been sent up to Khartoum on November 25th. Gordon sent her back on December 15th with the last post that ever left Khartoum. She took the last volume of his Journal, which concluded as follows :—"*Now mark this,* if the Expeditionary Force, and I ask for no more than 200 men, does not come in ten days, *the town may fall,* and I have done my best for the honor of our country. Good bye.—C. G. GORDON."

"Kartoum, 26/11/84.

"MY DEAR WATSON,—Thanks for your letter, which I received yesterday. The steamer* which brought it had to run the gauntlet of no end of rifle fire and of six guns. She was struck three times by shells, but only seven were wounded.

"I hope you and Mrs. Watson are well. I am not ill-treated, I consider, but the Cairo people up here—they are the ill-used. I will accept *nothing whatsoever* from Gladstone's Government. I will not

* The "Bordeen."

even let them them pay my expenses. I will get the King* to pay them. I will never put foot in England again, but will (D.V. if I get out) go to Brussels and so on to Congo.

"How is Miss Arnott, the lady who was with you? I greatly fear for Stewart, Power, and Kerbin, French Consul.

"With kindest regards, believe me yours sincerely, with kindest regards to Mrs. Watson.

"C. G. GORDON."

[THE LAST LETTER.]

"Kartoum, 14/12/84.

"MY DEAR WATSON,—I think the game is up, and send Mrs. Watson, you and Graham, my adieux. We may expect a catastrophe in the town, on or after ten days' time. This would not have happened (if it does happen) if our people had taken better precautions as to informing us of their movements, but this is 'spilt milk.' Good bye. Mind and let my brother (68, Elm Park Road, Chelsea) know what I owe you.

"Yours sincerely,

"C. G. GORDON."

THE STATUE OF WELLINGTON TO THE MINISTRY OF MR. GLADSTONE.

Aye! in your generation ye are wise
To thrust me from your sight. 'Twould anger you
That loyal men should point to me, and say,
"Thus in old time did England's sons reward
Great deeds of duty done, and honour won."—
But, when a living hero ye betray,
Ye well may mar memorials of the dead.
What have ye done for Gordon? Urged him forth
With one brave follower, to appease a land
Where a trained army had been crushed in fight,
Pledging yourselves his counsel to accept
In choosing chiefs for the revolted States;
Promptly he names Zebehr—you temporize,
And finally refuse;—pointing your shame,

* The King of the Belgians.

By asking why your noble dupe remains
In the beleaguered town, where treachery,
Almost as fatal as your own, had striven
To palsy the defence. For six long months,
'Neath Afric's torrid zone, almost alone—
But God was with him—has the modern Bayard
Contended with encircling hosts, and ye,
Through ribald scribes, have taunted him for failing
To do what you arrested him in doing.
Ye hoped, perhaps, that death would end his work,
And your great "man of words" might eulogize,
'Mid plausive crowds, the man ye sacrificed:
Berber and Tokar, Sinkat and Khartoum,
Were worthy themes for mournful eloquence.
But lo! a marvel! from the dubious gloom
That hung about the "man of action," came
Rumours that still he lived, still fought, still saved
Old England's banner from the shame ye courted.
Not then for him, but for yourselves ye trembled,
Lest public scorn should hurl you from your seats;
And when already the Nile flood is waning,
Ye fit out expeditions, and ye send
Ten thousand men to deal with the disorder,
Which Gordon, through six drear, unhealthy months,
Was left alone to cope with.
Ye do well
In your own generation, if ye strive
To maim the records of the past in England.
But heavy is your task. I haunt no more
Your idle hours, but memories yet remain:
Chatham, Pitt, Peel, Nelson and Disraeli,
With hosts of others dedicate to duty,
Are honoured in the land; and could ye rase
Our glorious Abbey with its scrolls of fame,
The page of History would live to tell
The height to which Old England once had risen,
Ere, in an evil hour, she trusted you.

September 22nd, 1884.

Soon after Gordon's death, General Earle had been killed at Kirbekan in a somewhat desperate fight (February 10th, 1885), and Sir Herbert Stewart—a Winchester boy, grand cricketer, brilliant soldier, and a Bayard all his life—died of his wound, on February 19th, 1885. Further troops, including the Grenadier Guards, were now sent out; the 1st class Reserves and many Militia Regiments were embodied; whilst on March 3rd, from New South Wales, arrived a valuable contingent, sent out at the Colony's expense, and to show their willingness to be a friend indeed if England is in need. This contingent was commanded by Colonel John Soames Richardson, an old officer of the 12th. He had been their Adjutant for many years: then on retirement took service in the Colonies.

On March 22nd, the troops under Sir John McNeill, V.C., were surprised by some Arabs at Suakim and severely punished. Then on April 21st, the country was alarmed over the Pendjeh incident in Afghanistan, it voted £11,000,000, and stopped all further movements in the Soudan.

> "Order is Heaven's first law; and this confest,
> Some are, and must be, greater than the rest,
> More rich, more wise; but who infers from hence
> That such are happier, shocks all common sense."

Mountstuart Elphinstone, after an absence of thirty-four years in India, reached England in May, 1829. Next to the unspeakable pleasure of finding himself at last again in his Fatherland, came his admiration (as he walked through the streets of Dover) at the extraordinary neatness of the houses and streets, and at the comfort of the people. Every cottage was neat and finished with geraniums in the window, and often a little garden in front. This extreme tidiness has often been remarked upon by the citizens of the United States on their first visit to England, so much so, that our so-called "earth hunger" is explained away as being no more than a sinister desire to make straight the untidy places of the earth—in fact the gadfly that drives John Bull from one nation to another, from one kingdom to another people.

It had been Elphinstone's fortune to accompany conquering armies, and to be entrusted with the duty of bringing order out of the confusion caused by each shock of War, and of re-distributing large territories on some durable political basis—in fact to do the tidying up.* So in later times, the same good intentions flourish; hence it is not without significance that we owe so much to Lord Kitchener for his fairy wand, first for war, then for tidiness. It was on September 2nd, 1898, that British and Egyptian troops under Kitchener—in number 23,000—defeated 50,000 dervishes under the Khalifa, and regained for us Khartoum. The sackcloth and ashes, which alas! had been worn for so long, were now set aside, and the soldier again put on his proper "Red" to appear at Gordon's shrine.

"The very first thing the conquerors did on entering Omdurman, was to set to work to clean and tidy the town. Anyone who knows only Khartoum of to-day, with its river-wall, broad straight level streets, its neat houses, avenues of trees, its mosquito brigade, and its anxious cleanliness, may well find it difficult to realize how pestiferous a town it was that Gordon and Baker knew: whose unhealthiness was such that men died like flies, and lived most of them in the spirit of 'Let us eat and drink, for to-morrow we die.'

"Old Khartoum was a town of mud houses, filthy tortuous unlighted lanes, unlovely holes full of garbage or stagnant water, ill-policed and ill-guarded; while to-day the tourist seeking picturesque dirt bitterly compares the new town to Brixton. One good thing, at least, the Mahdi did—he laid Khartoum desolate, and left a clear field for the re-construction of the new town.

"New Khartoum is typical of the general re-construction of the Soudan. There has been nothing showy or sensational, no heroic remedies or schemes of advertisement; there has just been a general British determination to make things tidy, to mend the broken, to hunt the lazy, to restrain the unjust, to help along the feeble—a list that contains not much beyond the ordinary duties of the day, just the things that lie nearest to hand. But it may be

* Lahore in the Sikh days was notorious as the filthiest capital in India. It is now one of the most striking Oriental cities.

Shiluk
White Nile
1908

To face page 463.

said truly, that the British officers and officials have gained the confidence of the people, and that most of them care for nothing else but the success of that branch of the Government's work over which they are set.

"Just these things, with some measure of patience and of love of justice, and no undue subservience to rules and forms, are all the secret of the Briton's power of government; and it is they that make the difference between the present state of government and the incompetent tyranny and robbery of the Egyptian Government before the days of Gordon.

"To put the matter shortly, the Englishman is guided by duty, the Egyptian was guided by reward. It is much to be feared that our latter-day world is tending to seek reward where its fathers sought duty. As Robert Louis Stevenson once said: 'The world must return some day to the word Duty, and be done with the word Reward. There are no rewards and plenty of duties. And the sooner a man sees that, and acts upon it like a gentleman or a fine old barbarian, the better for himself.'"

The following short poem to the "Shiluk" portrays the necessity to give the dwellers in our half-owned territory, the Upper White Nile, a certain amount of care:—

"THE SHILUK.

"Man—as God made him, strong and free,
No slave to bonds of modern ways;
Content to let the World go by
Nor reckon count of passing days.

"Rejoicing with the sun, but when
The sun is hid, makes no great moan,
Content to let his daily needs
Rest wholely with the Great Unknown.

"He knows no shame, but treads the earth
With supple limbs and graceful gait,
Clad as Dame Nature made him first,
Save beads and charms of savage state.

"He tends his herds and sows his corn,
And hunts the game with trusty spear;
His simple nature loveth peace,
But when in battle knows no fear.

"For countless ages he has dwelt
The master of the rolling plains,
Lord of the river, swift and broad—
But now his savage freedom wanes.

"For slowly like a gathering cloud,
Remorselessly, the White Man's hand
Advances, bearing doubtful 'light'
Into a so-called 'darkened' land.

"D. S. C.,
"Upper White Nile.

"Christmas, 1908."

General Sir Frederick Forestier-Walker, G.C.M.G., held command in Egypt after Sir Frederick Stephenson. These two Generals possessed a most charming *suavitur in modo*, which kept all trouble out of sight—hence their reserve of *fortiter in re* was seldom, if ever, called upon. Walker's death (as recent as July, 1910), as Governor of Gibraltar, was a grief to all who knew him—his charm was so great. To Sir Frederick Stephenson, may we wish many years as Constable of the Tower.

CHAPTER XLV.

"What a nation most hates, is another nation."

THE absence of well-defined boundaries has to answer for many a fight on our Indian Frontier, for the Waziris, the Swatis, Boners, Mohmands, Bajuris, Afridis, Zakha Kheels are all neighbours, and *ipso facto* trespassers. It is when trespassing descends to spoliation and outrage that the tug of war comes in, and a punitive expedition has to set matters right again.

In 1895, the Chitral Expedition under Sir Robert Low was an example. The Mihtar of Chitral had been treacherously murdered, and the assassins marched for Chitral Fort, where Lieutenant Gordon discreetly avoided all collision until Surgeon-Major Robertson, the British Agent, arrived from Gilgit with reinforcements, bnt not in sufficient force to stave off Sher Afzul, who besieged the place at once, sending off one Umra Khan to oppose Sir Robert Low, advancing from the south.

In the sorties from Chitral many lives were lost: Captains Ross and Baird, and later on Peebles and Battye, and forty of their men. The defence under Robertson, Townshend, Gurdon, Fowler and Edwardes, fully exemplified the old saying, "The strength of a fortress is the strength of the man behind it;" and thus it lasted for forty-six days, when a force under Colonel Kelly arrived and gave relief.

In 1897, the tribes previously mentioned all broke out. This meant a larger business, and General Sir W. Lockhart, a most distinguished Frontier officer, was placed in command (he died as Commander-in-Chief of India). General Penn-Symons (first to fall in the Boer War, at Talana Hill), General Yeatman Biggs (who died at his post, working hard to the last), and General Hammond, V.C., D.S.O., C.B., all held important commands.

Supreme political and military power was placed with General Lockhart. To assist him in his political work, Warburton and Udny, both skilled diplomatists, were included in his Staff. The former was the son of an Artillery officer, taken prisoner in Cabul, 1842, who escaped through the good offices of an Afghan princess. The latter became his mother, and it was to her he owed the power and influence he wielded over the tribes of the north-west frontier. He died in 1900.

The force assembled consisted of the Dorsets, Gordons, Derbys, 2nd Ghoorkas, 3rd and 13th Sikhs, besides some mountain batteries of the Royal Artillery. To drive the enemy from the Dargai Heights (a vantage point to them in many ways) became the first objective. The first attack was on October 18th, 1897, when the enemy, according to their tactics, yielded to their foe—only to resume at night the position they had lost by day. A rest was given to the troops on the 19th, for the General had resolved that these heights must be captured and retained. The fight was resumed on the 20th, when the Dargai incident came as a happy relief on the following morn, in response to their Colonel's (Mathias) call: "Highlanders! the General says the position *must* be taken at all costs—the Gordons *will* take it." Swiftly the Gordons, with the 3rd Sikhs in close support, crossed the fire-swept zone, and were on the ridge that had caused the trouble. The units (*i.e.*, companies) employed before had been powerless; bravery alone cannot make up for numbers. It was the cohesion of the two regiments, the Gordons and the Sikhs, which gave that adequate supply of men so necessary to stay disaster.

Another fight (giving the usual tactics of the tribesmen, which were to allow their enemy to proceed to its destination, preferring to harass it in camp, and to attack it strongly when it should commence its return march) occurred on November 16th. A force had been sent ahead to carry out some survey operations. It consisted of the Gordon Highlanders, the Dorsets, 15th Sikhs, 2nd Ghoorkas,

36th Sikhs, and two mountain batteries, Engineers, etc., under General Dempster. The going, the survey, and the rest in camp, were satisfactorily accomplished ; but when the retirement commenced, the enemy appeared here, there and everywhere. The conduct of this retreat was admirable. The brave Sikhs and Ghoorkas—especially the 36th Sikhs, under Colonel Haughton—deserve to be on England's shield as her "supporters." The Dorsets and the Gordons shone up in equal splendour. But as the force became split up into smaller fighting units, then it was the Subalterns appeared in their *role* of heaven-born commanders. Now the sadness comes : Captain Smith of the Derbys, and Lieutenant Crooke of the 12th (attached to Dorsets) killed, and similar heavy casualties throughout the rest. Young Wylie of the Ghoorkas, although warned by his Colonel to keep more under cover, next time he peeped to see, was instantly shot. Lockhart, that same night, paraded all to thank them ; for it was a telling day, requiring the utmost pluck from all. The brave Haughton was spared this fight, to be killed, however, on the 29th.

The war soon closed, with a death roll of 287, and 853 wounded. In the former was included General Sir Henry Havelock-Allen, V.C., the Adjutant to Franks of the 10th, the Staff Officer to his father in the many fights from Cawnpore to Lucknow, and who nursed him in his dying hours. Like an old war horse, he had come out from England to look once again on an Indian battle scene. It may be

> "He mourned his former vigour, lost so far
> To make him now spectator of a war,"

for, alas! he out-paced his escort, and, riding down a ravine alone, was shot on December 30th, 1897.

It was when all was over, and Lockhart had returned to Peshawur, that his carriage was dragged to the station by Afridis. Lockhart in his early days, when serving on the frontier, quite won the heart of the Afridis, and tells a story :—

"On one occasion, when he was rather depressed in spirits, his Afridi servant said to him, 'Sahib, why are you so depressed?'

Lockhart replied that it was because he was 'hard-up.' The Afridi enquired, 'Have you no friends who could assist you?' Lockhart replied that he had an old aunt at home who might leave him some money when she died. The Afridi then said, 'Sahib, you take me home with you. I'll do the trick' (passing his forefinger across his throat in a very significant manner—an Afridi's idea of a really kind act to a friend!) This man was the first to meet him at Bombay on his return from leave, to take service again under his old master."

This tale from Lockhart tempts one to allude once more to the sad murder of Lord Mayo. Shere Ali the murderer, an Afridi, had his home in the Tirah mountains, and had been selected for the service of mounted orderly to the Commissioner of Peshawur. Like all these hill tribes, Shere Ali (representing one clan) was at feud with some other; he one day found his arch-enemy in Peshawur, and roughly they quarrelled, and Shere Ali slew him. Across the border, or beyond Peshawur, this might have been considered "the usual" (to wit, the willingness of Lockhart's orderly to dispose of his aunt), but in Peshawur it meant murder, and Shere Ali received penal servitude for life. The pain of this imprisonment seemed beyond his control, hence this horrible murder of good Earl Mayo.

Referring to the Jowaki Expedition, it may be interesting to give an instance of an Afridi's cunning in stealing rifles. The Jowaki country being very rugged and steep, with deep *khuds* or cliffs with ravines, it was quite impossible for sentries on out-post duty to be visited at night, and the cold was very intense. It was, therefore, decided that in some inaccessible places the group system should be adopted, and to provide a small bell tent as shelter for the men not on sentry at the time. One cold wintry night, when the sentries called out "Sentry go!" not a single rifle could be found in the tent! An Afridi had crawled up by some unknown path, passed into the tent whilst the Corporal and four men were asleep, and had handed out the five rifles to an accomplice outside without being detected! (4th Rifle Brigade.)

In an Afridi village where our troops had rested, they found amongst other letters a Propaganda (which for veracity reads like the many political posters we have seen of the whole of China being dragged about in chains, or of 40,000 starving bricklayers all munching at one loaf):—

"You Mahommedans must take care lest you be deceived by the British, who are at present in distressed circumstances. For instance, Aden, a seaport which was in possession of the British, has been taken from them by the Sultan. The Suez Canal, through which the British forces could easily reach India in twenty days, has also been taken possession of by the Sultan, and has now been granted on lease to Russia. The British forces now require six months to reach India. The friendly alliance between the British and the Germans has also been disturbed, on account of some disagreement about trade, which must result in the two nations rising in arms against each other.

"The Sultan, the Germans, the Russians, and the French, are all in arms against the British at all seaports, and fighting is going on in Egypt, too, against them. In short, the British are disheartened nowadays. The Viceroy, and Generals who are to advance against you, have received distinct orders from London that the operations in the Kkyber and Tirah must be brought to an end in two weeks' time, as the troops are required in Egypt and at other seaports.

"In the case of the Mohmands and people of Gandab, who had killed ten thousand British troops, and had inflicted a heavy loss of rifles and property on them, the British, in their great dismay, concluded a settlement with them for twenty-four rifles only, whereas thousands of rifles and lakhs of rupees should have been demanded. This peace with the Mohmands is by way of deceit, and when the British get rid of their other difficulties, they will turn back and demand from the Mohmands the remaining rifles and compensation for their losses. They will say that, as the Mohmands have become British subjects by surrendering twenty-four rifles, they must make good the remaining loss too. The British are always giving out their troops will enter Khyber and Tirah on such and such dates, but they do not march on those dates, and remain where they are.

"This is deceitful on the part of the English, who wish to mislead Musselmans by a payment of five rupees, and seek for an opportunity to make an attack by surprise. I have thus informed you of the deeds and perplexities of the English."

It is now time to set seal to book, and to close our Roll of the Brave and Grand. Ere this is done, however, record must be made of a short campaign into regions hitherto untrodden by a soldier, where the troops had to march and fight (sixteen engagements) at altitudes varying from 10,000 to 18,500 feet, to face blizzards and snowstorms, and a temperature ranging down to 50 degrees of frost, before the goal was reached, and the veil removed from the hidden city of Lhassa. There the treaty of peace was signed by Colonel Younghusband and the Tibetan Regent in the Great Potala of the Lamas, in June, 1904—at a cost, however, of over 200 casualties, including twenty-four British officers. One life, indeed, was of no common value, for in Major Bretherton, who was swept away and drowned by the capsizing of a boat, the Expedition lost its chief Supply and Transport Officer, whose untiring energy had been its mainstay.

The Castle Yard of Windsor gives home and residence to Colonel M. Battye. His four brothers Quentin, William, Fred and Richmond, and little Richmond his nephew, have all been killed in our recent wars. Of the two latter, Richmond Battye was killed in the last Black Mountain fight, and his son Richmond fell fighting in the campaign just noted in Chitral.

Then the family of Peebles (with father and uncles all soldiers) —four brothers—followed suit. Allan Peebles, of the Devons, was killed at Chitral. He had invented a Maxim carriage, and with the Maxim guns was there on special service. The three remaining brothers were all wounded in South Africa, and two wear the D.S.O.:

Evelyn Peebles, D.S.O., Norfolks; Herbert Peebles, Army Service Corps; Arthur Peebles, D.S.O., 12th Suffolks (distinguished conduct at Bothaville, see Despatches).

Another hero of the 12th—for that was his regiment until the language and his service approved him for the I.S.C.—was Charles Grant. He heard that in far-off Manipur a horrid tragedy had occurred—the murder of Mr. Quentin, the High Commissioner, and his troops in escort; likewise of Mr. Grimwood, the Resident, and his Staff (in fact, the horrors of Cabul again). The gift of real initiative was in Grant, for in his letter home on April 16th, 1891: "On March 27th, thirty-five men of the 43rd Ghoorkas came into Tummu reporting there had been a great fight and massacre on the 25th. I wired all over Burmah and asked for leave to go and help Mrs. Grimwood and others (if possible) to escape, and got permission at 11 p.m. that night (the 27th). At 5 a.m. on the 28th, I started with fifty of my men and thirty Ghoorkas." Grant marched and fought each day until he reached Manipur—just like Lieutenant Edwardes in the old days to save his comrades at Lahore. The Queen at once conferred upon him the Victoria Cross "for the conspicuous bravery and devotion to his country," and in the same *Gazette* appeared: "Lieutenant Charles Grant to be Captain, dated May 10th. Captain Charles Grant to be Major (brevet), dated May 11th."

Then take Earl Roberts and Sir Charles Gough—each with a son, and all V.C.'s. Captain Roberts was killed whilst rescuing the guns at Colenso; all hoped that he knew, before he died, of the honour he had won.

Then take these four Marlborough boys—Coape-Smith, Thruston, Scott Moncrieff and Ernlé Money. Coape-Smith, as Commandant of the Camel Corps in the last Frontier dispute of 1908 with the Zakha Kheels, a campaign well planned and fought by Sir James Willocks (one who had reaped successes in other fields, to wit his relief of Kumasi in 1900), was the last to fall, shot by a Shinwari raider in the Khyber Pass. He was best known to his friends as "Bhiah"—a Marlborough boy, a Sandhurst

Cadet, brilliant in the gymnasium and at athletics, and as a horseman. Next when a Subaltern in the Hazara and Isaizi Campaign of 1891-92 (medal and clasp). In 1893 he accompanied Sir Harry Johnston to Central Africa, where he brought the Matipiri War to a satisfactory conclusion, which gave him mention in Despatches, medal and clasps, and a Brevet-Majority the day he became a Captain. He died in May, 1908, on the birthday of his mother—a woman of whom all could say:

> "No deed ungentle, or word unkind,
> The mildest manners, with bravest mind."

Arthur Thruston (born 1865) went from Marlborough to Sandhurst, from whence he was gazetted to the 43rd Light Infantry in 1882. His record is in safe keeping in the Crypt of St. Paul's:

In Memory of
ARTHUR BLYFORD THRUSTON,
Brevet-Major, Oxfordshire Light Infantry,
who served with distinction in the Dongola and Unyoro Expeditions,
and who, as a last service to his Country,
when the lives and supremacy of the English in Uganda
were in peril, went single-handed,
to endeavour by his personal intervention to recall to their duty
the mutineers of the Uganda Rifles,
by whom he was murdered at Fort Lubwas, East Africa,
October 19th, 1897.
Aetat 32.

Erected by his brother officers of the Oxfordshire Light Infantry,
as a mark of esteem and affection.

The crowning act of this brave man's life was in the manner of his death. His self-sacrifice is well shown in a letter from a brother officer:—"He was unwilling that his murder should be laid at the door of the private soldiers, who he knew were children in the hands of the native officers. The native officers had condemned him—let the chief of them be alone responsible." "If I am to be shot, shoot me yourself, Effendi—and do not tell the men to do it." Upon this Effendi raised his rifle to shoot; Major Thruston then directed it to his own forehead; Effendi shot—and the sacrifice of Major Thruston was complete.

The sad end to a short and fearless life, by the murder of Mr. Scott Moncrieff, is here given by his father:—

"TO THE EDITOR OF 'THE TIMES.'

"SIR,—As authentic news of the murder of my son has now reached me from the Sirdar, Sir Reginald Wingate, I beg to send through your columns the following brief account, which may be read by some of those who have felt interest and sympathy in hearing of it:

"'It must be remembered,' says the Sirdar, 'that the present generation of the Sudanese has been born and brought up in Mahdi-ism—the cult has been stifled, but we must expect occasional outbreaks from time to time, and it behoves us to crush them instantly before they attain any size.'

"It is clearly the duty of the civil administrators of the land (of whom my son was one) to endeavour by all means to effect a peaceful settlement of difficulties, and only to resort to arms when other efforts have failed; and it was in the endeavour to bring about such a settlement that my son lost his life.

"A certain landowner, Sheikh Abdel Kader Mohammed Imam, had proclaimed himself a prophet and collected a band of followers. He had refused to treat with the Mamur, or native administrator of the district; but on April 29th he sent a message to say that if my son and the Mamur would come and interview him unarmed and unattended, he would lay his grievances before them. He included in his invitation three of his brothers and the headman of his own tribe, who declined to go, and endeavoured to dissuade my son and the Mamur also. They, however, decided to trust the man, possibly attributing the refusal of the others to the enmity which in Moslem lands often exists between the nearest relatives.

"What followed is the evidence of men who were subsequently taken prisoners. My son and the Mamur left their small escort of police a mile away, and, dismounting from their camels, went into the enclosure where Abdel Kader and his followers offered them some hospitality. My son asked what their grievances were, to which the Sheikh replied that he had no grievances, but that what he was doing was for Allah. Then either he or some one behind him struck the Mamur with a sword or spear, while others attacked my son. They say he met his death fearlessly, smiling, and folding his arms while his lips moved as if in prayer.

"From subsequent events it appears that his action, followed by Major Dickenson's encounter, in which, I grieve to say, Captain Logan lost his life, was instrumental in checking what might have been a widespread revolt. Whether this be the case or not, it is some consolation to know that my only son did not flinch from what he judged to be his duty, and then that he met his death like a gallant gentleman and a Christian.

"His age was twenty-four; his school, Marlborough; and college, Corpus Christi, Oxford. He had entered the Sudan Civil Service in September, 1906.

"Yours faithfully,

"COLIN SCOTT MONCRIEFF,

"Late of the Egyptian Service."

Ernlé Money, Colonel of the 9th Bengal Lancers, was cruelly murdered in the mess tent, January 26th, 1895. The murderer had been reprimanded and some reward withheld. This prompted him to do the vile deed he did. Money was for five years a much-loved officer of the 12th (1867 to 1872).

This tempts me to quote an incident which happened to Colonel Nicholson, a kinsman of Nicholson the Great. Colonel Nicholson (Beloochi Regiment) had occasion to punish his orderly. This occurred a few days before a shikar trip into the jungle had been fixed. The night before they started, the senior native officer approached his Chief to report that this said orderly had vowed to take the Colonel's life. The Colonel thanked the man and said, "No matter." The morrow came—the two started. On arrival, the orderly pitched his master's tent and got the guns and ammunition in readiness for the first day's shoot. The Colonel, ere he retired to rest, sent for his orderly and charged him with what he had been told. The orderly said, "Yes! I admit it." The Colonel then replied, "I sleep there" (pointing to his tent); "you sleep there" (pointing to the ground beneath the fly), "and murder me if you dare!" The two men—the Colonel and his orderly—remained out shooting longer than a week, and the latter of his intention was for ever bitterly ashamed.

Then the two brothers Alexander, two hero explorers: Captain Claud Alexander (died 1904) and Boyd (shot 1910). The heroine who was to have been the latter's wife is now engaged in travelling to place her tribute on his tomb.

It is when noble families, like the Battyes, Broadfoots, Connollys, Peebles, etc., besides all other heroes, have been claimed by the God of War, that I see sympathy and simile in the last picture of Holman Hunt, viz., "The Sacred Fire in the Church of the Holy Sepulchre at Jerusalem." There beautifully delineated, we see the rush made by so many on this Sainte Samedi to obtain the holy spark, to carry back and so rekindle the Light of Life in their respective Churches. So let it be with our fallen heroes—their noble deeds must be the spark to give the light and fan the flame of Patriotism and Devotion.

Wars will never cease! For now an ill-advised Ultimatum (November 11th, 1899) from the Transvaal to England—the paramount Power in South Africa—was an insult no country could withstand; and England once again entered on a war in a light-hearted sort of way. "Home by Christmas!" said the many.

It has been said that many of the rifles that were picked up on Majuba Hill were found at the last moment, when the Boers were closing in, sighted to 800 yards. This shooting without method has always been a difficulty with men, which good Generals for a hundred years and more—Abercrombie, "No flint" Grey, etc., have struggled to prevent—perhaps

> "An effort only, and a noble aim—
> Still to be fought for, never to be won."

Are not our rifles full of bull's-eyes, if the men behind would but hold them straight? for to see and not be seen, to shoot and not be shot, are very early dictums, which should never be forgotten.

The previous campaign of Colley, and his defeats at Laing's Nek, Ingogo, and at last Majuba, all placed in evidence the good shooting and mobility of the Boers. "The Life of Colley," by General Sir William Butler, gave ample warning of our danger ahead, but "forewarned" was *not* "forearmed."

Time and space do not permit me to refer much to this campaign, beyond recording appreciation and indebtedness of all to Lord Wolseley, Commander-in-Chief, for the way in which he had nursed his country, thereby enabling her to send out 60,000 men to South Africa within six weeks without mishap of any kind. Six thousand also went from India. But this small army had to conquer two huge States, the Transvaal and the Cape. Our War Minister was not like Earl Chatham, who always asked his General: "How many men he should require for a certain expedition?" "Ten thousand," was the answer. "You shall have twelve thousand," said the Minister; "then if you do not succeed it is your fault."

However, we were not alone, for the young folk who were born in the days of Captain Cook and later, had grown into splendid children, and gave their manly aid; and in days still more recent, when old England thought her Navy running short, they all (our Colonies) replied with the one word, "Dread-naught!"

The next point of merit was the excellent service given to the country by our grand Militiamen (now our First Reserve). Mention must be made of the Royal South Down, a regiment commanded by Colonel Robert Wallace, one who had the knack, like the regimental band, to help his men along—a real propeller in his way—so when the chance of service came, one and all volunteered with him to go, and they went. For twelve months this regiment held their section on the line of communications, against the many rushes of the Boers, who were eager to get through. Wallace got his C.B., and Sergeant-Major Robert Gardiner his D.C.M.

"IN MEMORIAM" TO THE TWELFTH.

One letter from kind-hearted Arthur Watson (brother to Sir Charles Watson, the friend of Gordon—see page 456), Colonel commanding the 12th, is here given only to show his deep-rooted care for all, ere leaving England for the War:—

"Dover,

"6th Nov./99.

"Dear Gardiner,

"Very many thanks for your letter and good wishes to me and the Battalion. All are delighted at being sent out, albeit at the eleventh hour. We sail from Southampton on the 11th in the "Scot." I am arranging to get Militia washing for the married people left behind, and am not starting a special fund. My wife is going to look after their interests when we go. I will tell her of your kind offer.

"Best regards from all.

"Believe me, very sincerely yours,

"A. Watson."

One other letter, only two months later (from my friend Colonel Cubitt), gives the glory of Watson's fall and his 150 men, for he died with his sword in his hand, the word of command in his mouth, and with victory in his imagination:—

"We had a very severe handling at Colesberg, but the men behaved splendidly. It was 3 a.m., and very dark. The men made three desperate charges, losing heavily in each. The Colonel was shot dead in the last one. Brett made a gallant stand, but at last had to surrender. Williams and White were killed, and Brown mortally wounded, and Graham badly wounded, in the first charge. Carey was killed in the second charge, and Brett and Butler wounded. Colonel Watson was killed, and Thomson and Allen wounded, in the third charge. Only one officer, young Wood Martin, was not wounded." *

---

* Lieutenant-Colonel Arthur Watson, Captain and Adjutant F. A. P. Wilkins, Captain A. W. Brown, Lieutenant C. A. White, and Lieutenant S. J. Carey, killed at Colesberg, February 6th, 1900.

"The longer I live (says Fowell Buxton), the more certain I am that the great difference between the feeble and the powerful, the great and the insignificant, is energy, invincible determination, a purpose once fixed, and then death or victory!" If this is written of the individual, how much more so of the mass? The Suffolks had *their* purpose fixed, and they fought for it to the death. The victory came soon afterwards, for the enemy had to cease their operations, which were all-important at the time, viz., the invasion of Cape Colony.

On the Cornhill at Ipswich stands this soldier* dazed with grief and despair, "Since the flowers of the forest are weeded away;" but, at the same time, his attitude denotes that "Here I testify for County Suffolk, that her Centurions, her soldiers, her sailors, have not 'dropt into the grave unpitied and unknown.' God bless the Twelfth!"

It was during the early days of 1901, with the War Cloud hanging round, that all things became further darkened by the death of our beloved Queen Victoria, who in her sixty-two years of rulership had tempered man to meet these storms. "If for anything she loved her greatness, it was because she might exercise her goodness." So with intense sorrow, the nation learnt that on

"January 22nd, 1901,

"6.45 p.m.

"Her Majesty the Queen breathed her last at 6.30 p.m., surrounded by her children and her grandchildren.

(Sd) "James Reid, M.D.,
"Douglas Powell, M.D.,
"Thomas Barlow, M.D.

"Long live the King!"

The Queen is dead! Long live the King! The Commons soon assured His Majesty of our loyal attachment to His person, and further assured Him of our earnest conviction that His Reign will be distinguished under the blessing of Providence by an anxious

* The sculptor is Mr. Toft.

The Suffolk Memorial, Ipswich.

desire to maintain the laws of the kingdom, and to promote the happiness and liberty of His subjects."

This humble Address from the Commons needs no comment. Their earnest conviction was no empty phrase. His Majesty has promoted the happiness and liberty of his subjects, and has been named from one end of his Dominion to the other—and by all Monarchs and Republics—as "the Peacemaker." Long may he reign!* It was amongst his first acts to secure peace between South Africa and his own people, which happily soon came (1903). One of his other great humanities was also shown in his fondness to make provision for the sick and poor.

It is to his worthy helpmates, the Doctors, our concluding words are due, although they seek not their reputation at the cannon's mouth—for in their *métier* "they love mercy and delight to save." Thus it is hard to sing their praises, for so much of merit is concealed from the light of day, and has to pass unnoticed. But in this short chronicle of a century or more, no profession can boast of a nobler pedigree than our worthy friends the doctors; for we can place as their grand ancestor one Edward Jenner (born 1749, died 1823),

"One dear to the human race."

A slight proof of this epithet is that between the birth of Napoleon (1769) and the War after Amiens (1804), the death-roll from small-pox alone went down from 2443 to 580 a year; and in this interval poor Jenner had been damned from every pulpit as one of our three great plagues, viz., Evil, Bonaparte, Vaccination. Napoleon when asked "Who, in his opinion, was the greatest woman of his day?" replied, "The mother who has borne the largest family, especially in sons;" so perhaps he only saw in Jenner a preserver of his sex. However, he always thought of him as a grand *savant.* When the latter tried hard to mitigate the discomforts, if not the sufferings, of

---

* King Edward VII. died on May 6th, 1910. It was on March 17th, 1904, that *our* Duke of Cambridge—the soldiers' friend, and to memory endeared—died in his 86th year.

the prisoners of Verdun, he at first met with ill-success; but when two great friends of Jenner's—a Dr. and Mrs. Williams—fell into the mesh, he wrote direct to the Emperor asking for their release. This was about to be refused, when Josephine uttered the name of Jenner. "Ah!" said Napoleon, "we can refuse nothing to that name."

Abroad, Jenner was always a hero. In Russia, the children first vaccinated were called "little Vaccinoffs;" but in England, in spite of Lady Mary Montagu and a Royalty or two, who with their children gladly submitted to vaccine, Jenner, like the Prophet, was of no honour in his land.

England is a poor country to live in, but a grand country to die in; for at his death England rang with his praise. Take poor Chatterton, who found it so hard to exist even for a short seventeen years of life, that he had to pretend his brains belonged to another man named Rowley of centuries ago, for who would believe a boy could write a poem? Hence Chatterton became Rowley until his death by starvation, when he became Chatterton again, and his name is now blazened in fame. Then in Dr. Parkes, of Hygiene note, who from the Military Hospital of Fort Pitt, Chatham, produced a mine of wealth by his "Sanitary Reforms."

Perhaps the Military Medical will allow it also to pass as a compliment—the simple fact that on August 7th, 1815, when Napoleon left the shores of England on board the "Northumberland" (74 guns), he chose for his medical adviser, not a Frenchman, but a Naval doctor—a good all round Irishman—Mr. Barry O'Meara. O'Meara had been at the Battle of the Nile, little thinking that one day he would "fight this battle o'er again" with the great Napoleon in far-off St. Helena.

The march of Sir David Baird across the Desert has already been alluded to, as giving a niche in the Temple of Fame to Sir James McGregor; and what scores of other good medical men deserve the same notice in the long War of the early years of the last Century! But it is of a later date that one is pleased to record the name and doings of others just as celebrated: Robertson of Chitral commenced his career as Civil Surgeon, and like many another good doctor who tired of paper forms and returns, with wordy wars over trifles, took exception, and sought other ways and means. We have also read of Dr. Andrew Smith, Kennedy, Dempster, and McRae, who with energy, courage, and skill, carried on their operations, when crowds of fighting men broke throught their dhoolies and jumped their amputating tables. McRae was told, after the second Punjaub Campaign, "that no man in the whole army of 25,000 men had done the State better or more useful service." There the despatch abruptly ended—a C.B. or so would have seemed the proper finish. Then the story of Dr. John Murray, who when Assistant Surgeon to a troop of Horse Artillery, on seeing no field companion or amputating instruments about, made tracks to the nearest station, where boxes lay addressed to the Superintending Surgeon, many miles away; and there by means of a hatchet, he soon produced saws, knives, plasters, lint, and tourniquets, sufficient to save many lives—but at the cost of many pounds to himself when the Auditor came round. Scores of others one could mention—take one from the Mutiny: Sir Anthony Home, V.C.,* for in the first relief of Lucknow, his task was to pick up the wounded as Havelock fought through those loop-holed streets into the Residency. As each man dropped he was tended and guarded by Surgeon Home and his Staff, they improvised a splendid defence, until their turn came for relief. Havelock saved the women and children—Home saved the wounded. But for a finish, as an example of duty and devotion to their calling, Director-General Jameson, C.B., and his aide, Surgeon-General Skey Muir, C.B.: "For in no campaign have the sick and wounded been so well looked after as they have been in the Boer War of 1900" was the wording of the Royal Commission of 1901.

* His daughter, Miss Ethel Home, is now the talented and popular head mistress of Kensington High School.

POSTSCRIPT.

" Catch, then, O catch the transient hour ;
Improve each moment as it flies."

Many of the "Centurions" commenced their earthly pilgrimage at the early age of thirteen—with little or no money at their back—for instance Sir John Malcolm, who joined at thirteen, and was one of seventeen children (fifteen of whom grew up), so probably his allowance did not make of him a millionaire. His ingrained common-sense gave him much better help. It is said of him, when asked by a Director of the East India Company at his examination, "What would you do if you met Hyder Ali?" he replied "Out with my sword, sir, and cut off his head!"—like as Reynell Taylor, who, as a young child, met a noisy crowd of roughs, hissing and hooting our great Duke of Wellington as he was riding through the Park. On his return home, he related this adventure. "And what did you do?" said his father. "What did I do? Why, took off my hat, of course."

Many other such lives offer reasons for returning to that age as the minimum for obtaining a Commission, and let the maximum be forty-five years. This seems a mighty margin, but to quote, only three Peerages were given in the Peninsula, viz., to Beresford, Hill, and Graham. The latter became Lord Lynedoch. He joined the Army at forty-three years of age, born in 1748, was educated at home, completing same by travel. His two elder brothers died, which brought the family estates into his keeping; so he married and settled down as a country gentleman, beloved by neighbours and tenantry—known as a daring rider and sportsman, and a good classical scholar. After a happy married life of twenty years, his wife died in 1792. This depressed him greatly; he sought hard to dissipate his grief in travel, but failed. At last he obtained permission to enter the Army as a Volunteer under Lord Hood at the siege of Toulon, and then passed over to the military as aide-de-camp to Lord Mulgrave. Napoleon in his first career of arms at Toulon had observed Graham, and called him then "a daring old man." On his return to Scotland he raised the best part of a battalion (the 90th Regiment) and commanded it. He served in the disastrous Campaign of 1796; then became British Commissioner to the Austrian Army on the Rhine; then home again, only to be sent to the West Indies. In 1808 he went to Spain with Sir John Moore. His services during the retreat were of great importance. At Barossa he defeated the French Marshal Victor. Ciudad Rodrigo, Vittoria, St. Sebastian and Bidassoa were more of his battles. He received the thanks of Parliament, and was raised to the Peerage with a pension of £2000 a year. He died in 1843, aged ninety-five years. Sheridan, in his eloquent and affecting speech in his Vote of Thanks to the victor of Barossa, said "The hero of a race renowned of old," and "Never was there seated a loftier spirit in a braver heart."

Then Lord Hill, one of seven, had four brothers at Waterloo, whilst his cousins in the same number all became Church Militants—and in fighting qualities they all excelled; for surely Rowland and the General knew no limits in fighting for Saviour and for King.

Then if we take the life of John Nicholson, John Jacob, and throw in Henry Lawrence, Herbert Edwardes, Reynell Taylor and Hodson—it was by this small group, of whom perhaps the greatest was John Nicholson, that the peril of the great Mutiny was averted, and thanks to them that India had not either to be abandoned or re-conquered from its very seaboard.

Nicholson, as a mere youth, found himself a prisoner in the hands of the Afghans after the disastrous days and murders at Cabul. He was liberated, like the rest, by the advance of Nott and Pollock. He then devoted himself to professional reading. In 1845-46, the first years of the Punjaub, he formed a great alliance with Lawrence and Edwardes, and at the early age of twenty-five was entrusted with full political control between the Jhelum and the Indus. In the second Sikh War, 1848, he left his sick bed at Peshawur, and with sixty Pathan Horse and two Companies of Infantry, he made a forced march to seize the important fort of Attock. He also at the same time overawed a mutinous regiment. He was soon named "the mighty Nichalsain," and became a worship to the Sikhs. In

1850 he returned to Europe, and entered heart and soul into the Hungarian revolt, giving his full sympathy to Kossuth. He was in full political power at Peshawur when the Mutiny broke out. He soon reached Delhi. "Nicholson has come," wrote Hodson; "he is a host in himself." We have read of his Delhi work and sad death. "It is difficult to describe the man," wrote Herbert Edwardes; "he must be seen." Lord Dalhousie summed him up as a tower of strength—a born ruler of men. He shone equally as Judge, Commissioner or General.

Then Hodson, we know, joined the Army late in life, from the University. These instance are given, for perhaps the travel and fuller education of these men may have quickened them into abler action—just as Colonel Mountain often coveted a more mixed society, a more varied life. Remember! Wellington was Chief Secretary to Ireland when he won Roliça and Vimiera, and only thirty-nine years old. Give our officers a life of more variety—it may develop common sense, which is so becoming in every-day wear.

And what greater exponent have we of the immense value of travel, education, cultured society, and variety of life, than in Cecil Rhodes?—the British, Colonial, Imperial Statesman, and we may justly add Soldier, for had he not, when it became inevitable, fought the Matabeles, and carried out with a success, perhaps unique of its kind, a war entirely organized by his own directing genius. When invited by the Matabele chiefs to attend a Council, he rode unarmed, attended with two others, into the hornets' nest at the foot of the Matoppo Hills, and there encamped. He came away with Peace. He was fond of mentioning this scene as one which made life worth living. Surely, if Rhodes had been the Generalissimo of the British forces in the recent South African War, the "regrettable incidents" would not have been so many.

Then do not the Warrant and Non-Commissioned Officers of the Army and Navy form a nursery in themselves, promising a growth of men to meet requirements in higher things?—for as a class they are quite unsurpassed. Their duties are voluminous, impossible to define ; for like the second hand to a watch, they are the first beat, which makes the regiment or warship "go." To mention one, Sir Francis Bolton (born 1831) : he was promoted from the ranks into a Commission in the Gold Coast Artillery. He then was transferred into the 12th (the Suffolks), where he developed in a marvellous way the lime-light and his code of flashing signals, which the Army and the Navy soon adopted as their guide ; he next was entrusted with the management of the Waterworks of London. He was knighted in 1884. In the early days of Exhibitions, say 1880, he had a tower in South Kensington, from which he illumined with his lime-light the whole of London. I was there often with him, and at midnight or when the hour of closing came, we then repaired to the house of Admiral Sir Edward Inglefield in Prince's Gate, there to meet at supper the interesting people of the day—often Sir Henry Irving, Squire Bancroft, and other actors, straight from the Play. He died fairly young in 1887, from an affection of the throat, contracted, as he told me, from the lengthy hours passed in close contact with his lamp.

Then Lieutenant-Colonel Alexander Arthur of the 11th (the Devons), who, two days before his death (November, 1908), pictured to me most vividly some scenes of his early life. He was descended from a family of soldiers and sailors, who have had distinguished men among them. His grandfather was locked up in the same dungeon with David Baird at Seringapatam, where both were most cruelly treated. Arthur was born with prospects considerably above the average, and was intended for the Army from his cradle, but fates were against his entering in the way intended. He was one of a large family, and just as he was emerging from boyhood, a financial misfortune overtook his father. The contemplated fortune of the family was altered, among others his own. He was given to understand that he must give up the idea of the Army. As the Army was his great ambition from his early boyhood, this intimation was heart-breaking. Fortunately, up to that period (he was nearly seventeen years of age) he had received a fair education. He had heard that other gentlemen, overtaken by misfortune, had made their way to commissions through the ranks, and he determined to try. He was at school in Dublin at this time, but about to leave for home. He had received a little money to meet necessary requirements and to take him home; but instead of going home, he started for Liverpool, where he enlisted in the 11th Regiment, which had just arrived from Canada, and was then quartered at Devonport, where he joined. This was no wild escapade. It was as deliberate an act as any act of his every-day life. His regiment in those days was commanded by a distinguished Peninsular officer, who from the outset was most kind to him—who indeed never lost sight of him. They had a number of Peninsular men in the senior grades at that time, all perfect gentlemen and first-rate officers. He passed through the non-commissioned grades very rapidly, but never asked or received the slightest aid from a relative, until he was recommended for a commission. His commissioned service was chiefly in Staff appointments. At last a cruel accident rendered him powerless to walk, so he sought retirement from the Army, and soon became one of the leading men in London. He died aged eighty-seven, a Director of many Companies (Star Assurance), and Editor of a Military Gazette. His cheeriness and wit remained with him to the last.

Then our fine Corps of Commissionaires—who all possess their medals for good service rendered, but knowing that "Work is Salvation," they continue on in ways of utility unto the end. The 12th Regiment have one fine representative, Sergeant Thomas McKenzie, now on the Staff of the Royal Military College of Canada. I remember him a Sergeant as far back as 1864. He revisited his old Regiment on Minden Day, 1909.

On September 15th, 1871, seven officers met at No. 1, Albany Court Yard: Major Munnings, 24th Regiment; Major McCrea, 8th King's; Captain Ernest Lewis, 21st; Captain Nolloth, R.N.; Captain Gammell, Control Department; Captain A. Cockerell, 20th Hussars. They there and then rented a small tenement, 12, Vigo Street, Regent Street, for the purpose of carrying on a business as general dealers, with the object of supplying shareholders and their friends with articles of domestic consumption and general use, observing the rule that all such articles purchased must be paid for before being taken or sent out of the Store. (Capital to be £15,000, of 15,000 shares, £1 a share.) The business done during the first year (1871) was £128,000. The business done during the year ending February 1st, 1908, was £3,158,835. Surely not a bad Peace Manœuvre for the United Services!

ARMY AND NAVY CO-OPERATIVE SOCIETY, LTD.

A few Facts and Figures of Business Done, etc., Year Ending February 1st, 1908.

Business done	£3,158,835
Gross profit on trading and miscellaneous accounts	£487,571
Working expenses	£289,037
Nett profit on the year's business	£198,533
Assets	£1,991,527
Number of shareholders, life members and subscribers	=72,385
Number of employes in London—General staff ...	= 3,357
Manufacturing staff	= 1,943
	5,300
Daily average attendance of purchasers at Stores ...	10,225
Average daily cash takings ... ... (on an average £1 a head per visitor)	£10,162
Average number of letters received in London daily ...	6,601

Encourage all to read circumspectly, so that study may come as an easy after process—not to trust too much to memory, unless able at once to verify or rectify; to travel with eyes open, for language is the grand leverage to knowledge, which is so great a power. It was Von Moltke who could keep his silence in seven languages. Encourage all officers to do their clerical work (such as reports on positions, reconnaissance) and to keep diaries in German, French and Russian, or the languages which were included in examination test. In the Indian and Egyptian Army, all officers have not only to write, but to give their utterance in Persian, Arabic or the tongue of Hindostan. Further, insist on all officers knowing something of the history, manners, customs and religion of the country they soldier in.

Then take the charm and magical effect of music, to wit the following (favoured me by Colonel H. O'Donnell, of the 14th Regiment):—On May 23rd, 1793, at the attack on the entrenched camp at Famars, near Valenciennes (for the English were investing Dunkirk at the time), the success of which was due to one column —although three were in the fight, two had failed in their attack, and the third was *in extremis*—as the French advanced in grand array to their revolutionary and well-known air of "Ca Ira," the Colonel of the 14th (Sir Francis Doyle) dashed to the front, and seized with a happy inspiration, called on his drummers: "Come along, lads, we'll beat them to their own damned tune—strike up 'Ca Ira!'" The effect was irresistible; the lines re-formed, charged, and captured the Redoubts with seven guns and 200 prisoners. This famous tune has kept the 14th going, for it has been their quick-step ever since:—

"Ah! ça Ira, ça Ira, ça Ira;
Les aristocrates á la Lanterne.
Ah! ça Ira, ça Ira, ça Ira,
Malgré les mutins, tout réussira."

Again, in the following year at Tournay, the whistling of this tune stood the regiment well; for when cut off, and puzzled how to penetrate at dusk the enemy's lines, the whistling of "Ca Ira" gave instinct to the enemy that they must be friends, not foes, so the 14th successfully crept along.

Then, who can tell but what the "Marseillaise" has not a great deal to answer for on the side of good or wrong; for as a battalion of Volunteers, raised in Marseilles for service with Dumouriez against the Prussians (Valmy and Jemappes), had occasion to march through Paris to the seat of war, singing "Chant de guerre de l'armée du Rhin," they were seized upon by Marat and made to join the rabble on account of their soul-stirring song (August 10th, 1792, downfall of Monarchy), which soon became the battle cry for the miscreants of the day. "Music has charms," etc., but this was the wrong way about. Surely the battalion had better stuck to their colours and gone on to fight, but they preferred the rabble—hence the "Marseillaise."

Like an actor in the play, you are expected to know your part; or as a rower, you must pull your weight in the boat—and even then Victory is not assured: far from it. For was it not Garrick who, on being told that no more letters of "Junius" were to appear in the "Public Advertiser," had mentioned same to some noblemen about the Court. It was when "Junius" heard of this, he caused a letter to be sent to him at the theatre, just as he was going upon the stage to play one of his great parts. The letter was virulent and abusive, hinting to him that he might well be contented "Plausu sui gaudere theatri," and not interfere in politics. This letter so upset Garrick that our wonderful actor quite forgot his part.

Then carry a cheerful mien—the real splendour of which was found with Captain Inglefield (grandfather to Sir Edward) in H.M.S. "Centaur," 1782, on his voyage home from Jamaica after his many fights, for this was Gibraltar year. He knew his ship had been somewhat knocked about, but never doubted her ability to stand a gale of wind. This latter she soon encountered, when she rolled and rolled until she filled and sank. Of the company saved was Captain Inglefield, the master, one midshipman, and seven others, of whom David Woodward was one. It came to the sad lot of the latter to be in the same predicament again when chief mate of the "Enterprise," bound from Batavia to Manilla; and remembering what Captain Inglefield had done to cheer up his men, namely, to tell stories, and perhaps to sing a song, Woodward chose for his *repertoire*, night after night until his rescue came, Inglefield's narrative of the "Loss of Centaur," Holwell's account of the Black Hole of Calcutta, and the Scripture history of Joseph in the Pit. What a mystery is man, when the above gave solace to those perishing from famine in an open boat at sea! Surely it behoves us to

"Count our blessings every day;
Count them over and over again;
Count them over one by one,
And see what the Lord has done."

Finis.

APPENDIX.

With my best thanks to Messrs. Cox and Co., the following Roll of Colonels, *i.e.*, Generals and Lieutenant-Colonels of the 12th, is given :—

1754	...	General Henry Skelton.
1758	...	,, Robert Napier.
1767	...	,, Henry Clinton.
1779	...	,, William Picton.
1812	...	,, Sir Charles Hastings, Bart.
1824	...	,, The Honourable Robert Meade.
1852	...	Lt.-Gen. R. G. H. Clarges, C.B.
1857	...	,, C. A. F. Bentinck.
1864	...	,, H. Colvile.
1875	...	General J. Patton.
1888	...	,, J. M. Perceval, C.B.
1900	...	,, Hon. Sir P. R. B. Feilding, K.C.B.
1904	...	Major-Gen. (Hon. Lt.-Gen.) Hon. B. M. Ward.

The first five Generals on the roll had also been Lieutenant-Colonels in turn. Lord Cornwallis was Lieutenant-Colonel from 1761 to 1764.

From 1794 the Roll of Lieutenant-Colonels is as under :—

1794	...	H. H. Aston.
1799	...	Shaw.
1812	...	Harcourt.
1814	...	Stirke.

1817	...	Forsteen.
1826	...	Bayly.

Colours presented by Sir Richard Don, Gibraltar, June 28th, 1827.

1831	...	Gervas Turbervill.
1835	...	Joseph Jones (by purchase—Turbervill).
1843	...	John Patten (by purchase—Jones).
1847	...	John Patten.
		Stirling Freeman Glover.
1848/49	...	John Patten.
		William Bell (by purchase—Glover).

Colours presented by Mrs. Arbuthnot, Weedon, July 14th, 1849.

"July 14th, 1849.

"Went to Weedon to see Mrs. Arbuthnot present new Colours to the 12th Regiment of Foot. Colonel Arbuthnot made an excellent speech, followed by Gleig, whose address was also capital. Mrs. Arbuthnot's address I could not hear, but the whole thing went off remarkably well; the weather was beautiful. General Meade, the Colonel of the regiment, sent £100 to the regiment. The men had a capital dinner outside their barracks, which we all went to see. The sergeants had a ball. There was a *dejeuner* at 3 p.m., to which more than 150 people sat down, and everything was very pleasant. There was dancing afterwards, and I left at 8.15 p.m."

—From the diary of General Balders, C.B.

Surely a happy afternoon!—for the Chaplain-General, the Rev. G. Gleig, alone was a *tour de force*. He had fought with Wellington in the last of his big fights, and his book, "The Subaltern," was and is still the Subaltern's delight. He then took Holy Orders, and gave to military society, from its apex to its base, a tone which redounds to its credit in the four quarters of the globe.

Other worthies of this type one would like to mention, but time does not permit, although at present, at All Saints', Vevey, Switzerland, the comfort of our Church is so well administered by the Rev.

Douglas Harrison—a man to revere—late Lieutenant-Colonel of the 4th Bombay Rifles. And recently we have all lamented the loss of Dr. Maclagan, Archbishop of York—late Lieutenant Indian Army. But we must return to the Chaplain-General and to General the Hon. R. Meade, for surely these two were of merit enough to give a pleasant memory to this pretty ceremony—further to be enhanced, by the full consent of the Rev. W. St. Leger, and the wish of all, for the old Colours to be deposited in the Church of St. Mary's, Ipswich, accordingly on a later date, the Colours were returned, as they had been received, with religious celebration.

A short time after, a dignitary calling himself Archdeacon Ormerod, thought fit at his first Visitation to order the removal of the Colours, decreeing they were emblems of blood, and should have no habitation there.

The Regiment, to express their gratitude to the Rev. St. Leger, sent him a handsome gift. The last sentence in the Address refers to the noble protest the Incumbent made against the ignoble behaviour of this Archdeacon: "Presented by General the Honourable R. Meade, Lieutenant-Colonel Patten, the officers, non-commissioned officers and men of the 1st Battalion 12th (or East Suffolk) Regiment, to the Rev. William Nassau St. Leger, A.B., Incumbent of St. Mary-le-Tower, Ipswich, in grateful appreciation of his chivalrous and reverent reception of their old Colours into the chancel of the Church, and *his noble protest* against the violation of their consecrated repose." A letter from Colonel Patten accompanied the Presentation, fully thanking the Incumbent for his Christian act, and desiring that the episode of the Colours might be passed on to his successors, "Lest we forget" Captain Souter and Colours (sole survivor of 44th Regiment, Cabul), Ensigns Melville and Coghill, 24th Regiment (who perished swimming with their Colours, January 22nd, 1879). The soldier's *role* is set:

> "For the rights of fair England his broadsword he draws;
> Her King is his leader, her Church is his cause;
> His watchword is Honour! his pay is Renown!
> God strikes with the gallant that strikes for the Crown!"

1849/53	...	Randal Rumley (from 6th Foot).
		John Maxwell Perceval (by purchase—Bell.
1854/5	...	J. M. Perceval (Randal Lumley exchanges).
1855/60	...	J. M. Perceval.
		Thomas Brooke.
1860/1	...	John Maxwell Perceval. (Thomas Brooke retires upon half-pay).
1861/2	...	Henry Meade Hamilton. (J. M. Perceval retires upon half-pay).
1862/3	...	Henry Meade Hamilton.
		Edward George Hibbert.
1863/4	...	Henry Meade Hamilton. (E. G. Hibbert exchanges Grenadier Guards).
1864/8	...	Henry Meade Hamilton.
		Arthur Edward Valette Ponsonby.
1868/9	...	Henry Meade Hamilton. (A. E. V. Ponsonby deceased ; died of cholera, June 16th, 1868).
1869/70	...	Henry Meade Hamilton.
		John McNeill Walter, C.B.
1871/2	...	Henry Meade Hamilton.
		John McNeill Walter, C.B. (seconded).
1872/3	...	John McNeill Walter, C.B.
		John McKay. (Henry Meade Hamilton transferred to a Brigade depot).
1873/4	...	John McNeill Walter, C.B.
		John McKay.
		Edward Foster.
1875/8	...	John McKay.
		Edward Foster.
1879/80	...	George Fuller Walker.
		Frederick Bagnell.

1881	...	George Fuller Walker. (Frederick Bagnell retires).
		Colours presented by Duchess of Connaught, Southsea.
1882	...	Charles Jocelyn Cecil Sillery.
		William Cooke O'Shaughnessy.
		William Thomas Baker.
		(George Fuller Walker half-pay).
1883	...	William Cooke O'Shaughnessy.
		Hugh Pearce Pearson.
1884	...	William Cooke O'Shaughnessy.
		Hugh Pearce Pearson.
		Henry MacGregor Lowry.
		William Keough.
1885	...	William Cooke O'Shaughnessy.
		Hugh Pearce Pearson.
		William Keough.
		(H. McG. Lowry retires).
1886	...	William Keough.
		Richard Hebden O'Grady Haly.
		(W. C. O'Shaughnessy and H. P. Pearson half-pay).
1887	...	William Keough.
		Richard Hebden O'Grady Haly.
		James Edward Harris.
		(Charles Henry Gardiner to 45th).
1888/9	...	Richard Hebden O'Grady Haly.
		James Edward Harris.
		Arthur Tower.
		(William Keough half-pay).
1890	...	James Edward Harris.
		Arthur Tower.
		(E. H. O'Grady Haly half-pay).
1891	...	James Edward Harris.
		John Campbell Robertson Glasgow.
		(Tower 9/12/90).

1892/3	...	John Campbell Robertson Glasgow.
		Owen Williams. (Harris 1/4/92).
1894	...	Owen Williams. (J. C. R. Glasgow half-pay).
1895/6	...	Owen Williams.
		Richard Thomas Edward Dowse. (Glasgow).
1897	...	Richard Thomas Edward Dowse.
		Charles Richard Townley. (Williams).
1898	...	Richard Thomas Edward Dowse.
		Charles Richard Townley.
		Arthur John Watson. (Dowse).
1899	...	Arthur John Watson.
		C. R. Townley.
1900/1	...	C. R. Townley.
		George Frederick Campbell Mackenzie.
		(A. J. Watson killed in action, Colesberg, February 6th, 1900).
1902	...	C. R. Townley. (Half-pay 24/2/02).
		G. F. C. Mackenzie.
		Archibald Cyril Cubitt. (Townley).
1903	...	G. F. C. Mackenzie.
		A. C. Cubitt.
1904/6	...	A. C. Cubitt.
		Frank Graham (vice Mackenzie).
		(G. F. C. Mackenzie half-pay 7/1/04; died at Mentone, 1909, when on sick leave from India; A.A.G. 8th Division; fondly remembered by all).
1907/8	...	F. Graham.
		Vivian Waldegrave Hall Graham.
1909	...	V. W. H. Graham.
		Edward Montagu.
1910	...	E. Montagu.
		C. H. C. Van Straubenzee
		Massy-Lloyd.

It was my good fortune on embarking with troops for India, in 1887, to secure the services of Lieutenant E. Montagu as my Adjutant. Since then, his Staff College and other certificates, together with war service and the position he now holds, all speak of his great worthiness. In similar good keeping are the other battalions of the Twelfth (2nd Van Straubenzee, and 3rd Massy-Lloyd). The 1/12th, under Montagu, are now stationed at Malta (under orders for Egypt); and in a letter from a great friend, Captain Harry Jones, H.M.S. "Suffolk" (son of Jones, V.C., of Delhi), he said what pleasure it had been to meet and entertain the Suffolk soldier men—for at Malta the two "Suffolks" lay, the blue and the red. It was then the leadsman said, that there is no use in taking soundings, for there is no bottom to this sea of friendship on which the two "Suffolks" were afloat. H.M.S. "Suffolk" seemed a perfect lifeboat available for both.

It may interest, may I say encourage the young, *i.e.*, the Centurions coming on, to know that the first copy of this book was sent by kind permission to His Majesty King George V.

Christmas Day, 1910.

My thanks are due to the following for their great kindness in allowing me a few extracts from their books :—

Sir Frederick Maurice, K.C.B.,
Sir E. Fortescue,
General D. Hutchinson, C.S.I.,
Sir Charles Watson, R.E., G.C.M.G.,
Colonel H. Brock, 50th Queens,
Surgeon-General Skey Muir, C.B.,
Miss Waller,
The Honourable Mrs. H. S. Liddell (Records of the 10th Hussars),
Colonel H. O'Donnell and Officers of the 14th Regiment,
Colonel W. J. Boyes, 10th Regiment,
Colonel E. Montagu,
C. R. Andrae, Esq.,
Mr. Long, High Street, Portsmouth,
The Editor, *Edinburgh Review*,
The Editor, *Quarterly Review*,
Messrs. Kegan Paul, Trench, Trübner & Co.,
Wm. Heinemann, Esq.,
A friend in the Sudan,
W. Dupernex, Esq.,
Miss Wormald.

N.B.—If these pages should meet the eye of any who were children prisoners at the time of the Cabul disaster—or during the Mutiny at Lucknow or elsewhere—it would be most interesting to receive their recollections ; further, any incident or act, collective or individual, either on sea or shore, and worthy of insertion in the "Centurions of a Century," would be most thankfully received by the Editor for the next edition.

F. V. Hadlow and Sons,
Printers,
8, Prince Albert Street, Brighton.

www.ingramcontent.com/pod-product-compliance
Lightning Source LLC
LaVergne TN
LVHW040423011026
849194LV00022B/68